C0-ALK-838

ANIMAL SPECIES FOR DEVELOPMENTAL STUDIES

Volume 1
Invertebrates

ANIMAL SPECIES FOR DEVELOPMENTAL STUDIES

Volume 1
Invertebrates

Edited by
T. A. Dettlaff and S. G. Vassetzky
N. K. Kol'tsov Institute of Developmental Biology
Academy of Sciences of the USSR
Moscow, USSR

Translated into English by
G. G. Gause, Jr., and S. G. Vassetzky

Technical Editors
Frank Billett
and
L. A. Winchester

CONSULTANTS BUREAU • NEW YORK AND LONDON

Library of Congress Cataloging in Publication Data

Ob′′ekty biologii razvitiia. English.
Animal species for developmental studies / edited by T. A. Dettlaff and S. G. Vassetzky; technical editors Frank Billett and L. A. Winchester.
p. cm.
Translation of Ob′′ekty biologii razvitiia.
Includes bibliographical references.
Contents: V. 1. Invertebrates.
ISBN 0-306-11031-8
1. Embryology, Experimental. 2. Developmental biology. 3. Laboratory animals. I. Detlaf, Tat′iana Antonovna. II. Vassetzky, S. G. III. Title.
QL961.O2413 1989 89-22389
591.3′0724—dc20 CIP

This translation is published under an agreement with the Copyright Agency of the USSR (VAAP)

A Division of Plenum Publishing Corporation
233 Spring Street, New York, N.Y. 10013

Printed in the United States of America

PREFACE

This volume comprises normal tables (description of normal development) for protozoa and invertebrates widely used in developmental biology studies. The species chosen reflect their advantages for laboratory studies, the information available, and their availability for experimentation. Chapter 11, which contains the normal tables for the starfish *Asterina pectinifera*, was written specially for this edition, which is the invertebrate section of the revised and augmented translation of *Ob"ekty Biologii Razvitiya* published in Russian in 1975 as a volume in the series of monographs *Problemy Biologii Razvitiya* (*Problems of Developmental Biology*) by Nauka Publishers, Moscow.

The description of every species is preceded by an introduction in which the advantages of working with the particular animal are stated and the problems studied (with the main references) are outlined. Data are also provided on its taxonomic status and distribution of the animal, and conditions of keeping the adult animals in laboratory. Methods of obtaining gametes, methods of artificial fertilization, methods of rearing embryos and larvae, and tables of normal development are also given.

In this book an attempt is made to facilitate the study of temporal patterns of animal development by the introduction (for some animal species) of relative (dimensionless) characteristics of the timing of developmental stages comparable in different animal species and at different temperatures. Special emphasis on this approach is given in Chapters 5, 10, and 11, in which such data for invertebrates are provided for the first time. In these chapters, the relative time unit (τ_0) is defined as the interval between the appearance of successive cleavage furrows during first divisions. However, it remains to be seen in cytological studies whether this interval corresponds to the actual mitotic cycle. See also Chapter 1 of the vertebrate volume, written by T. A. Dettlaff.

As a whole, these two volumes should facilitate both the selection of experimental species for developmental studies and work with the chosen species. One should bear in mind that the use of normal tables is necessary for standardizing experimental materials and obtaining comparable results.

In conclusion, we would like to thank all those who kindly provided their permission to reproduce the normal tables, drawings, etc., as well as those who gave valuable advice.

T. A. Dettlaff
S. G. Vassetzky

FOREWORD TO THE ENGLISH EDITION

Quite obviously, our understanding of how animals develop at any particular period of history has depended to a large extent on the embryos on hand at the time and the methods available to examine them. The great early 17th Century treatises of Fabricius are devoted to the study of the chick embryo, readily available mammalian fetuses, and the embryos of a smooth shark and a snake. The species described were not only mostly familiar to Aristotle but also continued to provide standard material for embryologists in the two centuries following the death of Fabricius. The comparative and experimental studies of the 19th Century used a wider range of materials; noteworthy was the increasing use of amphibia and marine invertebrates. Contemporary studies show species dominance of a different kind, with choice of material dependent on the perceived usefulness for genetic, molecular biological, and essentially cell-based investigations. At the present time the embryos of a great many species are available for study, covering most of the major phyla; accessibility to this material depends on knowing how the embryos of a particular species may be obtained and maintained. The present two-volume English edition of the Russian book *Ob"ekty Biologii Razvitiya* (*Objects in Developmental Biology*) adequately relates to this need to know the practical requirements for studying a particular species; it also fulfills the additional, and essential service, of describing the development of the embryos of many animals in detail.

Although many of the species dealt with in these volumes will be familiar to those working outside the Soviet Union, others will be less so. Of particular interest are the chapters in the second volume on the stone loach (*Misgurnus fossilis*) and the sturgeon (*Acipenser güldenstädti*). Here we find embryological material that has been studied mainly in the Soviet Union. Apart from these rather special cases, the second volume also contains comprehensive treatments of these embryos, amphibian, avian, and mammal, which are used by developmental biologists the world over. The first volume is devoted to invertebrates. Both volumes contain descriptions of species that relate to practical needs; these include various fish species, the silkworm (*Bombyx mori*), and the honeybee (*Apis mellifera*). In these books our Soviet colleagues give us an insight into the breadth of their achievement in the field of animal development. They also provide us with a most useful and stimulating survey of what might be called the raw material of developmental biology. There are relatively few works of this kind, but they are essential. Without an adequate and informed description of the species available for developmental studies, we are not only hampered in our individual investigations but also that much poorer in our appreciation of both the diversity and essential unifying features of animal development.

As with the second volume of this series, during the preparation of the English edition the authors took the opportunity to update, and in some cases enlarge, their chapters. Thus the English edition contains new material and is larger than the original single-volume edition. Dr. Leslie Winchester undertook the initial reading and correction of the translation from the Russian; she was also involved in facilitating the transition from the format of the Russian publishers, Nauka, to that of Plenum. Together the English editors were concerned mainly with achieving some uniformity of presentation without detriment to the perceived style of the authors and the intentions of the Soviet editors.

Frank Billett
Southampton

CONTRIBUTORS

E. D. Bakulina, Institute of Developmental Biology, Academy of Sciences of the USSR, Moscow
L. V. Belousov, Biological Faculty, Moscow State University, Moscow
G. M. Burychenko, Institute of Developmental Biology, Academy of Sciences of the USSR, Moscow
G. A. Buznikov, Institute of Developmental Biology, Academy of Sciences of the USSR, Moscow
P. V. Davydov, Institute of Developmental Biology, Academy of Sciences of the USSR, Moscow
L. I. Gunderina, Institute of Cytology and Genetics, Siberian Branch, Academy of Sciences of the USSR, Novosibirsk
I. I. Kiknadze, Institute of Cytology and Genetics, Siberian Branch, Academy of Sciences of the USSR, Novosibirsk
V. V. Klimenko, Department of Ecological Genetics, Academy of Sciences of the Moldavian SSR, Kishinev
N. N. Kolesnikov, Institute of Cytology and Genetics, Siberian Branch, Academy of Sciences of the USSR, Novosibirsk
K. V. Kvitko, Biological Research Institute, Leningrad State University, Leningrad
O. E. Lopatin, Institute of Cytology and Genetics, Siberian Branch, Academy of Sciences of the USSR, Novosibirsk
V. N. Meshcheryakov, Biological Faculty, Moscow State University, Moscow
V. G. Mitrofanov, Institute of Developmental Biology, Academy of Sciences of the USSR, Moscow
E. N. Myasnyankina, Institute of Developmental Biology, Academy of Sciences of the USSR, Moscow
V. I. Podmarev, Institute of Developmental Biology, Academy of Sciences of the USSR, Moscow
E. V. Poluektova, Institute of Developmental Biology, Academy of Sciences of the USSR, Moscow
D. V. Shaskolsky, Institute of General Genetics, Academy of Sciences of the USSR, Moscow
O. I. Shubravyi, Institute of Developmental Biology, Academy of Sciences of the USSR, Moscow
S. G. Vassetzky, Institute of Developmental Biology, Academy of Sciences of the USSR, Moscow
A. L. Yudin, Institute of Cytology, Academy of Sciences of the USSR, Leningrad

CONTENTS

Chapter 3
Hydra and Hydroid Polyps
L. V. Belousov

Chapter 4
THE SLUDGEWORM *Tubifex*
V. N. Meshcheryakov

Chapter 5
THE COMMON POND SNAIL *Lymnaea stagnalis*
V. N. Meshcheryakov

Chapter 6
The Midge *Chironomus thummi*
I. I. Kiknadze, O. E. Lopatin,
N. N. Kolesnikov, and L. I. Gunderina

Chapter 7
The Fruit Fly *Drosophila*
E. V. Poluektova, V. G. Mitrofanov, G. M. Burychenko,
E. N. Myasnyankina, and E. D. Bakulina

Chapter 8
The Honeybee *Apis mellifera* L.
D. V. Shaskolsky

Chapter 9
The Silkworm *Bombyx mori*
V. V. Klimenko

Chapter 10
The Sea Urchins *Strongylocentrotus dröbachiensis*,
S. nudus, and *S. intermedius*
G. A. Buznikov and V. I. Podmarev

Chapter 1

Amoeba AND OTHER PROTOZOA

A. L. Yudin

1.1. INTRODUCTION: PROTOZOA IN DEVELOPMENTAL BIOLOGY

When evaluating the perspectives of using Protozoa in developmental biology one should remember that this group comprises organisms with a cellular level of organization. Morphogenesis, regeneration, differentiation, and control of genetic apparatus, all phenomena which are of great interest to developmental biologists, may be observed in Protozoa in a manner no less impressive than in higher multicellular organisms, although they take place at the level of an individual cell. Protozoa are often more suitable and less expensive than the cells of higher animals for studying these phenomena. One should not forget, however, that Protozoa are not just cells but individual living organisms, too.

Of the tremendous number of protozoan species, only a few are widely used laboratory models, but their number is increasing. Primarily they include certain ciliates, green flagellates, and amoebae. Slime moulds (myxomycetes) such as *Dictyostelium discoideum, Physarum polycephalum*, and related species have also become very popular in developmental biology. According to zoological taxonomy [38] all these organisms belong to the phylum Protista.

Several monographs and reviews deal specially with the biology of such protists and the methods of experimenting with them; these include works on *Paramecium* [35, 58, 59, 62, 63, 67], *Tetrahymena* [17, 18, 29], *Stentor* [61], *Blepharisma* [24], *Euglena* [9], myxomycetes [7, 10, 16, 43, 6, 15, 40], and flagellate amoebae [12, 21, 37, 44].

This chapter deals with a species which is of interest to developmental biologists. This is the large freshwater *Amoeba*. Special attention has been paid to them because several laboratories in the Soviet Union have experience working with them and hold collections of strains.

1.2. *Amoeba* AND RELATED SPECIES

This section deals only with the large free-living freshwater amoeba, of which a typical representative is *Amoeba proteus*. These amoebae can be cultured for long periods in the laboratory and are widely used both in teaching and in various investigations in cell biology. Amoebae are characteristically large and have a unique resistance to mechanical lesions. They are extremely suitable for various microsurgical interventions, including the transplantation of nuclei and cytoplasm between cells [64, 65]. These methodological opportunities have to a large extent determined one of the main trends of studies with amoebae. In brief, this can be defined as the study of various aspects of the interactions between the cell nucleus and cytoplasm – a long-standing problem in developmental biology.

1.3. TAXONOMY

Taxonomy of amoebae in general, and of the considered group in particular, continues to be one of the most difficult and vague sections of zoological taxonomy (see, for example, [8]). Identification is recognized as an extremely difficult problem; it can often only be done in the vital state under more or less standard conditions, because these organisms do not have a constant shape and have only a poor morphological differentiation. The main field characteristics used for identification include size and type of movement (including the shape of the cell and individual pseudopodia under different physiological conditions); other characteristics include the number, size, and shape of interphase nuclei and various intracellular inclusions such as crystals and granules [51]. The problem is compounded because amoebae show agamous reproduction.

Amoeba proteus and *Amoeba discoides* (the latter is not considered an independent species by some authors) as well as *Polychaos dubium* (Fig. 1.1) are large mononuclear amoebae belonging to the same family which are frequently used in cell and developmental biology, as is *Thecamoeba sphaeronucleolus*, a representative of a different family. Recent experimental studies make use of *Amoeba indica* [13] and *Amoeba amazonas* [19] whose status as an independent species is still doubtful. The gigantic multinuclear amoebae *Chaos* (= *Pelomyxa*) *carolinense* and *Chaos* (= *Pelomyxa*) *illinoisense* complete the list of popular laboratory species.

Several schemes have been put forward for classifying amoebae [8, 51]. Information on the morphology and physiology of these amoebae can be found in the two collections of papers [30, 31] and a monograph [66].

1.3.1. Strains and Mutants

At the present time most workers prefer to work with laboratory strains of amoebae of known origin, rather than with material isolated from nature. This is partly because the establishment of a permanent culture of large amoebae from a specimen found in nature is by no means always successful, this partly related to the difficulties of identifying newly isolated amoebae. Furthermore, the use of laboratory strains makes comparison of the results obtained in different laboratories more reliable. At present several laboratory strains of *Amoeba proteus* of different origins are available. They are mentioned in current literature and kept in laboratories in various countries including the USSR. A large collection has been built up and maintained at the Institute of Cytology (Academy of Sciences of the USSR, Leningrad). The hereditary differences between some of these strains in a number of morphological, physiological, and biochemical traits have been described [33, 66]. Various treatments of cells may induce inherited phenotypic instability with

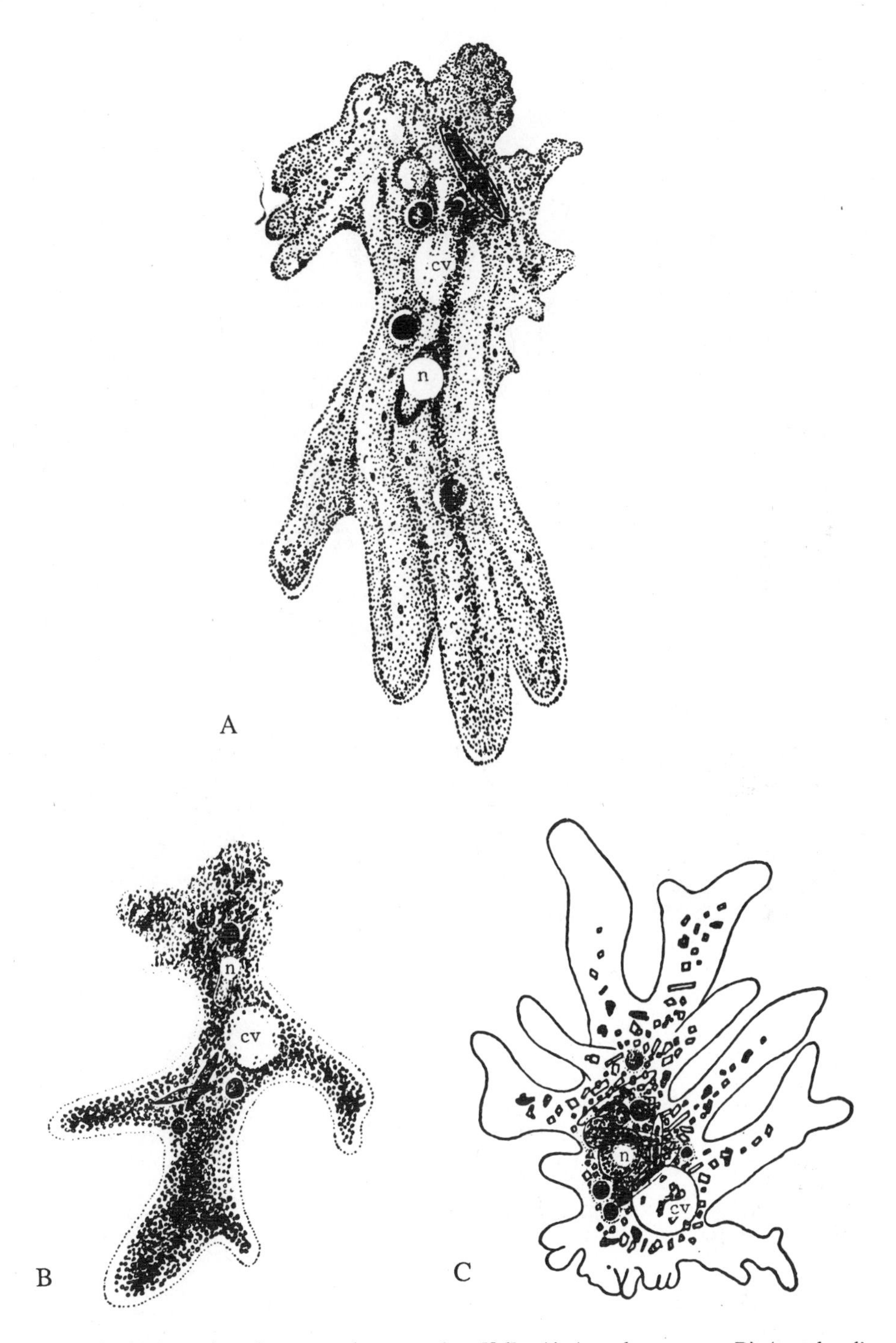

Fig. 1.1. Several species of mononuclear amoebae [36]. A) *Amoeba proteus*; B) *Amoeba discoides*; C) *Polychaos dubium*. n, nucleus; cv, contractile vacuole.

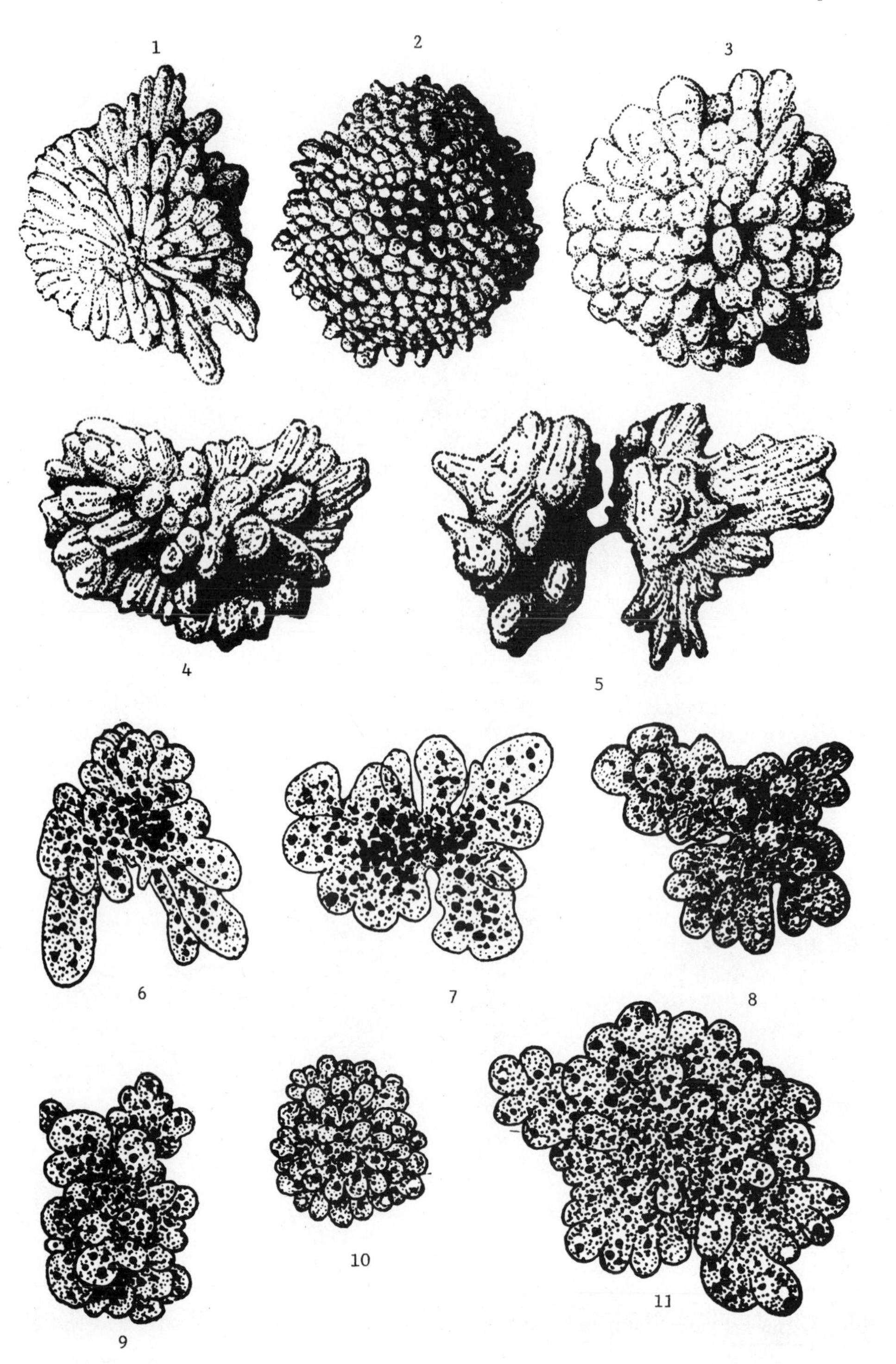
1
2
3
4
5
6
7
8
9
10
11

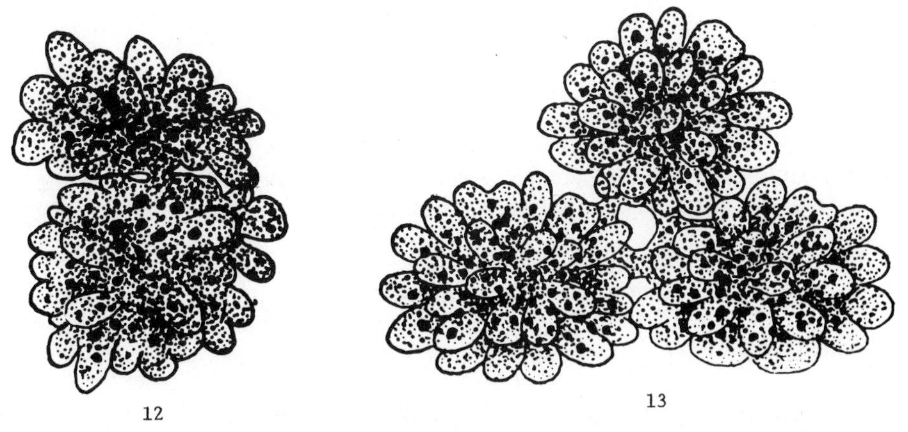

Fig. 1.2. Changes in appearance during the division of *Amoeba proteus* (1-5) and *Chaos carolinense* (6-13) [36]. 1) Shortly before the formation of the "division sphere"; 2) a later stage of the "division sphere"; 3) before elongation; 4) further elongation; 5) division almost complete; 6) early prophase; 7, 8) late prophase; 9) metaphase; 10) early anaphase; 11) late anaphase; 12, 13) late telophase up to the stage of resting nuclei (12 – cytoplasmic division yielding two daughter cells; 13 – cytoplasmic division yielding three daughter cells).

respect to many of these traits (for details, see [66]). Strains of *Amoeba proteus* may be either incompatible in transplantation, that is, they yield nonviable chimeric cells after the transplantation of nuclei or cytoplasm between the strains, or they may be compatible (for details, see [34, 66]). Endosymbionts are found in the cytoplasm of a large proportion of strains [32].

Mutants of *A. proteus* have been obtained using chemical mutagens [45, 49, 23]. Clones with increased content of DNA in the nucleus have been obtained by intracellular colchicine injections [3]; similar experiments have been performed with *Thecamoeba sphaeronucleolus* [14]. A similar "polyploidization" can be induced in *A. proteus* by low temperatures [65]. In some clones with increased content of nuclear DNA spontaneous depolyploidization was observed within a different time after they had been obtained.

Species such as *Amoeba discoides, A. indica, A. amazonas, Polychaos dubium*, and *Chaos carolinense* in most laboratories are represented by stocks of just one strain.

1.4. HABITATS

Amoebae live in shallow reservoirs with a slow current, abundant aquatic plants, and away from direct sunlight. A water temperature of 18-20°C is optimal for their reproduction. Samples of water and sludge taken from such reservoirs are inspected in the laboratory, and the amoebae are usually found in the layer adjacent to the bottom and on the surface of aquatic plant leaves. The specimens are removed and may be cultured using one of the techniques described below. Techniques of sampling, examining, and isolating the amoebae have been described by Page [51].

1.5. CULTURE CONDITIONS

Amoebae are usually cultured either in glass jars or in plastic containers, but it should be remembered that certain types of glass are toxic for them. The size and shape of the jars does not matter, depending solely on what the experimenter considers suitable. Mass cultures can be kept in beakers, Koch dishes, etc., while microtiter plates can be used for cloning. Cloning in glass capillaries has also been described [50].

Numerous inorganic media for growing amoebae have been described. The composition of salt solutions normally employed for this purpose is given below.

The medium of Chalkley [11] is used in several modifications. One has the following composition [41]:

	g/liter distilled water
NaCl	0.08
$NaHCO_3$	0.004
KCl	0.002
$Na_2HPO_4 \cdot 1H_2O$	0.001
$CaHPO_4$	traces

The Prescott medium [55] contains the following ingredients:

	g/liter distilled water
NaCl	0.01
$CaCl_2$	0.01
KCl	0.006
$MgSO_4$	0.002

The Prinsgeim medium [12] contains the following ingredients:

	g/liter distilled water
$Ca(NO_3)_2 \cdot H_2O$	0.2
KCl	0.026
$Na_2HPO_4 \cdot 2H_2O$	0.02
$MgSO_4 \cdot 7H_2O$	0.02
$FeSO_4 \cdot 7H_2O$	0.002

It appears that amoebae may be cultured equally well in any of these solutions.

A description of certain other older techniques for growing amoebae can be found in manuals on cultivating invertebrates [22]. In accordance with these recipes, several untreated or boiled grains of rice or wheat are placed in a jar with the inorganic medium, and the mixture is then inoculated with protozoans such as *Colpidium, Chilomonas*, or any others which serve as food for the amoebae. After a certain time several dozens, even hundreds, of amoebae are released in this medium. A modification of such a culture has been described among others by Lorch and Danielli [41]. Such "pond" cultures are frequently employed for keeping stocks in collections because they require the minimum amount of equipment and attention. Transfers should be made approximately once monthly. These cultures, however, have a number of serious drawbacks: the development of such cultures is usually phasic in nature; amoebae from parallel cultures and even from the same culture may be in different physiological conditions; the cloning and mass growing of amoebae in such cultures is difficult and sometimes even impossible; and it is not easy to get rid of the accompanying organisms when required.

The procedure suggested by Prescott and James [54, 55], which has been further improved by Griffin [26], produces far better results. According to this method, amoebae are kept in an inorganic medium (e.g., the Prescott medium) and thoroughly washed ciliates are periodically added for feeding. The most commonly used ciliate is *Tetrahymena pyriformis*. *Tetrahymena* is specially grown for this purpose in axenic (bacteria-free) cultures made in 2% proteose-peptone or yeast extract. To prepare the yeast extract for *Tetrahymena* growth 70 g of fresh baker's yeast are boiled in 1 liter of tap water for an hour; the extract, freed from the remnants of yeast cells, is diluted with an equal volume of 1/45 M phosphate buffer, pH 6.5-7.0, dispensed into flasks and autoclaved. The medium is inoculated under sterile conditions using *Tetrahymena* from axenic cultures and incubated at room temperature for 2-3 days. Thereafter the *Tetrahymena* are collected, concentrated, and cleansed of the remaining culture medium by three centrifugations in conical centrifuge tubes at about 800*g*; a fresh supply of the Prescott medium is used to wash the *Tetrahymena*, which are then diluted with this medium to the required density for feeding amoebae.

The following procedure for feeding produces good results. The Prescott medium in the amoeba culture is changed daily; it can be decanted directly from the culture flask, since the majority of amoebae are tightly attached to the walls. Feeding with *Tetrahymena* is performed every other day after the medium has been changed, so that the amoebae have ingested almost all the *Tetrahymena* by the following day. Other feeding conditions may also be used, and are specifically employed to grow large quantities of amoebae [25, 27]. In all cases, however, overfeeding the cultures should be avoided since, for some unknown reason, this results in an irreversible deterioration of their state and eventually in the death of the culture.

Containers with amoebae should be kept at 17-25°C away from direct light. From time to time, as the container bottom and walls become contaminated, the culture should be transferred to a clean vessel.

Methods of axenic cultivation of amoebae have still to be developed.

1.6. REPRODUCTION

The content of DNA in the *Amoeba proteus* nucleus measured by cytofluorimetry 1 h after division is, in most strains, 14.6 ± 0.6 pg [2]. By the end of the cell cycle this amount increases 2.2-2.9 times and more [42]. There are circumstantial arguments in favor of the polyploid nature of this species which has a very large number (500-1000) of small chromosomes [39, 48].

Large mononuclear amoebae multiply by a simple division, each parent cell yielding two daughter cells (Fig. 1.2); the nucleus is divided by a typical mitosis of the partially closed type [28]. The mean generation time of *A. proteus* varies depending on the strain, feeding conditions, and cultivation temperature [56, 57, 60]. The highest reproduction rate recorded for this species is one division every 24 h at 25°C, provided food is constantly present in the culture. Usually the generation time at 18°C is 40-50 h. Phase G_1 is virtually absent in the cell cycle; the S-phase lasts 6-8 h for a 36- to 48-h cycle at 23°C; the G_2 phase occupies the main part of the cycle [53]. According to other data [46], at 18-20°C when the generation time is 48-54 h, the S-phase lasts 12-14 h. During this time there are two peaks of incorporation of labelled precursor into DNA with the additional one following the main peak.

Characteristics of the mitotic cycle in other species of mononuclear amoebae may differ significantly from those given above, but the G_1 period is always absent.

The content of DNA in one nucleus of the multinuclear amoeba *Chaos carolinense*, measured by cytofluorimetry 1 h after division, is 9.6 pg; the mean generation time at 25°C is 48 h [1]. According to other data [52], the mean time of cell doubling in the culture was 1.7 days and, assuming that three daughter cells on average are formed after each division, the cell cycle was 2.7 days [52].

There is no reliable information on the sexual processes in amoebae of the listed species, at least not under culture conditions.

REFERENCES

1. S. Yu. Afon'kin, "DNA content in the nuclei of *Chaos chaos*, a multinuclear amoeba," in: *Modern Problems in Protozoology* [in Russian], Vilnius (1982).
2. S. Yu. Afon'kin, "DNA content in the nuclei of various strains of *Amoeba proteus* measured by cytofluorimetry," *Tsitologiya* **25**, 771–776 (1983).
3. S. Yu. Afon'kin, "Producing *Amoeba proteus* clones with increased nuclear DNA content," *Acta Protozool.* (1984).
4. S. Yu. Afon'kin, "Spontaneous 'depolyploidization' of cells in *Amoeba* clones with increased nuclear DNA content," *Arch. Protistenk.*, B. **131**, 101–112 (1986).
5. S. Yu. Afon'kin and L. V. Kalinina, "Low temperature induces polyploidization in dividing *Amoeba*," *Acta Protozool.* **26** (1987).
6. H. C. Aldrich and J. W. Daniel (eds.), *Cell Biology of Physarum and Didymium*, Academic Press, New York–London (1952).
7. J. T. Bonner, *The Cellular Slime Molds* (2nd edn.), Princeton University Press, Princeton, New Jersey (1967).
8. E. C. Bovee and T. L. Jahn, "Taxonomy and phylogeny," in: *The Biology of Amoeba*, K. W. Jeon (ed.), Academic Press, New York–London (1973).
9. D. E. Buetow (ed.), *The Biology of Euglena*, Academic Press, New York–London (1982).
10. P. Cappuccinelli and J. M. Ashworth (eds.), *Development and Differentiation in the Cellular Slime Molds*, Elsevier/North Holland, Amsterdam (1977).
11. H. W. Chalkley, "Stock cultures of *Amoeba*," *Science* **71**, 442 (1930).
12. C. Chapman-Andresen, "Pinocytosis of inorganic salts by *Amoeba proteus* (*Chaos diffluens*)," *C. R. Trav. Lab. Carlsberg* **31**, 77–92 (1958).
13. S. Chatterjee and M. V. N. Rao, "Nucleocytoplasmic interactions in the *Amoeba* interspecific hybrids," *Exp. Cell Res.* **84**, 235–238 (1974).
14. J. Comandon and de P. Fonbrune, "Action de la colchicine sur *Amoeba sphaeronucleus*. Obtention de variétés géantes," *C. R. Soc. Biol. Paris* **136**, 410–411 (1942).
15. W. F. Dove, S. Hatano, J. Del, F. B. Haugli, and D. E. Wohlfarth-Botterman, *The Molecular Biology of Physarum polycephalum*, Plenum Press, New York (1986).

16. W. F. Dowe and H. P. Rusch (eds.), *Growth and Differentiation in Physarum polycephalum*, Princeton University Press, Princeton, New Jersey (1980).
17. A. M. Elliott (ed.), *Biology of Tetrahymena*, Dowden, Hutchinson, and Ross, Inc., Stroudsburg, Pennsylvania (1973).
18. L. P. Everhart, Jr.,"Methods with *Tetrahymena*," in: *Methods in Cell Physiology*, D. M. Prescott (ed.), Vol. 5, Academic Press, New York–London (1972).
19. C. J. Flickinger, "Maintenance and regeneration of cytoplasmic organelles in hybrid amoebae formed by nuclear transplantation," *Exp. Cell Res.* **80**, 31–46 (1973).
20. C. Fulton, "Amoebo-flagellates as research partners: The laboratory biology of *Naegleria* and *Tetramitus*," in: *Methods in Cell Physiology*, D. M. Prescott (ed.), Vol. 4, Academic Press, New York–London (1970).
21. J. G. Gall, *The Molecular Biology of Ciliated Protozoa*, Academic Press, New York (1986).
22. P. S. Galtsoff, F. E. Lutz, P. S. Welch, and J. G. Needham, *Culture Methods for Invertebrate Animals*, Comstock Publishing, Ithaca, New York (1937).
23. A. Gangopadhyay and S. Chatterjee, "A cell-cycle-phase-specific mutant of *Amoeba*," *J. Cell Sci.* **68**, 95–111 (1984).
24. A. C. Giese, *Blepharisma. The Biology of a Light-Sensitive Protozoan*, Stanford University Press, Stanford, California (1973).
25. L. Goldstein and C. Ko, "A method for the mass culturing of large free-living amoebae," in: *Methods in Cell Biology*, Vol. 13, Academic Press, New York–London (1976).
26. J. L. Griffin, "An improved mass culture method for the large, free-living amoebae," *Exp. Cell Res.* **21**, 170–178 (1960).
27. J. L. Griffin, "Culture: Maintenance, large yields, and problems of approaching axenic culture," in: *The Biology of Amoeba*, K. W. Jeon (ed.), Academic Press, New York–London (1973).
28. D. B. Gromov, "Ultrastructure of mitosis in *Amoeba proteus*," *Protoplasma* **126**, 130–139 (1985).
29. D. L. Hill, *The Biochemistry and Physiology of Tetrahymena*, Academic Press, New York–London (1972).
30. H. I. Hirshfield (ed.), "The biology of the Amoeba," *Ann. N. Y. Acad. Sci.* **78**, 401–704 (1959).
31. K. W. Jeon (ed.), *The Biology of Amoeba*, Academic Press, New York–London (1973).
32. K. W. Jeon, "Symbiosis of bacteria with amoeba," in: *Cellular Interactions in Symbiotic and Parasitic Relationships*, Ohio State University Press, Columbus, Ohio (1973).
33. K. W. Jeon and I. J. Lorch, "Strain specificity in *Amoeba proteus*," in: *The Biology of Amoeba*, K. W. Jeon (ed.), Academic Press, New York–London (1973).
34. K. W. Jeon and I. J. Lorch, "Compatibility among cell components in large free-living amoebae," *Int. Rev. Cytol.*, Suppl. **9**, 45–62 (1979).
35. A. Jurand and C. G. Selman, *The Anatomy of Paramecium aurelia*, St. Martin's Press, New York (1969).
36. R. R. Kudo, *Protozoology*, C. C. Thomas, Springfield (1966).

37. M. Levandowsky and M. M. Hutner. (eds.), *Biochemistry and Physiology of Protozoa*, Academic Press, New York, Vols. 1 and 2 (1979), Vol. 3 (1980), Vol. 4 (1981),
38. N. D. Levine, J. O. Corliss, F. E. G. Cox, G. Deroux, J. Grain, B. M. Honigberg, G. F. Leedale, A. R. Loeblich, III, J. Lom, D. Lynn, E. G. Merinfeld, F. C. Page, G. Poljansky, V. Sprague, J. Vavra, and F. G. Wallace, "A newly revised classification of the Protozoa," *J. Protozool.* **27**, 37–58.
39. W. Liesche, "Die Kern- und Fortpflanzungsverhaltnisse von *Amoeba proteus* (Pall.)," *Arch. Protistenk.* **91**, 135–169 (1938).
40. W. F. Loomis (ed.), *The Development of Dictyostelium discoideum*, Academic Press, New York–London (1982).
41. I. J. Lorch and J. F. Danielli, "Nuclear transplantation in amoebae. I. Some species characters of *A. proteus* and *A. discoides*," *Q. J. Microscop. Sci.* **94**, 445–460 (1953).
42. E. E. Makhlin, M. V. Kudryavtseva, and B. N. Kudryavtsev, "Peculiarities of changes in DNA content of *Amoeba proteus* nuclei during interphase. A cytofluorimetric study," *Exp. Cell Res.* **118**, 143-150.
43. G. W. Martin and C. J. Alexopoulus, *The Myxomycetes*, University of Iowa Press, Iowa City, Iowa (1969).
44. D. L. Nanney, *Experimental Ciliatology: An Introduction to Genetic and Developmental Analysis in Ciliates*, Wiley, New York (1980).
45. G. V. Nikolajeva, L. V. Kalinina, and J. Sikora, "Transparent *Amoeba proteus* originated from the strain C," *Acta Protozool.* **19**, 339–344 (1980).
46. M. J. Ord, "The synthesis of DNA through the cell cycle of *Amoeba proteus*," *J. Cell Sci.* **3**, 483–492 (1968).
47. M. J. Ord, "Mutations induced in *Amoeba proteus* by the carcinogen N-methyl-N-nitrosourethane," *J. Cell Sci.* **7**, 531–548 (1970).
48. M. J. Ord, "*Amoeba proteus* as a cell model in toxicology," in: *Mechanisms of Toxicity*, Macmillan, New York (1971).
49. M. J. Ord, "Chemical mutagenesis," in: *The Biology of Amoeba*, K. W. Jeon (ed.), Academic Press, New York–London (1973).
50. M. J. Ord, "A capillary technique for cloning *Amoeba* from single cells," *Cytobios* **21**. 57–69 (1979).
51. F. C. Page, *A New Key to Freshwater and Soil Gymnamoebae with Instructions for Culture*, Freshwater Biol. Assoc., Ambleside (1988).
52. D. M. Prescott, "Mass and clone culturing of *A. proteus* and *Chaos chaos*," *C. R. Trav. Lab. Carlsberg, Ser. Chim.* **30**, 1–12 (1956).
53. D. M. Prescott, "The cell cycle in *Amoeba*," in: *The Biology of Amoeba*, K. W. Jeon (ed.), Academic Press, New York–London (1973).
54. D. M. Prescott and R. F. Carrier, "Experimental procedures and cultural methods for *Euplotes eurystomus* and *Amoeba proteus*," in: *Methods in Cell Physiology*, Vol. 1, D. M. Prescott (ed.), Academic Press, New York–London (1964).
55. D. M. Prescott and T. W. James, "Culturing of *A. proteus* on *Tetrahymena*," *Exp. Cell Res.* **8**, 256–258 (1955).
56. A. Rogerson, "Generation times and reproductive rates of *Amoeba proteus* (Leidy) as influenced by temperature and food concentration," *Can. J. Zool.* **58**, 543–548 (1980).

57. R. A. Smith, "Growth temperature acclimation by *Amoeba proteus*: Effects on cytoplasmic organelle morphology," *Protoplasma* **101**, 23–25 (1979).
58. T. M. Sonneborn, "Methods in the general biology and genetics of *Paramecium aurelia*," *J. Exp. Zool.*, **113**, 87–143 (1950).
59. T. M. Sonneborn, "Methods in *Paramecium* research," in: *Methods in Cell Physiology*, D. M. Prescott (ed.), Vol. 4, Academic Press, New York–London (1970).
60. V. A. Sopina, "The multiplication rate of amoebae related to the cultivation temperature," *J. Therm. Biol.* **1**, 199–204 (1976).
61. V. Tartar, *The Biology of Stentor*, Pergamon Press, New York–Oxford (1961).
62. W. J. van Wagtendonk (ed.), *Paramecium: A Current Study*, American Elsevier, New York (1974).
63. R. Wichterman, *The Biology of Paramecium*, Plenum Press, New York–London (1986).
64. A. L. Yudin, "Nuclear transplantation in *Amoeba*," in: *Methods in Developmental Biology* , T. A. Dettlaff, V. Ya. Brodsky, and G. G. Gause (eds.), Nauka. Moscow (1974).
65. A. L. Yudin, "Amputation and transplantation of cytoplasm in *Amoeba*," in: *Methods in Developmental Biology* [in Russian], T. A. Dettlaff, V. Ya. Brodsky, and G. G. Gause (eds.), Nauka, Moscow (1974).
66. A. L. Yudin, *Nucleocytoplasmic Relationships and Cell Heredity in Amoebae* [in Russian], Nauka, Leningrad (1982).
67. H. D. Görtz (ed.), *Paramecium*, Springer-Verlag, Berlin (1988).

Chapter 2

THE GREEN FLAGELLATE *Chlamydomonas*

K. V. Kvitko

2.1. INTRODUCTION

Green flagellates belonging to the genus *Chlamydomonas* are widely used as a model in experimental studies. *Chlamydomonas* is one of the more suitable unicellular eukaryotic organisms for studying cell differentiation with modern techniques. This is due to the relative simplicity of the structure allied with the presence of a wide range of cell organelles, as well as the availability of a considerable body of information on the physiology, biochemistry, and genetics of this organism. Another advantage is that well-developed culture procedures are available.

2.2. TAXONOMY

Chlamydomonas are unicellular eukaryotes, a group in which the division into animals and plants is controversial. Zoologists include them in the phylum Protozoa, type Sarcomastigophora, class Flagellata, subclass Phytomastigina, order Phytomonadina. Botanists believe that phytomonads hold the rank of phylum Chlorophyta, and the group Volvocacea is considered a class while chlamydomonads comprise an order. However, both botanists and zoologists agree that the genus *Chlamydomonas* is polymorphic and includes several hundred species. Detailed analysis of the taxonomy and species diversity of *Chlamydomonas* has been conducted by Ettl [18, 19]. Cultures of more than fifty species are maintained in the collections of protozoans and algae in Texas [67], Cambridge [15], Prague [3], and Leningrad [25, 26]. They include marine, fresh water, and soil species.

Most common laboratory strains belong to *Chlamydomonas reinhardii* 137C. The strains were isolated by Smith in 1946 and are now being kept as two stocks: 137C (+) and (–) sex and 137F (+) and (–) sex [34]. The strains are compatible, but the cultures of the 137F strain are burdened by meiotic lethality. These strains have been subjected to detailed genetic, biochemical, and biophysical studies; therefore when describing *Chlamydomonas* culture, mutant selection, etc., we shall be referring to the descendants of 137C strain, which are capable of three types of nutrition: heterotrophic, mixotrophic, and phototrophic. Investigators interested in obligate phototrophs should refer to the review on the biology of *C. eugametos* [24, 42, 45, 46]. [For recent bibliography see *Chlamydomonas Newsletter*, M. Adams (ed.).]

2.3. CELL STRUCTURE AND DIFFERENTIATION DURING THE LIFE CYCLE

The structure of *Chlamydomonas* cells and gametes is presented in Fig. 2.1. Each cell has two flagella each with a basal body; the cup-shaped chromatophore accounts for more than half the cell volume and includes pyrenoid granules, lamellae, the space occupied by ribosomes, and the eye (stigma). The nucleus, with a distinct nucleolus, is submerged into the chloroplast cup, and mitochondria form a complex network of strands covering the chloroplast and periflagellar space [64]. Two pulsating vacuoles are present near the bases of the flagella.

Biogenesis of organelles and physiological changes during the cell cycle proceed in strict sequence: the formation of the cell wall and of the chloroplast membranes, DNA synthesis in the chloroplast, induction of gametogenesis, and increase in the level of chlorophyll and carboxydismutase; these events follow each other during the light phase of the cycle (i.e., that part of the 24-h cycle occurring in daylight or its equivalent). There is a corresponding sequence of periods characterized by a lack of sensitivity to inhibitors such as rifampicin, bromouracil, spectinomycin, ethidium bromide, diurone, chloramphenicol, and *p*-chloromercuribenzoate. During the dark phase (i.e., during the hours of darkness), the sequence of events is as follows: tubulin accumulation, dissolution of the flagella, biogenesis of new basal bodies, synthesis of nuclear DNA, cytokinesis (mitosis and chloroplast division), formation of daughter cells [29].

The vegetative cells are haploid and contain 19 ± 1 chromosomes [71]. DNA has been detected in the chloroplast and mitochondria [11, 57]. Vegetative reproduction of *Chlamydomonas* belongs to the palintomic type, each cell yielding 2, 4, 8, or even 16 or more daughter cells (zoospores). The duration of one sporulation period at an optimal temperature is close to 24 h. *C. reinhardii* shows the isogamic type of sex differentiation, that is, cells and gametes of the opposite sex (pairing types) are similar to each other [20]. Whether they belong to the (+) or (–) pairing type depends on the state of the *mt* locus in the VIth linkage group. The ultrastructure of the mating machinery or "sexosome" is different for (+) and (–) types [22].

The induction of gametogenesis is a necessary prerequisite for the transition from vegetative to sexual reproduction [14]. The main factor inducing gametogenesis is a nitrogen deficit in the culture medium. Under standard culture conditions the gametogenic division is completed 19 h after the beginning of induction. However, stationary cultures may yield gametes after only 2 h [63]. Morphological differences between gametes, zoospores, and vegetative cells are not very prominent, but their biochemical composition is drastically different. In gametes the level of cytoplasmic and chloroplastic ribosomes is reduced, the total content of DNA per cell remaining constant while the proportion of chloroplast DNA decreases from 14% to 7%; then the relative amount of nucleolar DNA is increased from 1% to 4%. Under normal conditions the pear-shaped prozygote (two nuclei, four flagella, and two chloroplasts) is formed 5-10 min after the mating bridge. The fusion of the nuclei and a corresponding decrease of the number of flagella from 4 to 2 is observed 3 h later; the chloroplasts fuse after 6 hours and the chloroplast structure is rearranged after 10 h of zygotic life [9]. Later the zygote begins to increase in size; fertilization and the first 24 h of zygotic development proceed upon illumination and thereafter the cell passes into a resting state, becomes

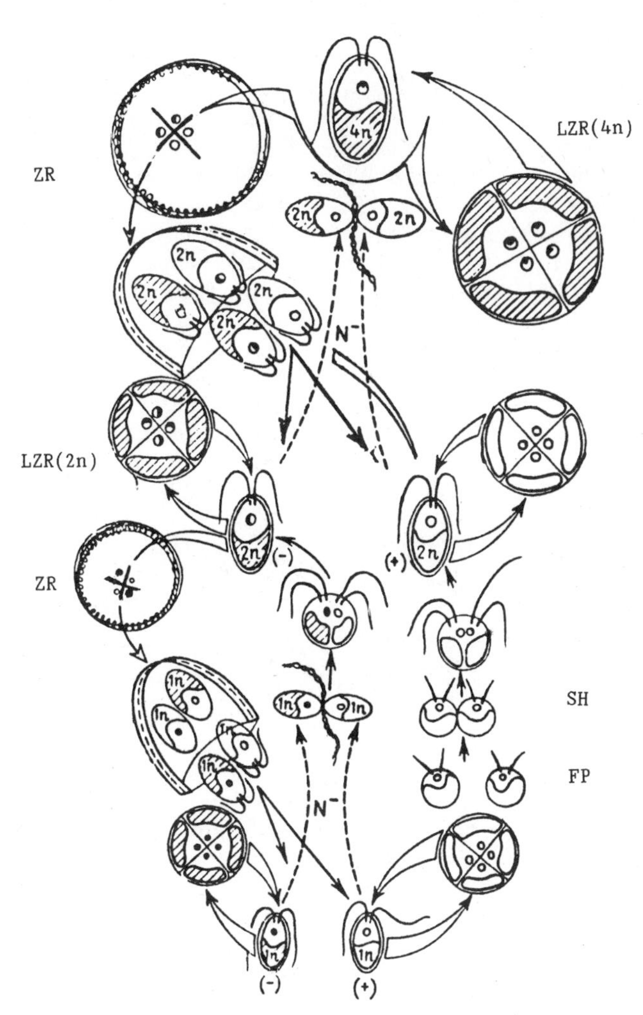

Fig. 2.1. Life cycle of *Chlamydomonas* (137C). Changes in ploidy from the haploid state (1*n*) to the diploid state (2*n*) after the fusion of (+) and (–) gametes, and to the tetraploid state (4*n*) upon the loss of zygotic reduction (LZR) in (2*n*) diploids and (4*n*) tetraploids. Decrease in the level of ploidy by a factor of two (thin arrows) occurs as a consequence of zygotic reduction (ZR). Vegetative reproduction is shown by open arrows. Dashed arrows indicate gametogenesis upon depletion of nitrogen (N^-). Short arrows indicate the formation of diploids of (+) mating type achieved by the formation of protoplasts (FP) followed by somatic hybridization (SH). Hatched nucleus and chloroplast indicate the transmission of traits determined by the nucleus.

encapsulated, and requires no further illumination during this period. Induction of meiotic division is accomplished after at least 5 days in the resting state by the transfer of the culture onto a fresh medium followed by illumination. Meiosis continues for one day yielding 4, 8, or 16 haploid zoospores (vegetative cells). The loss of zygotic meiosis plus protoplast fusion in the presence of polyethylene glycol, which was achieved by Matagne *et al.* [51] using *Chlamydomonas* mutants devoid of a cell wall, results in diploid clones, crosses between which yield fertile 4n zygotes (Fig. 2.1). Triploids are unstable in chlamydomonads [36].

2.4. INTRASPECIES DIVERSITY - MUTANTS

Since the wild type strains of *C. reinhardii* are facultative heterotrophs, one may choose conditions for selecting and cultivating various mutants. Such mutants include nonphotosynthetic forms, some of which are sensitive to light and die under such conditions, and others which are autotrophic forms and obligate phototrophs. Observations of motility and phototaxis type have identified mutants which have paralyzed, elongated, or truncated flagella as well as forms with negative phototaxis. Nuclear and extranuclear mutations resistant to antibiotics and inhibitors have been found as well. Mapping with the use of tetrad analysis demonstrates the presence of 16 linkage groups of mutations inherited according to Mendelian principles [34, 44]; a group of mutations inherited uniparentally has also been described [21, 58, 59] as well as the group of mutations "yellow" which is defined by the trait "yellow in the dark, green under illumination" [60]. Maps have been published by Sager [61] and Harris [28].

In the Peterhof genetic collection of green algae [35, 40] in the Biological Research Institute a collection of mutants isolated from the strains 137C (+) and (–) is being maintained. These include forms with centromeric markers for 16 linkage groups, which include various autotrophies, acetate dependence, and impaired mobility; another group includes extrachromosomal markers which are resistant to antibiotics [21]. Yet another group consists of strains with markers whose localization is still unknown; these impair biogenesis of chloroplast membranes (pigment–proteolipid components of photosystem I, photosystem II, and the light-collecting complex). These mutants have been described in a review [34] and a number of papers [1, 13, 27, 32, 37, 38, 69, 70].

The most complete collection of mutants derived from the strain 137C is now being maintained at Duke University in the Chlamydomonas Genetics Center. (Information can be obtained from Dr. Elizabeth Harris, Senior Research Associate, Department of Botany, Chlamydomonas Genetics Center, Duke University, Durham, NC 27706, USA.)

2.5. METHODS USED TO GROW CELLS AND MODIFY THEIR DIFFERENTIATION

2.5.1. Media

The L_2 media of Levine [42, 68], HSC of Sueoka [66], MC [66], and TMB of Surzycki [72] are used for the phototrophic growth of vegetative cells. The solution of trace elements prepared according to Hatner [67] is used for all these

TABLE 2.1. Composition of Media for Phototrophic Cultivation of *Chlamydomonas*

Stock solutions and their components	Extent of stock dilution and final salt concentrations (g/liter)						
	L_2	HSC	MC	MM	TB	TMB	C
Beyerinck solution	1:10	–	–	1:20	1:20	1:20	1:10
NH_4Cl (NH_4NO_3 in L_2 and C)	0.30	0.50	0.05	0.40	0.40	0.40	0.7
$MgSO_4 \cdot 7H_2O$	0.02	0.02	0.02	0.10	0.10	0.10	0.1
$CaCl_2 \cdot 2H_2O$	0.01	0.01	0.01	0.05	0.05	0.05	0.05
Phosphate buffer	1:10	–	–	1:20	1:100	1:1000	1:10
K_2HPO_4	0.72	1.44	0.72	0.72	1.43	0.93	0.72
KH_2PO_4	0.36	0.72	0.36	0.36	0.73	0.63	0.36
Tris-buffer (1:100)							
Tris-base	–	–	–	–	2.42	2.42	–
HCl, ml/liter	–	–	–	–	1.44	1.5	–
$MgSO_4$ solution (1 M), ml/liter	–	–	–	1.0	–	1.0	–
Trace-element solution, ml/liter	1.0	1.0	1.0	1.0	1.0	1.0	1.0
pH before sterilization	6.8	6.8	6.8	6.7-6.8	7.0	7.0	6.8+ 2 g Na acetate

media (Table 2.1). The stock solution of trace elements should be prepared at least two weeks before use, and the solution may then be kept for more than one year. The following quantities of salts are dissolved sequentially in 550 ml of water:

$ZnSO_4 \cdot 7H_2O$	22.0
H_3BO_3	11.4
$MnCl_2 \cdot 4H_2O$	5.1
$FeSO_4 \cdot 7H_2O$	5.0
$CoCl_2 \cdot 6H_2O$	1.6
$CuSO_4 \cdot 5H_2O$	1.6
$(NH_4)_6Mo_7O_{24} \cdot 4H_2O$	1.1

50 grams of Na_2EDTA is then dissolved in 250 ml water by heating. The first solution is heated to 100°C, mixed with the EDTA solution, and brought to the boil. The pH of the resulting solution is adjusted to 6.5-6.8 by titration at 70-90°C adding no more than 100 ml of 20% KOH solution. The pH-meter should be calibrated using a standard buffer at 75°C. The volume is adjusted to 1 liter. After 2 weeks of storage in a loosely corked flask the green solution turns purple. After removing the precipitate, the solution should be stored in a refrigerator.

Of the media described in Table 2.1, those with high salt concentration (HSC), magnesium excess (MM) and tris-buffer (TB and TMB) have been proposed for mass synchronous cultures. Higher concentrations of nitrogen and buffer stabilize the conditions in liquid media. To prepare solid media (agar content 1.5%) for maintaining wild-type strains, an MC solution is used [41]. Adding organic substances, such as sodium acetate (up to 2 g/liter), dry yeast extract (2 g/liter), peptone (5-10 g/liter) or yeast autolysate (5-10 ml/liter) allows all the above-mentioned media to be used for the heterotrophic and mixotrophic growing of vegetative cells, not only of wild type strains but also of mutant ones.

Treatment of biologically active substances has been repeatedly used on the studies on organelle biosynthesis in *Chlamydomonas* cells [4, 56, 73].

Gametogenesis is induced by transferring cells to completely nitrogen-free media, since even traces of nitrogen in water may interfere with the transition of cells to gametogenesis. The well-known results of the Moewus experiments [52] on gametogenesis regulation using chemical agents could not be reproduced (see [14]); consequentially these findings are treated with reservation. In the simplest experimental technique, cells are transferred from solid media into either water or a mineral medium devoid of nitrogen. Cells grown in liquid cultures should first be removed from the medium by centrifugation. Mating gametes may be obtained by simply mixing their suspensions in a nitrogen-free medium. Transferring gametes onto a medium containing nitrogen induces their dedifferentiation and transformation into vegetative cells [33]. Growing in liquid media contributes towards flagella formation. Medium C enables 1.5-2 grams of cell wet weight per liter to be obtained under mixotrophic conditions when the salt concentration was doubled (2 × C) [5].

2.5.2. Culture Conditions

Growth is optimal at 25-28°C in darkness or at 21°C under illumination. Keeping a collection in test tubes containing a solidified acetate yeast autolysate medium requires moderate illumination (1000 lx). In screw-cap test tubes the colonies retain viability on agar slopes for many months; usually tubes containing cultures resulting from 2 or 3 consecutive transfers are stored. If test tubes with cotton wool plugs are used, it is recommended that the medium be diluted with the same amount of water, and non-sloped agar should be prick-inoculated, while the height of the medium slab should be doubled. Under these conditions any drying up of the medium is less harmful for *Chlamydomonas* colonies. Cultures of most *Chlamydomonas* strains may be grown in the dark and stored at 15-20°C.

Cultures actively grown for experimentation are maintained on Petri dishes at 2000 lx, transfers being made twice weekly. Experiments are usually performed in a liquid medium at either an illumination of 2000-4000 lx or in darkness. Cultures with densities around 10^4-10^5 cells per 1 ml growth, in small cultures (5-50 ml with a liquid layer no thicker than 2-4 cm), do not need to be stirred either in darkness or under illumination. Under such conditions, owing to motility and chemotaxis, the algae migrate to those regions of the vessel where the conditions are favorable for them to multiply. At higher densities (10^5-10^6 cells per 1 ml) additional aeration is needed, which is achieved by either shaking or bubbling through air.

Culture, particularly under heterotrophic and mixotrophic conditions, requires aseptic conditions, that is, conventional microbiological routines for sterilization media, glassware and pipettes, as well as sterile conditions for transferring axenic cultures [50]. Cleansing the cultures of contaminants is achieved by cloning, using solidified media enriched with organic components, in which colonies of contaminating microorganisms may be seen.

2.5.3. Synchronization Methods

General principles for synchronizing the growth of green flagellates are given in reviews [30, 53]. For *Chlamydomonas* the important factors include strain homogeneity, the stabilization of phototrophic conditions, and periodic light/darkness change. Two regimes have been recommended: 12 h of illumination and 12 h of darkness [72] or 12 h of illumination and 4 h in the dark. Synchronization starts around noon by inoculating cells from the stationery culture liquid to a cell density of $5 \cdot 10^4$ cells/ml. Illumination should be 6000-20,000 lx at the level of the vessel from all directions. In an established synchronous culture the divisions take place in the dark, and maximal culture densities prior to dilution are usually $(5\text{-}24) \cdot 10^6$ cells/ml. Dilution is performed at the end of the dark period by a factor of 4, 8, or 16, according to the increase in the number of zoospores.

Using this type of synchronous culture, the sequence of the main stages of differentiation during the vegetative cycle [29], gametogenesis [31], and during zygotic and meiotic stages [11] has been studied. This technique of synchronization is limited since it can only be applied to phototrophic cultures. The phenomenon of gamete dedifferentiation [33] opens up the opportunity for developing a universal method of synchronization, since the transformation of gametes into vegetative cells after nitrogen has been added does not require light and may be used for obligate heterotrophs, light-sensitive cultures, and other mutants.

2.6. USE OF MUTANTS FOR ANALYZING CELL DIFFERENTIATION

Detecting genotype factors in wild-type cells using mutations and their combinations allows biogenesis of the cell structures to be analyzed. An example is provided by analysis of the differentiation and function of flagella in *Chlamydomonas* [54]. The combination (achieved by mating) of cells with different flagellum structure leads to normal motility being restored as the result of gene interaction [65]. Normalizing biogenesis becomes manifest during the regeneration of flagella in prozygotes (4 flagella) or zygotes (2 flagella). Such an analysis cannot be performed for strains devoid of flagella, since mating is impossible in this case. A detailed analysis is given in the review [39].

Mutants with an impaired photosynthetic function have been successfully used for the genetic analysis of the mechanism of photosynthesis and chloroplast differentiation [44]. Arginine-requiring mutants have been used for analyzing arginine biosynthetic pathways [74], and mutants with a defective cell wall have been used for studying the biogenesis of its various components [16]. The opportunities offered by the genetic collection for analyzing cell differentiation are far from being exhausted [7, 8].

There are quite diverse methods for inducing mutants [41, 48] and for detecting various types of mutants. Similar techniques are being used in hybridological analysis. For revealing mutants with a lower content of chloroplast ribosomal RNA, fluorescence after acridine orange staining may be used [23]. For acetate-dependent mutants, the main technique used for observing their phenotype involves inoculation onto both acetate-containing and acetate-free media. Concomitantly, differences in chlorophyll fluorescence and relative rate of $^{14}CO_2$ incorporation [41] may be employed. Pigmentation differences can be seen with the naked eye and may be objectively recorded by the absorption spectra of individual cells [6] or by absorption spectra of the suspension [37]. Death (shown by bleaching) on overexposure to light and other changes of growth characteristics (e.g., pigmentation) are revealed when cultures grown under corresponding conditions are compared. Methods for detecting autotrophs are identical to those commonly used in microbiology. Motility types are identified in liquid cultures. The trait "paralyzed flagella" may be revealed on a solid mineral medium by the cells' inability to migrate to a drop of water placed near the colony. Resistance to toxic chemicals is best revealed using solidified media, since it is difficult to select proper inhibitory concentrations in liquid media. In a suspension these concentrations are usually an order of magnitude lower than on solidified media, while the concentration ratio enabling mutants to be identified may be inconstant.

The method of replicas is used for mass tests [62]. Problems may be caused by the tendency of the *Chlamydomonas* cells to form dense colonies which either remain completely on the master plate or are completely transferred onto a velvet printer, and then only onto one of the printed plates. A certain softening of cell clusters may be achieved if colonies are transferred onto a plate with a nitrogen-free medium.

If the studied trait is lost by the strain, then the initial forms have to be found among the clones of the original culture stored for various periods of time. In the case of reversion as a result of suppression, one may attempt to obtain the initial mutant type by crossing with the wild type.

Crossing methods have been described in earlier works [43, 68]. Tetrad analysis assumes that the zoospores of zygote progeny are isolated. The isolation procedure is performed using a Petri dish under a dissecting microscope and a blunted glass needle. For this purpose N. N. Alexandrova and Chunaev *et al.* [13] have used a micromanipulator in combination with a microscope; this enabled errors owing to low magnification during micromanipulation to be avoided and contributed to better results. Similar techniques of zoospore isolation are used if the inheritance of a given trait is followed in a series of vegetative divisions (cell pedigree analysis).

In a number of cases inheritance analysis may be achieved without isolating zoospores in tetrads (octads, etc.). By plating zygotes onto a fresh medium after killing the vegetative cells with chloroform, one may obtain progeny from each zygote and the types of descendants may be identified on the basis of visible markers and by using replicas.

Diploid vegetative clones which are not obtained very frequently allow the complementation of various mutations to be examined [13, 10, 49, 65, 74, 36]. Details of other experiments may be found in the monograph of Lewin [47], half of which deals with the *Chlamydomonas* strain 137C.

The study of mutants with defective biogenesis of flagella in a similar species *Chlamydomonas moewusii* (= *C. eugametos*) revealed similar trends in the hereditary variation of forms belonging to one genus [46]. These species have been found to be similar both in the spectrum of mutations for 16 genes studied and in their localization: 11 loci were found to belong to known linkage groups. The peculiarities of flagellum regeneration [39, 54] in *Chlamydomonas reinhardii* 137C are identical to those found in *C. eugametos* [2]. Removing the flagella without damaging the whole cell is followed by regeneration and unscheduled tubulin synthesis [75]. Normally tubulin synthesis in *C. reinhardii* 137C follows a circadian rhythm but precedes the nuclear division; in other words it coincides with the reproduction of cytoskeleton. In Alexandrov's experiments [2] a delay in flagellum development after a heat shock was observed. The duration of this delay showed a correlation with temperature during the shock treatment (1 h at 36°C, 15 h at 42.3°C). This delay may be explained by postulating dedifferentiation of cell organelles, which is correspondingly greater under more drastic conditions. Such cell "rejuvenation" allows the repair of damaged structures in the natural course of development.

The simplest model of "flagellum regeneration" may be used for analyzing the nature of cell ontogenesis in protists. The combined efforts of geneticists, morphologists, and embryologists in studying *Chlamydomonas* are perhaps related to the fact that the cellular cycle of this organism shows a certain similarity to the differentiation of macro- and microgametes in multicellular organisms; on the other hand, owing to palintomy, the cycle also resembles cleavage of the fertilized egg.

REFERENCES

1. L. S. Abros'kina, L. M. Vorob'eva, and K. V. Kvitko, "Chlorophyll luminescence in mutants of *Chlorella* and *Chlamydomonas*," *Fiziol. Rast.* **26**, 383 (1979).
2. V. Ya. Alexandrov, "Stimulation of flagellum recovery in *Chlamydomonas eugametos* after heat injury," *Arch. Protistenk.* **124**, 345 (1981).
3. M. Baslerová and J. Dvoráková, *Algarum, Hepaticarum, Muscorumque in Culturis Collectio.*, Nakladat. Českosl. Akad. ved., Praha (1962).
4. W. Behn and C. G. Arnold, "Localization of extranuclear genes by investigations of the ultrastructure in *Chlamydomonas reinhardii*," *Arch. Microbiol.* **92**, 83 (1973).
5. E. P. Bers, "Effect of cultivation conditions on productivity and certain physiological traits of *Chlamydomonas reinhardii* cells," in: *Experimental Algology* [in Russian], Peterhof Biological Institute, Leningrad (1977).
6. P. Kh. Boyadzhiev, A. F. Smirnov, and K. V. Kvitko, "Microspectrophotometry of *Chlamydomonas* pigment mutants," in: *Control of Biosynthesis in Microorganisms* [in Russian], Nauka, Krasnoyarsk (1973).
7. V. G. Bruce, "Mutants of the biological clock in *Chlamydomonas reinhardii*," *Genetics* **70**, 537–548 (1972).
8. V. G. Bruce and N. C. Bruce, "Circadian clock-controlled growth cycle in *Chlamydomonas reinhardii*," in: *International Cell Biology 1980–1981*, H. G. Schweiger (ed.), Verlag B., New York (1981).
9. T. Cavalier-Smith, "Electron microscopy of zygospore formation in *Chlamydomonas reinhardii*," *Protoplasma* **87**, 297 (1976).
10. V. I. Chemerilova and K. V. Kvitko, "Study of mutations modifying pigmentation in *Chlamydomonas reinhardii* strains at various ploidity," *Genetika* **11**, 44–49 (1976).
11. K. S. Chiang, "Replication, transmission, and recombination of cytoplasmic DNA in *Chlamydomonas reinhardii*," in: *Autonomy and Biogenesis of Mitochondria and Chloroplasts*, North Holland Publishing Co., Amsterdam (1971).

12. A. S. Chunaev, "Genetics of photosynthesis in *Chlamydomonas reinhardii*," *Usp. Sovrem. Genet.* **12**, 63–92 (1984).
13. A. S. Chunaev, V. G. Ladygin, T. A. Gavrilenko, L. P. Krela, and G. A. Kornyushenko, "Inheritance of the trait 'absence of chlorophyll b' and the variability of the light-collecting complex in the meiotic progeny C-48 of *Chlamydomonas reinhardii*," *Genetika* **17**, 2013–2024 (1981).
14. A. W. Coleman, "Sexuality," in: *Physiology and Biochemistry of Algae*, Academic Press, New York–London (1962).
15. Culture Collection of Algae and Protozoa, List of Strains, Cambridge University, Cambridge (1976).
16. D. R. Davies and K. Roberts, "Genetics of cell wall synthesis in *Chlamydomonas reinhardii*," in: *The Genetics of Algae*, R. A. Lewin (ed.), Oxford (1976).
17. V. A. Dogel, *Invertebrate Zoology* [in Russian], Vysshaya Shkola, Moscow (1981).
18. H. Ettl, "*Chlamydomonas* als geeigneter Modellorganismus für vergleichende cytomorphologische Untersuchungen," *Algol. Stud.* **5**, 259 (1971).
19. H. Ettl, "Die Gattung *Chlamydomonas* Ehrenberg," *Beih. Nova Hedwigia* **49**, 1–22 (1976).
20. J. Friedman, A. L. Colwin, and L. H. Colwin, "Fine structural aspects of fertilization in *Chlamydomonas reinhardii*," *J. Cell Sci.* **3**, 115–128 (1968).
21. N. Gillham, "The uniparental inheritance in *Chlamydomonas*," *Am. Nat.* **103**, 355–387 (1969).
22. U. W. Goodenough, "Sexual microbiology: mating reactions of *Chlamydomonas reinhardii, Tetrahymena termophila* and *Saccharomyces cerevisiae*," in: *Eukaryotic Microbial Cell*, G. W. Gooday et al. (eds.), Cambridge University Press, Cambridge (1980).
23. U. W. Goodenough and R. P. Levine, "Chloroplast structure and function in ac-20, a mutant strain of *Chlamydomonas reinhardii*. III. Chloroplast ribosomes and membrane organization," *J. Cell Biol.* **44**, 547 (1970).
24. C. S. Gowans, "Genetics of *Chlamydomonas moewusii* and *Chlamydomonas eugametes*," in: *The Genetics of Algae*, R. A. Lewin (ed.), Oxford (1976).
25. B. V. Gromov, "The collection of algae cultures of Leningrad University Biological Institute," *Tr. Peterhof. Biol. Inst. Leningr. Gos. Univ.* **19**, 125 (1965).
26. B. V. Gromov and N. N. Titova, "Culture collection of algae at the Microbiology Laboratory, Biological Institute, Leningrad State University," in: *Cultivation of Collection Strains of Algae* [in Russian], Leningrad University Press, Leningrad (1983).
27. I. Gyurjian, G. Erdös, and A. H. Nagy, "Polypeptide composition of thylacoid membrane in pigment-deficient mutants of *Chlamydomonas reinhardii*," in: European Meeting on Molecular Genetics and Biology of Unicellular Algae, Liege (1980).
28. E. H. Harris, "Nuclear gene loci of *Chlamydomonas reinhardii*," in: *Genetic Maps*, S. O'Brien (ed.), Vol. 5, National Cancer Inst. (1982).
29. S. H. Howell, W. J. Blaschko, and C. W. Drew, "Inhibitor effects during the cell cycle in *Chlamydomonas reinhardii*. Determination of transition points in asynchronous cultures," *J. Cell Biol.* **67**, 126 (1975).
30. T. W. James, "Induced division synchrony in the flagellates," in: *Synchrony in Cell Division and Growth*, Interscience, New York–London (1964).

31. R. F. Jones, "Physiology and biochemical aspects of growth and gametogenesis in *Chlamydomonas reinhardii*," *Ann. N.Y. Acad. Sci.* **175**, 648–659 (1970).
32. N. V. Karapetyan, M. G. Rakhimberdieva, N. G. Bukhov, and I. Gyurjan, "Characterization of photosystems of *Chlamydomonas reinhardii* mutants differing in their fluorescence yield," *Photosynthetica* **14**, 48–54 (1980).
33. J. R. Kates, K. S. Chiang, and R. F. Jones, "Studies on DNA replication during synchronized vegetative growth and gametic differentiation in *Chlamydomonas reinhardii*," *Exp. Cell Res.* **49**, 121–135 (1968).
34. K. B. Kvitko, "Biology and genetics of *Chlamydomonas reinhardii* 137C," in: *Experimental Algology* [in Russian], Leningrad (1977).
35. K. V. Kvitko and T. N. Borshevskaya, "Peterhof collection of pigmentation mutants of green algae," in: *Methods Used to Study the Structure of Photosynthetic Apparatus* [in Russian], Pushchino-on-Oka (1972).
36. K. V. Kvitko and V. I. Chemerilova, "Adaptive significance of polyploidy in algae, chlamydomonads taken as an example," in: *Evolutionary Genetics* [in Russian], Leningrad University Press, Leningrad (1982).
37. K. V. Kvitko, P. Kh. Boyadzhiev, A. S. Chunaev, B. T. Mukhamadiev, A. A. Baranov, and V. S. Saakov, "Genotypic and phenotypic variation of pigment–lipoprotein complex of the green algae mutants. 2. Study of the absorption spectra of mutants with altered response to illumination in *Chlamydomonas reinhardii* 137C," in: *Experimental Algology* [in Russian], Leningrad (1977).
38. K. V. Kvitko, V. V. Tugarinov, Ph. T. Ho, A. S. Chunaev, E. E. Temper, and B. T. Mukhamediev, "Mutational analysis as a method of studying genotype structure of green algae," in: *Genetic Aspects of Photosynthesis*, Yu. S. Nasyrov (ed.), Junk (1975).
39. K. V. Kvitko, Vl. Vl. Matveev, and A. C. Chunaev, "Motility and behavior of *Chlamydomonas* and their changes induced by mutations," in: *Motility and Behavior of Unicellular Animals* [in Russian], Nauka, Leningrad (1978).
40. K. V. Kvitko, T. N. Borshchevskaya, A. S. Chunaev, and V. V. Tugarinov, "Peterhof's genetical collection of strains of *Chlorella, Scenedesmus, Chlamydomonas*," in: *Cultivation of Collection Strains of Algae* [in Russian], Leningrad University Press, Leningrad (1983).
41. R. P. Levine, "Preparation and properties of mutant strains of *Chlamydomonas reinhardii*," in: *Methods in Enzymology*, Vol. 23, Part A. Academic Press, New York–London (1971).
42. R. P. Levine and W. T. Ebersold, "Gene recombination in *Chlamydomonas reinhardii*," *Cold Spring Harbor Symp. Quant. Biol.* **23**, 395–410 (1958).
43. R. P. Levine and W. T. Ebersold, "Genetics and cytology of *Chlamydomonas*," *Annu. Rev. Microbiol.* **14**, 197–216 (1960).
44. R. P. Levine and U. Goodenough, "The genetics of photosynthesis and of the chloroplast in *Chlamydomonas reinhardii*," *Annu. Rev. Genet.* **4**, 397–407 (1970).
45. R. A. Lewin, "The genetics of *Chlamydomonas moewusii* Gerloff," *J. Genet.* **51**, 543 (1953).
46. R. A. Lewin, "Genetic control of flagellar activity in *Chlamydomonas moewusii* (Chlorophyta, Volvocales)," *Phycologia* **13**, 45–55 (1974).
47. R. A. Lewin (ed.), *The Genetics of Algae*. Bot. Monographs, II, Oxford (1976).

48. R. Loppes, "Ethyl methane sulfonate: an effective mutagen in *Chlamydomonas reinhardii*," *Mol. Gen. Genet.* **102**, 299--231 (1968).
49. R. Loppes, R. Motagne, and P. J. Strijkert, "Complementation of the Arg 7 locus in *Chlamydomonas reinhardii*," *Heredity* **28**, 239–251 (1972).
50. Y. Maynell and E. Maynell, *Experimental Microbiology* [Russian translation], Mir, Moscow (1967).
51. R. F. Matagne, R. Deltour, and L. Ledoux, "Somatic fusion between cell wall mutants of *Chlamydomonas reinhardii*," *Nature (London)* **278**, 344–346 (1979).
52. F. Moewus, "Carotinoid derivative als geschlechtsbestimmende Stoffe von Algen," *Biol. Zbl.* **60**, 143–166 (1940).
53. G. M. Padilla and J. R. Cook, "The development of techniques for synchronizing flagellates," in: *Synchrony in Cell Division and Growth*, Interscience, New York–London (1972).
54. J. Randell and D. Starling, "Genetic determinants of flagellum phenotype in *Chlamydomonas reinhardii*," in: *The Genetics of the Spermatozoon*," Bogtrykkereit Vorum, Edinburgh–N.T.–Copenhagen.
55. R. A. Lewin (ed.), reprinted in *The Genetics of Algae*, Oxford (1976).
56. R. W. Rubin and P. Filner, "Adenosine-3',5'-cyclic monophosphate in *Chlamydomonas reinhardii*: influence on flagellar function and regeneration," *J. Cell Biol.* **56**, 628–635 (1973).
57. R. S. Ryan, D. Grant, Chiang Kwen-Sheng, and H. Swift, "Isolation of mitochondria and characterization of the mitochondrial DNA of *Chlamydomonas reinhardii*," *J. Cell Biol.* **59**, Pt. 2, 297a (1973a).
58. R. Sager, "Mendelian and non-Mendelian inheritance of streptomycin resistance in *Chlamydomonas*," *Proc. Natl. Acad. Sci. USA* **40**, 356–370 (1954).
59. R. Sager and Z. Ramanis, "A genetic map of non-Mendelian genes in *Chlamydomonas*," *Proc. Natl. Acad. Sci. USA* **65**, 593–600 (1970).
60. R. Sager and J. Tsubo, "Mutagenic effects of streptomycin in *Chlamydomonas*," *Arch. Mikrobiol.* **42**, 159–175 (1962).
61. R. Sager, *Cytoplasmic Genes and Organelles,* Academic Press, New York (1972).
62. I. A. Sakharov and K. V. Kvitko, *Genetics of Microorganisms* [in Russian], Leningr. Gos. Univ., Leningrad (1967).
63. E. T. Schmeisser, D. M. Baumgartel, and S. H. Howell, "Gametic differentiation in *Chlamydomonas reinhardii*: cell cycle dependency and rates in attainment of mating competency," *Dev. Biol.* **31**, 31–37 (1973).
64. F. Schötz, H. Bathelt, C. G. Arnold, and O. Schimmer, "Die Architectur und Organization der *Chlamydomonas*-celle. Ergebnisse der Elektronenmikroskopie von Serienschnitten und der daraus resultierenden dreidimensionalen Reconstruktion," *Protoplasma* **75**, 229–254 (1972).
65. D. Starling, "Complementation tests in closely linked flagellar genes in *Chlamydomonas reinhardii, Genet. Res.* **14**, 343–347 (1969).
66. R. C. Starr, "Algae cultures – sources and methods of cultivation," in: *Methods in Enzymology*, Vol. 23, Part A, Academic Press, New York–London (1971).
67. R. C. Starr, "The culture collection of algae at University of Texas at Austin," *J. Phycol.* **14**, Suppl., 47–100 (1978).
68. A. V. Stolbova, "Genetic analysis of pigment mutations of *Chlamydomonas reinhardii*," *Genetika* **7**, 90–94 (1971).

69. A. V. Stolbova, "Genetic analysis of pigment mutations in *Chlamydomonas reinhardii*," *Genetika* **8**, 123–128 (1972).
70. A. V. Stolbova, "Genetic analysis of light-sensitive mutants of *Chlamydomonas reinhardii*," in: *Genetic Aspects of Photosynthesis*, Yu. S. Nasyrov and Z. Sesták (eds.), Junk (1975).
71. R. Storms and P. J. Hastings, "A fine structure analysis of meiotic pairing in *Chlamydomonas reinhardii*," *Exp. Cell Res.* **104**, 39–46 (1977).
72. S. Surzycki, "Synchronously grown cultures of *Chlamydomonas reinhardii*," in: *Methods in Enzymology*, Vol. 23, Part A, Academic Press, New York–London (1971).
73. S. Surzycki, U. W. Goodenough, R. P. Levine, and J. J. Armstrong, "Nuclear and chloroplast control of chloroplast structure and function in *Chlamydomonas reinhardii*," in: *Control of Organelle Development, 24th Symposium in Experimental Biology*, Cambridge University Press, Cambridge (1970).
74. J. S. Sussenbach and P. J. Strikert, "Arginine metabolism in *Chlamydomonas reinhardii*," *Eur. J. Biochem.* **8**, 403–412 (1969).
75. D. P. Weeks and P. S. Collis, "Induction and synthesis of tubulin during the cell cycle and life cycle of *Chlamydomonas reinhardii*," *Dev. Biol.* **69**, 400–407 (1979).

Chapter 3

Hydra AND HYDROID POLYPS

L. V. Belousov

3.1. INTRODUCTION

Since the times of A. Trembley (first half of the 18th Century) (see [7]) freshwater hydra have been used in laboratories as a suitable species for studying the mechanisms of regeneration and asexual reproduction (for a review of earlier papers, see [6]). Since the middle of the 20th Century there have been many detailed investigations on the mechanisms of cell differentiation and the control of morphogenesis. It is impossible to review so much work in his short chapter (for references see [21]). Investigations make use of the adult, intact, regenerating, and budding hydra as well as mutant strains, their chimaeras, and cell cultures.

The main centers dealing with hydra biology include the Department of Developmental and Cell Biology, Irvine University, Irvine, California 92717 (USA) and the Zoologisches Institut der Universitat Zurich-Irchel, Winterthurerstrasse 190, CH-8057, Zurich, Switzerland.

This contribution does not pretend to provide a comprehensive description of *Hydra* as an experimental animal used in developmental biology. Specifically, we do not present any information regarding the mutant strains of hydra. The following information should be regarded as introductory, necessary for those investigators using *Hydra* for the first time. More detailed information can be found in the references at the end of the chapter or obtained from the scientific centers mentioned above.

Sexual reproduction and, in particular, embryogenesis of the hydra are less accessible to research. Nevertheless, *Hydra* is a suitable species for studying the early stages of gametogenesis.

Along with freshwater hydra, various marine coelenterates, particularly hydroid polyps, are being increasingly used in research. They are used in studies dealing with the mechanisms of cell differentiation, regeneration, and morphogenesis. Furthermore, they may be used as test systems for screening various physiologically active substances [9] or as indicator organisms of sea water pollution [17].

3.2. SPECIES USED

Freshwater hydra are usually solitary but sometimes form temporary colonies. They belong to the genus *Hydra* (family Hydridae, order Hydrida, class Hydrozoa, type Coelenterata). Kanayev and Naumov described the following species of the genus *Hydra* in the reservoirs of the Soviet Union: *H. oligactis, H. brauezi, H. baicalensis, H. vulgaris, H. attenuata, H. circumcincta, H. oxycnida, H. oxycnidoides*. Among other species *H. littoralis* occurring in America should be mentioned.

Hydra (Pelmatohydra) *oligactis* and *H. attenuata* are very widely used along with *H. littoralis* which is used in laboratories in the United States and Britain.

The most characteristic external feature of *H. oligactis* is a clear division of the body into trunk and stem parts, while in *H. attenuata* and *H. littoralis* there is no distinct separation of the body on the two regions.

Another feature useful for species diagnosis is the order of tentacle origin in the growing buds. In *H. hymanae* buds all the five tentacles arise quite simultaneously; in *H. oligactis* two lateral tentacles arise early and assume a slender shape before the other tentacles appear; in *H. attenuata* the tentacles arise in a slightly staggered sequence [10].

Of marine hydroid polyps the various *Obelia* species (= Gonothyrea, Laomedea) should be mentioned: *O. loveni, L. flexuosa, O. longissima, O. geniculata* (family Campanulariidae, order Leptolida, class Hydrozoa), and *Dynamena* (= Sertularia) *pumila* (family Sertulariidae, same order and class) [14]. All these species are colonial.

3.3. DISTRIBUTION AND COLLECTION

Hydra can be found during the summer in any sufficiently clean body of still or slow-moving water. Weeds and other objects (dead branches, stones, etc.) are collected from the water where the hydra are either known or expected to occur and placed into glass jars filled with water. After a while the hydra move to the illuminated side of the vessel or attach themselves to various objects placed into the jar (plants, branches, etc.) and spread out, thus becoming clearly visible. Alternatively the hydra can be taken directly from the water. Individual plants may be picked out of the water from the shore, stones, bridge, raft, or boat, or examined from close range in the upper water layer. Parts of plants covered with hydra are taken and placed into small jars of water and transported to the laboratory. Minimal experience in recognizing hydra in their natural environment is usually sufficient to make locating them a simple matter.

It should be remembered that hydra do not, by any means, live in every body of water. Generally it is recommended that cultures grown in the laboratory for many years be used, especially as they are better adapted to artificial conditions.

Asexual budding of hydra in nature proceeds throughout the summer. Sexual maturation and the formation of gonads in nature takes place predominantly in autumn, although in *H. vulgaris* gonads may appear during the summer as well.

The marine hydroid polyps belonging to the species mentioned above occur widely in the littoral areas of most seas in cold and temperate climatic zones. In the Soviet Union the best area to collect them is in the littoral zone of the White Sea, where they live in brown algae, stones, and on piles (*O. longissima*). Vegetative reproduction (budding) under natural conditions is at its most intense in June or early July. In late July the gametes mature, while vegetative reproduction is partially or completely arrested. It is resumed in the autumn; under artificial conditions budding may occur in winter too.

3.4. CULTURE CONDITIONS FOR *Hydra*

If one does not require a *Hydra* culture with a controlled rate of increase, one may follow the recommendations described below [6].

The animals are kept in glass jars varying in size from half a liter to large aquariums, with moderate amounts of *Elodea* which *Hydra* prefers for attachment. Sand on the bottom is not necessary. Tap water stored in the laboratory for at least 24 h or, even better, pond water is used. The water should be clean and fresh; if it does not become foul it can be kept for weeks without change. Using a pipette, the jars should be regularly cleansed of various dead residues. The animals should be kept away from direct sunlight and at room temperature. They should be fed on small crustaceans (1-2 *Daphnia* per *Hydra* per day) or mosquito larvae.

Under these conditions the animals usually appear healthy and budding occurs, although sometimes, for no apparent reason, they go into decline (they stop feeding and do not respond to touch; later their tentacles undergo resorption and the animals die). If this happens the water should be changed and the animals kept at a lower room temperature. They should be given food after they have resumed the capacity to feed.

Decline may be the result of infestation by parasites such as infusorians (e.g., *Tricodina, Cerona*). If this is the case, the culture should be cleansed of these parasites. The infusorians may be removed from the body of *Hydra* with a brush; alternatively this can be done by "rinsing" the animal in water. The parasites can also be removed using a pipette; alternatively the animals can be placed into a medium lethal to the parasites but which only temporarily and reversibly affect hydra, for example lukewarm water or a weak solution of Nile Blue sulfate.

After such manipulations the *Hydra* should be transferred to clean water and checked to see if the parasites have gone. The other enemies of *Hydra*, such as Turbellaria, *Microstomum* or molluscs (*Helix* and others), as well as the larvae of Chaoborinae [6], should not be present in aquariums containing *Hydra* cultures.

3.4.1. Conditions For Obtaining Cultures with a Controlled Reproduction Rate

Loomis and Lenhoff [10, 13] recommend the following conditions for growing *Hydra*. The culture should be set up on flat 15-mm culture dishes containing one of the following culture media (Table 3.1): CN, KNC, or M. CN is similar to that previously described [11] as L-solution; it is useful in experiments such as those involving glutathione, in which K^+ affects the animals negatively. Solution KNC is suitable for all brown *Hydra*, whereas M-solution prepared with either NaC1 or $NaHCO_3$ works for both the green and the brown *Hydra* [10].

Special precautions should be taken against heavy metals in the water for preparing the above solutions. It is recommended [10] to pass the distilled (deionized) water produced by reverse osmosis through a tank of mixed bed resin.

TABLE 3.1. Composition of Stock and Regular Strength Culture Solutions [10]

Culture solution	pH	Components of stock solutions	Dilution (to prepare culture solutions)	Final concentration
CN	7.6	(a) 0.1 M $NaHCO_3$	1 in 1000	10^{-4} M
		(b) 1.0 M $CaCl_2$	1 in 1000	10^{-3} M
KNC	7.6	(a) 0.1 M $NaHCO_3$ 0.1 M KCl	1 in 1000	10^{-3} M
		(b) 1.0 M $CaCl_2$	1 in 1000	10^{-3} M
M	7.8	(a) 1.166 M Tris-HCl buffer	6 in 1000	10^{-3} M
		0.166 M $NaHCO_3$		10^{-3} M
		0.016 M KCl		10^{-4} M
		0.016 M $MgCl_2$		10^{-4} M
		(b) 1.0 M $CaCl_2$	1 in 1000	10^{-3} M

a) Three liters of stock M solution (a) may be made by mixing 1000 ml 0.5 M Tris-HCl buffer pH 7.6-7.8, 500 ml 1.0 M $NaHCO_3$, 50 ml 1.0 M KCl, 50 ml 1.0 M $MgCl_2$, and 1400 ml distilled water.

b) Prepare Tris-HCl buffer from a 1 M solution of Tris adjusted to pH 7.6-7.8 with a mixture of equal volumes of water and concentrated HCl. Dilute the solution to almost 2 liters, check the pH, and readjust it as required; then bring it to a final volume of 2 liters. Refrigerate this stock buffer when it is not being used.

If one plans to use *Hydra* only occasionally and does not need a constant culture, one may detoxify the heavy metals present in tap water with Na_2EDTA (final concentration about 20 mg per liter of culture medium).

In [8] and [16] some other modifications of culture media are recommended (in no case above 10 mOsm).

The *Hydra* are fed using the larvae of the brine shrimp *Artemia*. To obtain a permanent culture of larvae, 1/4 teaspoon of dry *Artemia* eggs (about 0.7 g) should be placed onto the surface of 500 ml of NaC1 solution (3.5 g/liter) in a flat dish. The incubation should be at room temperature. After 48 h the hatched larvae should be collected using a net with fine cells washed in the L-medium and these larvae added to the *Hydra* cultures. Otto and Campbell [15] recommend that between 12 and 15 larvae be added daily to 60 ml plastic Petri dishes containing *Hydra* cultures.

Twenty four hours after feeding, the L-medium should be decanted (while the *Hydra* are still attached to the walls) and changed with a fresh portion.

Under these conditions the exponential increase in the number of *H. littoralis* and *H. vulgaris* has been observed over a number of days. If rapid growth is not necessary the *Hydra* may be left in a medium without feeding for several weeks at room temperature or for several months in a refrigerator. Within 48 h of returning to room temperature and beginning to feed, the *Hydra* start budding again. Sexual differentiation under these conditions is described below.

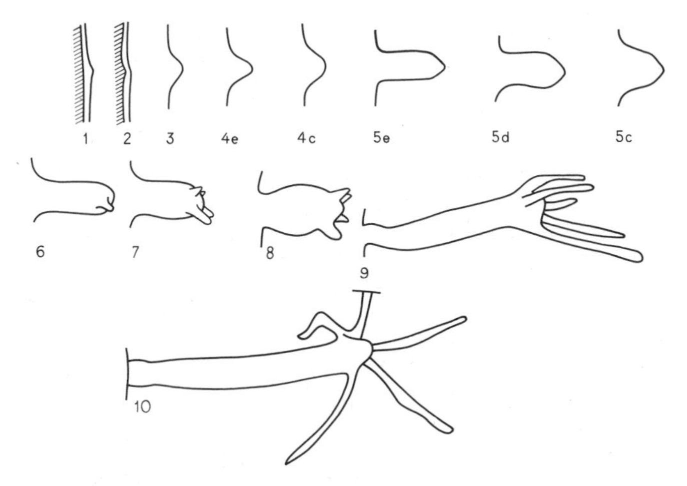

Fig. 3.1. Stages of bud development in *Hydra attenuata*. Buds are shown in profile with stage numbers below the sketches. Stages 1 and 2 were not followed. In some stages, buds are in various states of contraction: c) contracted; e) elongated; d) distended with fluid. Crosshatching in stages 1 and 2 represents endoderm. All sketches are drawn to the same scale [15].

For culturing large numbers of *Hydra*, Lenhoff recommends two methods: a simple tray method requiring no special equipment, and a special vertical-plates method [10].

3.5. DEVELOPMENT OF BUDS IN FRESHWATER *Hydra*

The fed *Hydra* undergo intensive budding. Buds are laid down as tubercules in the lower part of the trunk. In *H. attenuata* the first buds are formed opposite each other, while subsequent buds are, as a rule, located spirally; the angle between the consecutive buds is about 120°.

The consecutive stages of bud development according to Otto and Campbell [15] are described below (Fig. 3.1).

Stage 1: Condensation. The ectoderm thickens in the region of the future bud; this can be seen best when the animal contracts.

Stage 2: Peak. A small mass of endoderm penetrates the thickened ectoderm.

Stage 3: Mound. Both the ectoderm and endoderm of the bud protrude from the wall of the parent in a simple low mound. The endoderm intrusion has become blunter.

Stage 4: Shield. The profile of the contracted bud (Fig. 3.1, 4c), is shaped like a shield. A peak occurs at the distal end of the bud. The bud has elongated slightly since Stage 3. The length of the bud from top to base is less than the diameter of the parent. The extended bud (Fig. 3.1, 4e) does not have a shield shape.

Stage 5: Bulb. The bud has a bulbous shape when it is distended with fluid (Fig. 3.1, 5d), with the proximal part being narrower than the distal part, where the tissue appears to be stretched. When the bud contracts (Fig. 3.1, 5c), it has a similar shape to that of Stage 4. However, it is longer than in Stage 4; its length is at least equal to the parent's diameter.

Stage 6: Rudiment. Tentacle rudiments which are shorter in length than their breadth mark this stage. In *H. attenuata* there are generally two initial rudiments opposite one another on the lower surface of the bud.

At this and the preceding developmental stage a mouth appears in the bud for the first time and a feeding response to glutathione can be noted.

Stage 7: Tentacle. The initial tentacle rudiments are longer than their breadth. Three additional tentacle rudiments are usually present, two on the upper side of the bud and one on the lower side between the two initial rudiments. The base of the bud is slightly constricted when the bud is extended.

Stage 8: Constriction. The proximal region of the bud immediately adjacent to the parent has constricted and its diameter is less than two-thirds that of the bud gastric region.

Stage 9: Basal disc. The ectodermal cells in the basal region of the bud are translucent as a result of basal disc differentiation. This basal region is greatly constricted, but the tissues are still connected to those of the parent.

Stage 10: Independent. The bud remains on the parent, but no tissue connection between the bud and its parent can be seen. Basal disc differentiation appears complete.

The same authors [15] give the following data on the timing of *H. attenuata* developmental stages at 20-23°C:

Stages	Time, h
1-2	Cannot be accurately determined
2-3	3
3-4	6
4-5	6
5-6	10
6-7	8
7-8	9
8-9	20
9-10	8
10 (prior to the separation of the bud from its mother)	Varying duration up to several days

3.6. CONDITIONS LEADING TO SEXUAL REPRODUCTION

No two authors express the same opinion regarding this problem. Kanayev [6] has reviewed the vast but rather controversial data on the stimulation of sexual reproduction following temperature and feeding changes.

Tardent [18, 19], on the basis of a 2-year observation of *H. attenuata* cultures, came to the conclusion that there is a spontaneous cycling of appearance and disappearance of sexuality, which is unaffected by temperature, feeding conditions, or the presence of phosphates. The individual polyp does not change sex for a long period of time and as a rule its asexual progeny is of the same sex. Spontaneous and generally irreversible sex changes are possible.

Loomis and Lenhoff [13] observed the formation of gonads in all dense cultures of *H. littoralis* when they were fed with *Artemia* larvae and cleaned daily. If feeding is carried out every 30 min at a temperature of 21°C, gonads appear 10 days after the beginning of culture. To stop gonads from appearing, the water must be changed frequently and the culture should be kept in sparse aeration, which also inhibits the transition to sexual reproduction.

Aisenstadt [1] observed long-term sexual reproduction in *H. oligactis* proceeding for about 6 months, beginning from late September after the animals were changed over from their usual feed of *Daphnia* and *Cyclops* to pieces of sludgeworm (*Tubifex*). Pieces of the sludgeworm were given with forceps to each *Hydra* 2-3 times a week. The animals were kept at room temperature.

Davidson [4] stimulated transition to sexual reproduction in *H. hymanae* by transferring the animals from 24°C into a medium with a temperature of 15°C or 4°C for 5-25 days. The latter case brings particularly good results: 50% of the embryos are hatched by the 45th day and there is a high survival rate.

3.7. GAMETOGENESIS

3.7.1. Spermatogenesis

The testes initially appear at the distal end of the body and later reach the budding zone. They are conical protuberances on the body surface (Fig. 3.2) and vary in number from a few up to 60. In cases of hermaphroditism, which can even be observed in the *Hydra* that were initially dioecious, the testes are located distally to the oocytes, which develop in the budding zone.

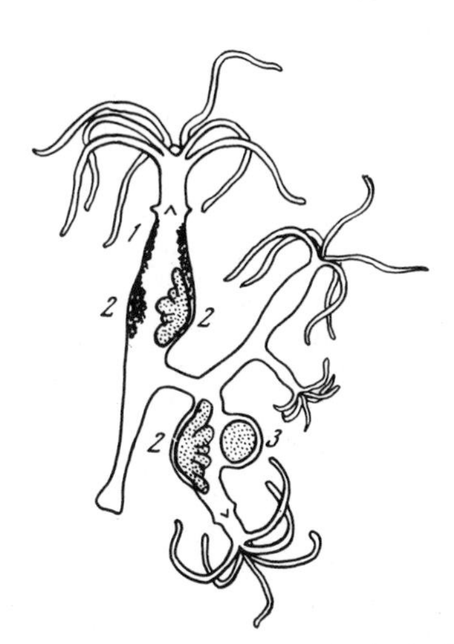

Fig. 3.2. Gametic development in the budding *Hydra*. 1) Testes; 2) oocytes at different stages of development; 3) mature egg.

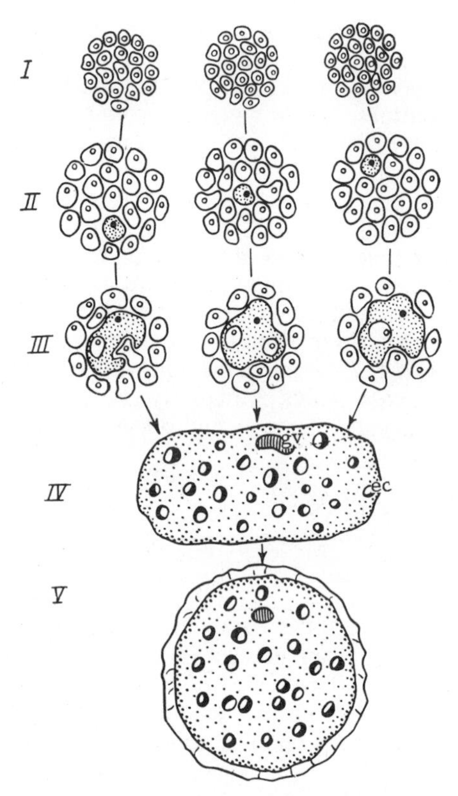

Fig. 3.3. Oogenesis in *Hydra* [23]. I) Stage of i-cell proliferation in the germinal zone; II) stage of organelle appearance in the cytoplasm of growing oogonia; III) phagocytosis of oogonia II by the grown oocyte; IV) fusion of oocytes; V) egg. gv, germinal vesicle; ec, endocytes.

Each testis consists of several cysts with an extended shape, formed by myoepithelial cells. During sperm maturation within the cyst in cell clusters, which represent the progeny of one or several interstitial cells, the spermatogonia occupying the basal cyst region can be distinguished; furthermore, primary and secondary spermatocytes can be seen as well as the more distally located spermatids and spermatozoa. Mature spermatozoa are released in batches into the surrounding water. The data concerning their viability in water is controversial; Kanayev [6], for example, states that spermatozoa stay viable for 1-3 days, while Zihler [23] has shown that, under optimal conditions (9-12°C, pH 7.3), the spermatozoa of *H. attenuata* remain viable for only about 4.5 h; he has emphasized that normal insemination requires the synchronous maturation of gametes in both males and females.

3.7.2. Oogenesis

The *Hydra* oocytes, which are frequently called "ovaries," are gigantic cells which show phagocytic activity. In females or hermaphrodites which have started sexual reproduction, a whitish spot appears in the budding zone; this spot corresponds to a thickening of the epidermal layer formed by the proliferating interstitial cells (Fig. 3.3, I). The cells within the thickening rapidly increase in size. In histological sections it can be seen that they form grape-like accumulations of syncytially connected oogonia (Fig. 3.3). As in the testis, these accumulations are separated by strands of myoepithelial cells. Each grape-like cluster is formed by the descendants of one or several interstitial cells. One cell within the cluster grows faster than the others and absorbs the surrounding germ cells by phagocytosis (Fig. 3.3, III). According to a number of authors [6, 23], actively phagocytosing oocytes may fuse with each other.

According to the most recent interpretation only the fast-growing cell of a cluster is the oocyte. All other cells begin to synthesize yolk precursors intensively before they are phagocytized. Hence, they are similar to the ovarian nurse cells of the other species [22].

The oocytes that have stopped growing acquire a spherical shape and become connected with the body surface through a pedicle consisting of epithelial muscle cells surrounding the whole egg and forming a characteristic follicle. In certain species this structure has a diameter of 1.5 mm. The number of eggs in one animal rarely reaches 10; they develop sequentially.

Before maturing the germinal vesicle (nucleus of the oocyte) is displaced to the distal cell wall. Meiotic divisions are followed by "ovulation" involving a rupture in the layer of epidermal cells surrounding the egg which, however, stays connected to its mother through a pedicle until the formation of the embryotheca (see below).

3.8. DEVELOPMENT OF THE *Hydra* EMBRYO

Cleavage is complete and initially uniform but uniformity is no longer present after the third cleavage division. The blastula has a large cavity. Gastrulation is of the mixed type involving delamination with multipolar immigration. The embryonic development of *Hydra* is as yet insufficiently studied and there are no accurate data on its duration. McConnell (cited from [6]) has shown that the duration of *H. oligactis* embryonic development under laboratory conditions varies from 35 to 70 days.

A thick chitin envelope with spikes, called the embryotheca, is formed around the embryo. At this stage the embryo usually separates from the mother's body (*Hydra* deposits eggs upon the substrate by body contraction) or remain attached to the mother through a system of filaments (Tannreuter, cited from [7]).

At the moment of hatching (see [5]) the embryo misleadingly looks like an undifferentiated syncytium, inside the embryotheca. In fact, the ectodermal and endodermal layers are already separated by the basal membrane, but the gastric cavity is not completely formed. Its formation begins from the oral end and spreads toward the aboral end. The first tentacles are often located without any apparent order. Under natural conditions Gruzova observed *Hydra* embryos hatching in mid-March [5].

3.9. KEEPING MARINE HYDROID POLYPS UNDER LABORATORY CONDITIONS

It is recommended that hydroid polyps be kept in aquariums containing natural or artificial sea water, with a closed cleaning cycle involving aeration and the stabi-

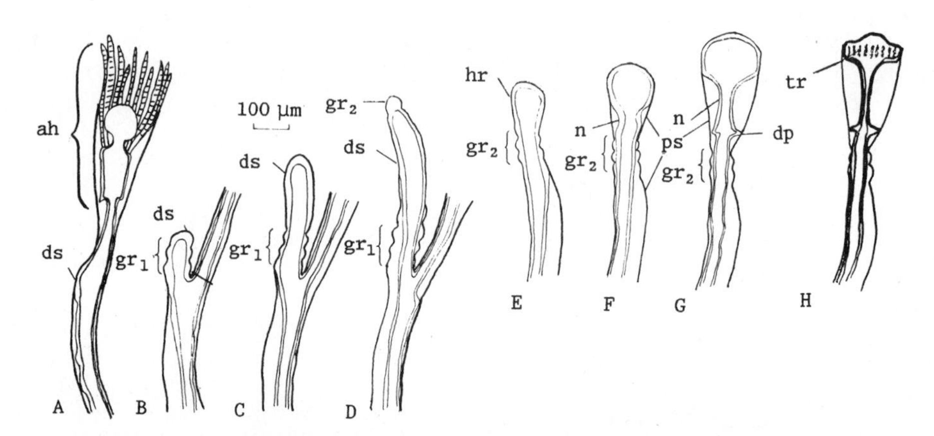

Fig. 3.4. A–H) Successive stages of stem and hydranth development in *Obelia loveni* (sketches from time-lapse film, modified from [2]). ah, adult hydranth; ds, daughter stem; gr_1, first series of grooves; gr_2, second series of grooves; hr, hydranth rudiment; n, hydranth "neck"; dp, diaphragm; ps, perisarc; tr, tentacle rudiments.

lization of an optimal temperature. After extracting colonies of hydroids from the sea, it is recommended that portions be cut, each bearing several hydranths, and tied with a thread to Plexiglas plates pretreated with abrasive tissue or chloroform to make them rougher (the chloroform should be thoroughly washed away). The plates are submerged and suspended on the walls of the aquarium.

A few days later the small colonies become attached to the plates, so the threads can then be removed. The hydroid polyps may be fed on *Artemia* larvae. According to Crowell and Rusk [3], colonies of *Campanularia* (= *Obelia*?) *flexuosa* grown in this way doubled their weight at 17°C in 3.5 days, while at 20-22°C the doubling time took 6 days.

Karlsen has reported on the long-term culture and growth of *Dynamena pumila* colonies in stationary sea water in Petri dishes provided that the animals were regularly fed.

3.10. DEVELOPMENTAL STAGES OF THE VEGETATIVE BUDS OF *Obelia loveni*

Stage 1 (Fig. 3.4A). The ectoderm of the future polyp or hydranth is adjacent to the perisarc, forming a plateau (ds) on the convex site of the maternal coenosarc. The ectoderm is not in contact with the perisarc in other regions.

Stage 2 (Fig. 3.4B). First series of circular grooves. The perisarc of the hydranth rudiment forms 1-4 circular grooves; these grooves appear sequentially, 30-40 min after each other.

Stage 3 (Fig. 3.4C). Smooth growth. The rudiment (ds) grows in length without adding any more new circular grooves.

Stage 4 (Fig. 3.4D). The second series of circular grooves (gr_2). After the smooth growth period the perisarc of the rudiment forms up to four more circular grooves.

Stage 5 (Fig. 3.4E). A vesicle-like rudiment of the hydranth (hr). That part of the rudiment located more distally than the last groove becomes wider while continuing to lie close to the perisarc.

Stage 6 (Fig. 3.4F). The hydranth rudiment with a zone of proximal contraction. The rudiment acquires a cone-like shape, its proximal part becomes thinner (n) and undergoes periodic contractions (every 5-10 min) while separating from the perisarc (ps), and again comes into closer contact with it.

Stage 7 (Fig. 3.4G). Early mushroom-like rudiment of the hydranth. The rudiment undergoes extensive elongation and differentiates to form the diaphragm (dp), a thinned proximal zone permanently separated from the perisarc (n), and an expanded distal region.

Stage 8 (Fig. 3.4H). Late mushroom-like rudiment of the hydranth. The thinned proximal zone elongates and the broadened distal part becomes shorter. Rudiments of the tentacles appear as numerous dense column-like regions (tr) around the whole perimeter of the hydranth rudiment.

Stage 9 (Fig. 3.4A). The adult hydranth, with its hypostome and tentacles, is located distally from the hydrotheca throughout its length (ah). The perisarc, which is a transparent outer skeleton made of a chitin-like substance, surrounds the entire body of the colony. The hydrotheca is a portion of the perisarc formed by the hydranth rudiment and is open distally.

Intervals between developmental stages (at 20°C) are as follows:

Stage	Time (h)
2 (two grooves)-3 (smooth length increase is equal to the length of the first groove zone)	4
3-4 (the second series consists of three grooves)	4
4-5	2
5-6	1-1.5
6-7	2-3
7-9	3-4

It is difficult to determine the interval between stages 1 and 2, since the development of a new hydranth rudiment may halt at the plateau stage. It is also difficult to determine the intervals between stages 7, 8 and 8, 9 because the tentacles form gradually.

3.11. DEVELOPMENTAL STAGES OF THE VEGETATIVE BUDS OF *Dynamena pumila*

Stage 1. Separation of the central rudiment (Fig. 3.5A). The central rudiment (cr) of the three-membered apical assembly has a convex roof. Its lateral parts are completely fused with the inner walls of the lateral rudiments (lr).

Stage 2. The central shoot (Fig. 3.5B,C). The central rudiment grows apically and becomes separated completely from the lateral rudiments. It has a rounded top, the width of which is less than the basal width. At this stage the central rudiment may look like a thick club (Fig. 3.5B), or may assume a more elon-

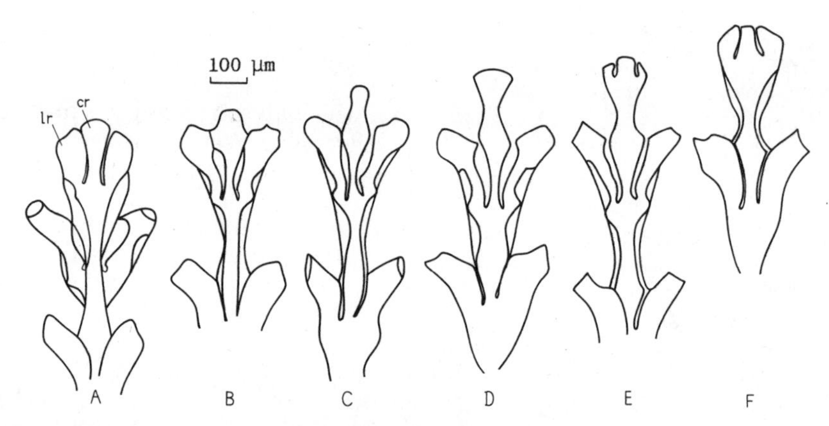

Fig. 3.5. Developmental stages of *Dynamena pumila* vegetative shoots. For explanation see text. cr, central rudiment; lr, lateral rudiment.

gated shape (Fig. 3.5C). In most *D. pumila* colonies both these configurations alternate uniformly and are located at adjacent levels.

Stage 3. The triangle (Fig. 3.5D). The upper part of the central rudiment assumes a triangular shape. Its greatest transverse length is at least equal to the width of the central rudiment base.

Stage 4a. Early three-membered whorl. (Fig. 3.5E). The central rudiment is divided by two longitudinal furrows into three regions. The length of the furrows does not exceed the width of the lateral rudiments they separate.

Stage 4b. Advanced three-membered whorl (Fig. 3.5F). The length of the longitudinal furrows is greater than the width of the lateral rudiments. The central rudiment has a flat roof which is not separated by any depressions from the roofs of the lateral rudiments.

The time intervals between individual stages at 15°C are as follows:

Stage	Time (h)
1-2	3-4
2-3	2-3
3-4a	4
4a-1	8 and more (during this period shoot growth may stop)

REFERENCES

1. T. B. Aizenstadt, "Study of oogenesis in the hydra. 1. The ultrastructure of interstitial cells at different stages of their transformation into oocytes," *Ontogenez* **5**, 18–30 (1974).

2. L. V. Belousov, L. A. Badenko, A. L. Katchurin, and L. F. Kurilo, "Cell movements in morphogenesis of hydroid polyps," *J. Embryol. Exp. Morphol.* **27**, 317 (1972).
3. S. Crowell and M. Rusk, "Growth of *Campanularia* colonies," *Biol. Bull.* **99**, 357 (1950).
4. J. Davidson, "*Hydra hymanae*: regulation of the life cycle by time and temperature," *Science* **194**, 618 (1976).
5. M. H. Gruzova, "New data concerning *Hydra vulgaris* (Pall.) development," *Dokl. Akad. Nauk SSSR* **109**, 670 (1956).
6. I. I. Kanayev, *The Hydra. Essays on the Biology of the Freshwater Polyps* [in Russian], Nauka, Moscow–Leningrad (1952).
7. I. I. Kanayev, *Abraham Trembley* [in Russian], Nauka, Leningrad (1972).
8. W. Kemmer and H. C. Schaller, "Analysis of morphogenetic mutants of *Hydra*," *Wilhelm Roux's Arch. Dev. Biol.* **190**, 191 (1981).
9. Yu. A. Labas, L. V. Belousov, L. A. Badenko, and V. N. Letunov, "Concerning pulsating growth in multicellular organisms," *Dokl. Akad. Nauk SSSR* **257**, 1247–1250 (1981).
10. H. M. Lenhoff (ed.), *Hydra: Research Methods*, Plenum Press, New York–London (1983).
11. W. F. Loomis, "Cultivation of *Hydra* under controlled conditions," *Science* **117**, 565–566 (1953).
12. W. F. Loomis, "Sexual differentiation in *Hydra*: control by carbon dioxide tension," *Science* **126**, 753–759 (1957).
13. W. F. Loomis and H. M. Lenhoff, "Growth and sexual differentiation of *Hydra* in mass culture," *J. Exp. Zool.* **132**, 555–573 (1956).
14. D. V. Naumov, *Hydroids and Hydromedusae* [in Russian], Nauka, Moscow–Leningrad (1960).
15. J. J. Otto and R. D. Campbell, "Budding in *Hydra attenuata*: Bud stages and fate map," *J. Exp. Zool.* **200**, 417–428 (1977).
16. D. I. Rubin and H. R. Bode, "The aberrant, a morphological mutant of *Hydra attenuata*, has altered inhibition properties," *Dev. Biol.* **89**, 316 (1982).
17. A. R. D. Stebbing, "Hormesis – stimulation of colony growth in *Campanularia flexuosa* (Hydrozoa) by copper, cadmium, and other toxicants," *Aquat. Tox.* **1**, 227–238 (1981).
18. P. Tardent, "Zur Sexualbiologie von *Hydra attenuata*," *Rev. Suisse Zool.* **73**, 357–381 (1966).
19. P. Tardent, "Experiments about sex determination in the *Hydra attenuata* Pall.," *Dev. Biol.* **17**, 483–511 (1967).
20. P. Tardent, "Coelenterata, Cnidaria," in: *Morphogenese der Tiere*, F. Seidel (ed.), Lieferung I: A-I, 68–415, VEB Gustav Fisher Verlag, Jena (1978).
21. P. Tardent and R. Tardent (eds.), *Developmental and Cellular Biology of Coelenterates*, Elsevier/North-Holland Biomedical Press, Amsterdam–New York–Oxford (1980).
22. P. Tardent, "The differentiation of germ cells in Cnidaria," in: *The Origin and Evolution of Sex*, Alan R. Liss, Inc., New York (1985).
23. J. Zihler, "Zur Gametogenese und Befruchtungsbiologie von *Hydra*," *Wilhelm Roux's Arch. Entwicklungsmech. Org.* **169**, 239–267 (1972).

Chapter 4

THE SLUDGEWORM *Tubifex*

V. N. Meshcheryakov

4.1. INTRODUCTION

Tubifex tubifex Müll. (Annelida, Oligochaeta) serves as a classic model in experimental and biochemical embryology. The development of this worm is an impressive example of ooplasmic segregation which occurs prior to the first cleavage as a formation of pole plasms rich in mitochondria and endoplasmic reticulum [16]. Owing to the strictly determined heteroquadrant spiral cleavage, the material of these plasms is distributed over definite individual blastomeres [43, 44]. Therefore *Tubifex* has for a long time been used for studying the problem of the early localization of anlage, using methods such as centrifugation, blastomere isolation, and cytochemistry [12, 17, 31]. As different blastomeres may easily be identified during early cleavage, the biochemical characteristics of certain cells or groups of cells may be studied in relation to subsequent differentiation [4, 69, 70].

The simplicity with which worms can be kept under laboratory conditions, plus the opportunity of obtaining eggs virtually all year round, add to the advantages of *Tubifex* as a laboratory experimental model [29]. Furthermore, *Tubifex* eggs may be cultured throughout development outside the cocoon in a simple salt solution [31]. The egg membranes of *Tubifex* are permeable to a wide range of substances, including, specifically, mitochondrial and mitotic poisons. This provides an opportunity for experimental studies of mitosis [32]. Early *Tubifex* embryos have been used for studying the effect of magnetic and electromagnetic fields on the chromosomal apparatus and chronology of mitoses [5-8]. Methods of studying the structure of *Tubifex* eggs using the transmission and scanning electron microscope have been developed, as well as methods for isolating meiotic spindles [56, 57]. *Tubifex* has been extensively used by Japanese scientists to study the function of microfilaments in oocyte maturation, ooplasmic segregation, and early cleavage. Cytoskeleton poisons, ionophoresis, and protein synthesis inhibitors are employed as tools in these studies [49, 50, 52-54, 56-62]. Along with methods enabling nucleate and non-nucleate egg fragments to be obtained, these studies open up the way to the elucidation of mechanisms underlying the intrinsic asynchrony of the cleav-

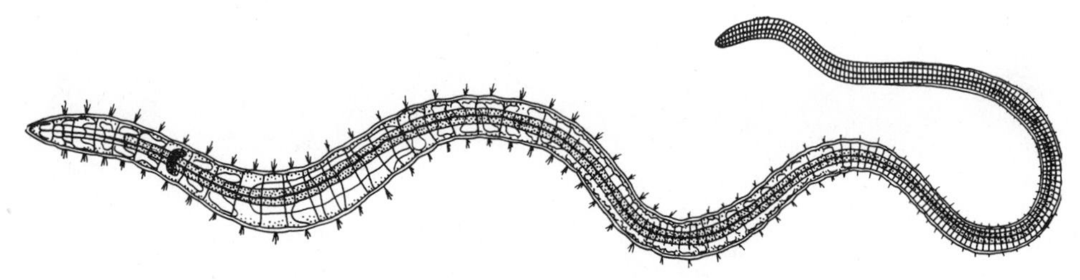

Fig. 4.1. General view of *Tubifex tubifex* [35].

age divisions characteristic of *Tubifex*. For example, evidence has been collected for the existence of the rhythmic pattern of actin polymerization/depolymerization in the egg cortex; this pattern does not depend on the nucleus and determines the rhythm of oocyte maturation and cleavage divisions [55]. Although no systematic studies of macromolecular synthesis in *Tubifex* development are known, it should be mentioned that techniques have been developed [55] which allow protein synthesis on semithin sections to be studied using autoradiography.

Adult worms may be used for studying regeneration [11, 27, 59]. Since sludgeworms are generally resistant to oxygen deficit, the biochemistry of their energy metabolism and the physiology of their respiration provide another developing area of research [1, 2, 13, 14, 21, 47, 48].

The drawbacks of *Tubifex* as an experimental animal include the small number of eggs in the cocoon, the not-very-high synchrony of their development, the rather high proportion of spontaneous developmental abnormalities, and the fact that they can be easily infected by fungi and other pathogens.

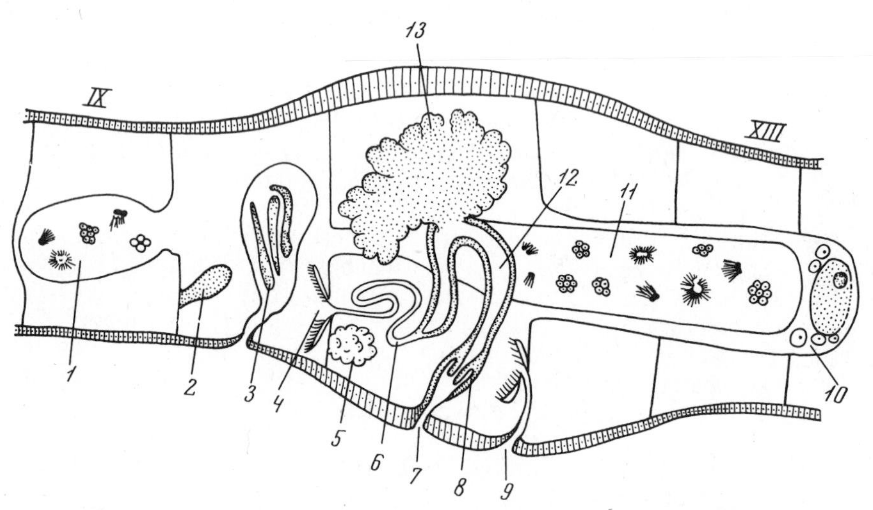

Fig. 4.2. Diagram of the reproductive system in *Tubifex tubifex* [11]. 1) Anterior seminal sac; 2) testis; 3) spermatheca containing spermatozeugmata; 4) seminal funnel; 5) ovary; 6) seminal duct; 7) male genital pore; 8) penis in the penial sac; 9) female genital pore; 10) egg sac; 11) posterior seminal sac; 12) atrium; 13) prostate. Roman numerals refer to segment numbers.

4.2. TAXONOMY AND DISTRIBUTION

The genus *Tubifex* (Oligochaeta, Naidomorpha, Tubificidae) includes about seven species living in the USSR [11], of which *T. tubifex* Müll. (= *T. rivulorum* Lam.) is the most common and can be found almost worldwide (Fig. 4.1). It is impossible to identify freshwater oligochaetes precisely without a detailed examination of the reproductive system (Fig. 4.2) and bristles. *T. tubifex* almost invariably comprises the only species of this genus found in material commercially sold in zoological shops as fish food. The salient features for the identification of *T. tubifex* are given below [11]. Penial bristles are absent. The penis is not included in the chitin tube, and the distal notch of the ventral bristles is no shorter than the proximal one. Long dorsal hairlike bristles are present, the length of these not exceeding twice the body width. Live worms have a light-red pigmentation of various shades ranging from yellowish to brownish. Their length is 20-100 mm and the number of segments varies from 36 to 120.

Another species, *Tubifex hattai* Nomura, is used in embryological studies in Japan. The early development of this species is better known in many respects than that of *T. tubifex.*

4.3. ECOLOGY AND CULTURE

Tubifex occurs in predominantly freshwater environments such as slow rivers, streams, ditches, and stagnant reservoirs. It prefers shallow places with slimy, sometimes sandy beds. The species is particularly abundant in polluted reservoirs. It is resistant to various chemicals and is tolerant of low or fluctuating oxygen levels. *Tubifex* feeds by passing mud through the intestines. No specialized respiratory organs are present; respiration takes place through the skin and is helped by rhythmic movements of the body's caudal section. The head region and approximately half the body are submerged in the river bed; particles of the substratum together with mucous secretions from the skin form a tube around the worm.

Worms obtained commercially or caught using a metal sieve can, if necessary, be kept wet for several days in a cool place. The thickness of the worm layer should not exceed 1 cm and the worms should be rinsed once a day. If the layer of worms is thicker, the animals are rapidly poisoned by the products of their own metabolism. If a constant culture is desired, the worms should be washed several times, and the longest specimens, possessing a marked spindle-like thickening in the anterior third of the body (where gametes are formed), should be selected; the large oocytes can often be seen through the transparent wall of the body.

River mud passed through a mesh, or alternatively a mixture of mud and sand, should be boiled with water for an hour and then placed into medium-sized dishes (15 to 20 cm diameter); the thickness of the mud layer should be 3-4 mm, with water carefully added until its level is 3-4 cm above the surface of the mud. About 100-200 worms are placed into each dish, which is then loosely covered by glass. If the substrate consists only of mud, no additional feeding is required for 1-2 months; after this period the mud should be changed.

Japanese scientists cultivate the worms on a 1.5-mm-thick layer of sterilized river sand. To feed the animals, a small quantity of wet yeast is added every 3 days [24]. There is no need to aerate the water, but Inase [24] believes that aeration aids reproduction. The water may be changed once a week or at longer intervals. Dead worms should be removed. A film of bacterial growth on the water surface is one indication that the water should be changed. The worms are best kept at 18-20°C; higher temperatures are less desirable than lower ones. Oviposition usually begins within several days of the worms being placed into cultivation vessels.

Experiments aimed at elucidating the optimal culture conditions for *T. tubifex* have been performed [28]. It is recommended that the animals be kept at a temperature below 24°C and at a population density no greater than 1500 animals per square meter; they can be fed on lettuce.

4.4. REPRODUCTION

Reproduction in *T. tubifex* takes place during most of the year, although recent data [64] indicate that reproduction is interrupted during the summer months (June to September). Oogenesis and spermatogenesis begin before the end of September, while oviposition takes place between mid-January and May.

The structure of the reproductive system, oogenesis, and the formation of the cocoon of the Japanese species have been described in great detail [18, 19, 26, 66], and there is a detailed account of the structure of *T. tubifex* spermathecae [15]. These worms are hermaphrodite, but self-fertilization is avoided in the course of copulation as clusters of spermatozoa are deposited into spermathecae of opposite partners (see Fig. 4.2 and page 42). Oogenesis is of the nutrimentary type. During the period of sexual reproduction the worms possess a girdle or "saddle" – a modified region of the skin epithelium which secretes material necessary for forming the cocoon envelope. A protein-containing liquid is secreted under the envelope and the cocoon is gradually pushed toward the head region. When it passes over segment XI, it picks up oocytes; the spermatozeugmata enter the cocoon as it passes over segment X. Finally, the cocoon passes over the head and is shed. Mating and oviposition may be separated by a considerable time interval.

It has been established that *Tubifex* is capable of parthenogenetic reproduction, which takes place under conditions of arrested spermatogenesis [47].

Under the above culture conditions, the whitish cocoons are deposited directly onto the mud surface where they can be easily detected. When the worms are cultured on sand the collection of cocoons is conveniently performed every 3-4 h. The likelihood of quick oviposition may be increased if one initially selects worms with 6-8 mature eggs in the ovisacs located in the first segments immediately behind the girdle (see Fig. 4.2). Freshly laid cocoons may be identified by the relatively low contamination of the external membrane by mud particles.

According to our observations, a single worm can lay 4-5 cocoons in 1 month. According to Kosiorek [28], one worm lays about 14 cocoons on average in 100 days, and the complete life cycle of *T. tubifex* (from zygote to the first oviposition by mature worms) at 24°C lasts 50-57 days, while embryogenesis (up to hatching from the cocoon) lasts 10-12 days. The first cocoons are laid by worms at least 40 mm in length.

4.5. GAMETES AND FERTILIZATION

When the eggs are laid into the cocoon at metaphase I, they are not yet fertilized. The cocoon is roughly elliptical in shape and has one tubular appendage at each end. These appendages later serve for the emergence of hatched juvenile worms (see Fig. 4.8, 35). The thin elastic transparent membrane of the cocoon is covered by an opaque sticky layer of external mucus with adhering particles of mud and detritis. Between 1 and 20 eggs may be found inside the cocoon, whose length varies from 1.2 to 2 mm. The cocoon is filled with a protein-containing fluid.

The *Tubifex* egg (see Fig. 4.5, 1/1) belongs to the isolecithal type. The equatorial diameter is 0.2-0.5 mm. The egg is opaque, yellowish, or rose-white, and contains a large amount of yolk. The mature eggs in the ovisacs (Fig. 4.2) are surrounded by a tightly apposed single-layered vitelline membrane consisting of a loose fibrillar network. The plasma membrane contains numerous microvilli. Cortical granules or their analogs cannot be detected in the egg cortex. In addition to yolk granules and lipid droplets the ooplasm contains mitochondria, ribosomes, endoplasmic reticulum, and multivesicular bodies [50].

Immediately after deposition the egg is somewhat irregular in shape but gradually becomes ellipsoid, the short axis coinciding with the animal-vegetal axis. A small region located at the animal pole during this stage is relatively poor in yolk. The vitelline membrane is in close contact with the egg surface. The *Tubifex* egg is protected from the environment by not only the vitelline membrane, but also by protein-containing fluid, the cocoon envelope, and the external layer of mud.

The cocoon envelope is resistant to a variety of enzymes and appears to consist of scleroproteins [26]. The envelope is permeable to inorganic ions, water, and mitochondrial poisons such as dinitrophenol and dyes; colloids of the gum arabic type cannot, however, pass through this membrane [8, 26, 42]. It should be pointed out, however, that both the cocoon membrane and the protein-containing fluid of the cocoon protect the egg to some extent from the above substances. The vitelline membrane is also permeable to colchicine, theophylline, various quinones and narcotics, cytochalasine B, cycloheximide, ionophore A 23187, and ^{3}H-amino acids [22, 23, 32, 33, 52, 53, 55]. Transfer through the membrane proceeds faster at higher temperatures.

The spermatozoa are deposited in the spermatheca during mating and enter the cocoon in the form of spermatozeugmata, long cylindrical arrow-shaped complexes of spermatozoa attaining 1.2 mm in length (Fig. 4.2). After deposition the spermatozeugmata disintegrate and the spermatozoa disperse in the protein-containing fluid. It has been established [3] that a spermatozeugma consists of an axial cylinder and a cortical layer. The spermatozoa typical of the Oligochaeta are located in the axial cylinder. Each consists of an acrosome, a long cylindrical nucleus 0.1-0.3 μm in diameter, a middle section, and a flagellum. The cortical layer is formed by helically arranged aberrant spermatozoa glued to each other by their caudal parts (tails). The tail ends are surrounded by a layer of mucopolysaccharides and are capable of wave-like movements. The mucopolysaccharide layer appears to play a protective role.

Insemination in the Japanese species, *T. hattai*, takes place roughly 30 min after oviposition. One spermatozoon enters the egg close to the vegetal pole where the fertilization cone is formed [20] (see Fig. 4.5, 1/1). This is where the vitelline membrane begins to detach from the egg membrane and forms the perivitelline space; its initial width is approximately 0.25 μm, but increases to 0.5-1.0 μm 40-50 min after oviposition. The vitelline membrane acquires a three-layered structure, the surface microvilli gradually disappear, and microfilaments appear in the cortex for the first time. The cortical layer becomes 0.5 μm thick on average, although later its thickness may undergo local variations. It does not contain any membranous organelles. Golgi elements and sac-like vesicular structures appear for the first time after fertilization [50]. The so-called meiotic deformations appear during the course of the maturation divisions (*see* the description of developmental stages). Cortical microfilaments are involved in these deforming movements [49]. The microfilaments also play a part in ooplasmic segregation and they appear to be closely connected with the meiotic deformations, segregational cytoplasm movements, and the cytokinesis of cleavage divisions.

The formation of pole plasms, one of the most remarkable features of *Tubifex* early development, proceeds in two stages [58]. Initially the cytoplasm consists predominantly of Golgi complexes, multivesicular bodies, and mitochondria but is devoid of yolk, while lipid droplets move away from the egg center to the periphery, forming a subcortical layer under the entire egg surface. Thereafter, this layer moves parallel to the cortex toward the animal and vegetal poles and forms lens-shaped accumulations at the poles by the metaphase of the first cleavage division. Usually, the animal pole plasm protrudes markedly from the egg surface and the organelles become compacted in the pole plasms. By this time the organelles are connected to each other by microfilaments. Mechanisms underlying the cytoplasmic movements at these two stages of segregation appear to be different [58].

Very occasionally the author has been able to perform artificial insemination of *Tubifex* eggs. This was achieved as follows. The ovisacs with oocytes are usually located near the spindle-like thickening of the worm body (see Fig. 4.2) and are easily discernible through the translucent body wall. The nonanesthetized worm is placed into a solution (composition described below), and the part of the body with oocytes is cut from the head and tail regions using a surgical knife. The body wall and ovisac membrane are then cut laterally from each side (see Fig. 4.2) and the oocytes are carefully pressed out into the salt solution. Only the largest oocytes showing germinal vesicle breakdown are selected. The oocytes with a germinal vesicle may be easily identified because the vesicle is located near the surface of the egg.

The spermathecae are a pair of small oval sacs located on the ventral side of the body at the border between the spindle-like thickening and the head region (see Fig. 4.2). They usually contain easily discernible spermatozeugmata, appearing as several whitish cylindrical bodies. The body region with the spermathecae is cut away from the worm in a salt solution, and the spermathecae are excised and suspended together with the spermatozeugmata in the salt solution. A suspension of generally immobile spermatozoa obtained from 2-4 spermathecae is then added to the oocytes along with 2-5 ml of the salt solution. A protein-containing fluid is needed to render the spermatozoa mobile.

To conclude this section, a procedure for the fixation of early *Tubifex* embryos for transmission electron microscopy is described [58].

The eggs taken from the cocoons (see embryo culture) are passed through three changes (of 2 h each) of 3% glutaraldehyde solution in 0.5 M phosphate buffer (pH 7.4) containing 0.2% tannin at room temperature. They are then placed overnight in a 0.05 M sucrose solution made in a 0.05 M phosphate buffer at 0-4°C; the egg is then cut with a safety razor into several parts and post-fixed for 1.5 h in a 1% OsO_4 solution made in a 0.5 M phosphate buffer in a cold room. The eggs are then dehydrated in alcohol, the vitelline membrane is removed with watchmaker's forceps, and the eggs are embedded in Epon 812. For general cytological purposes, semithin Epon sections may be stained with 1% toluidine blue.

4.6. CULTURE OF EMBRYOS

To observe the development in the cocoons the external mucus should be carefully removed with needles. The cocoon wall is not quite transparent, and strong dark-field illumination is needed for accurate staging. According to Penners [46], the cocoons should be cultured in soft aquarium water to avoid the deposition of calcium salts on the walls; aeration should be adequate and the water changed daily. It should also be pointed out that those cocoons whose mucus has been removed are often attacked by fungi. Cocoons with a moderate number of loosely located embryos are the most suitable for observation.

It often happens that not all embryos in a cocoon develop normally. Some embryos may stop developing while various abnormalities, including twin abnormalities, form in others. The proportion of normal embryos at hatching averages 74% [5]. Apparently, the fluid present in the cocoon is not assimilated by the embryo and serves rather as a mechanical protection and salt reservoir.

Salt solutions suitable for culturing an embryo throughout its developmental period have been described. The composition of such a solution for *T. tubifex* is as follows [30]:

$MgSO_4$	20 mM	50 parts
$CaCl_2$	20 mM	130 parts
NaCl	20 mM	16 parts
KCl	20 mM	4 parts

Embryos of the species common in Japan, *Tubifex hattai*, are cultured in a solution containing 5 mM NaCl, 1.4 mM KCl, 8.1 mM $CaCl_2$, 2 mM $MgSO_4$, pH 7.1 [23]. The success of the culture depends predominantly on the care with which an embryo is removed from the cocoon. It should be remembered that large *Tubifex* eggs are extremely sensitive to mechanical treatments, and this frequently results in the plasma membrane being disorganized. As a result the perivitelline space acquires a characteristic bluish opalescence. This happens when this space is overfilled with mitochondria which comprise the main bulk of the active subcortical plasm. In these cases the vitelline membrane undergoes a uniform detachment from the egg surface at a distance of 50-60 μm. This is a reliable indication that the egg is dying. Just as often mechanical trauma results in an abnormal cleavage, but here the rhythmicity of divisions is less affected. The eggs are relatively less sensitive between cell divisions and in the more advanced stages of development (24 blastomeres and later). The stage with prominent pole plasms is very suitable for taking the fertilized egg out of the cocoon prior to cleavage. The cocoon is placed onto the paraffin-coated bottom of a cuvette filled with a salt solution. Then, using a pair of sharp forceps, the cocoon wall is incised in two places and the region between the incisions removed; alternatively one of the cocoon tips (about one-third of the overall size) is carefully dissected together with several eggs, using an ophthalmological surgical knife. Afterward the cocoon is turned tip down (the hole should be sufficiently large) and the eggs are allowed to sink to the bottom. Touching the eggs with instruments is not recommended. The removal of eggs from the cocoon is performed under a dissecting microscope.

Successful culturing requires working with a dim light; bright light has an adverse effect on the isolated embryos. When high light intensity is required for ex-

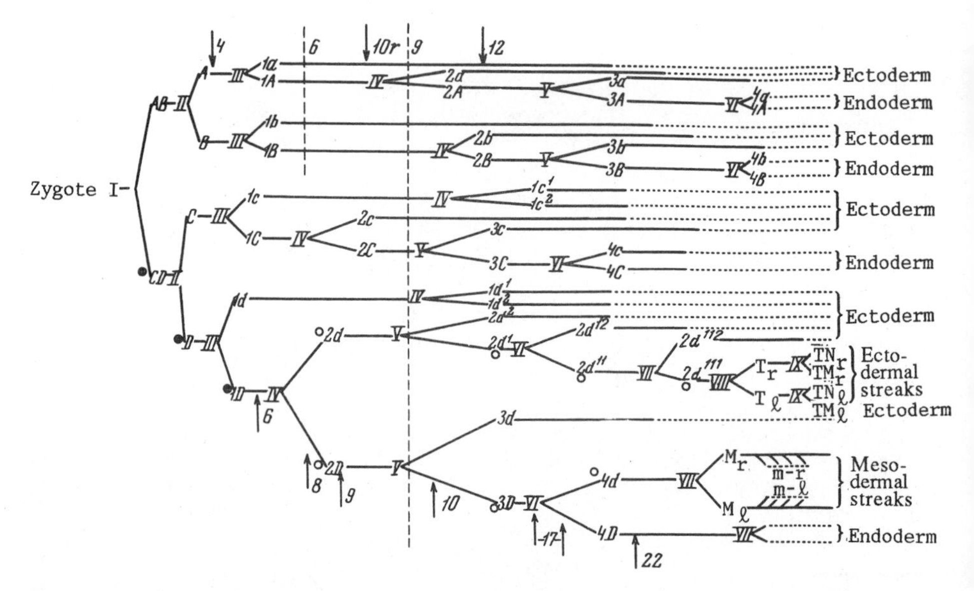

Fig. 4.3. Diagram illustrating cleavage in *Tubifex tubifex* [43]. M_l, M_r, left and right mesoblasts; T_l, T_r, left and right primary ectoteloblasts; TN_l, TN_r, left and right neuroteloblasts; TM_l, TM_r, left and right myoteloblasts; m-l, m-r, elements of the mesodermal streaks. Roman numerals refer to division numbers of different cell lines, Arabic numerals with arrows indicate the number of blastomeres.

amining the embryos under a microscope, heat filters should be used. Another problem is associated with the possible contamination of the cultures by fungi, bacteria, and infusorians; the latter are the most dangerous. They can be avoided if, prior to the extraction of embryos, the cocoons are thoroughly washed on cheesecloth, using first water and then a sterile salt solution. Subsequent surgical intervention should be performed using sterilized instruments. If, despite these measures, contamination still occurs, the culture solution should be changed at frequent intervals. The embryos are suitably cultured in small beakers containing 3-5 embryos per 5-10 ml of solution. Under favorable conditions of culture about 74% [22] or 85% [22] of normal embryos survive. Henzen [16] pointed out that there was no significant difference in the time embryos take to develop inside or outside the cocoons. Even so, it is not practical to work with isolated embryos unless there are particular reasons for doing so. Isolating embryos is not suitable for experiments aimed at elucidating the long-term results of the action of various substances which may pass through the cocoon envelope. It would be beneficial to take another cocoon at the same developmental stage as a control, especially since high synchrony in *T. tubifex* development is rather rare; usually no more than 60-70% of embryos develop synchronously at the early stages.

The temperature range of normal *T. tubifex* development is around 10-25°C. Bearing in mind that drastic temperature changes (between 5-10°C) result in various developmental abnormalities [46], such temperature fluctuations should be avoided.

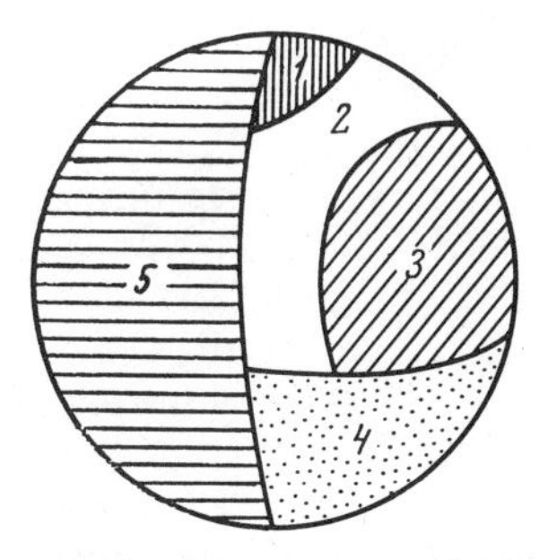

Fig. 4.4. Fate map of *Tubifex* blastula [10], view from the left. 1) Head and ectoderm; 2) epidermal ectoderm; 3) ectodermal embryonic streaks; 4) mesoderm; 5) midgut. Animal pole is on the right.

4.7. NORMAL DEVELOPMENT

Tubifex development belongs to the mosaic type. Penners [46] studied in detail the cleavage and fate of the derivatives of different blastomeres (Fig. 4.3). The fate map of primary rudiments for the blastula (prior to epiboly) is given in Fig. 4.4 [10, 16]. The main organ systems in *Tubifex* are formed of the following cells. The midgut and esophagus are formed from the yolk-rich derivatives of 3A-3C and 4D macromeres. The head ectoderm originates predominantly from the micromeres of the first quartet. The entire circulatory system, the connective tissue envelopes of the inner organs, the longitudinal muscles and nephridia, all originate from the coelomic mesoderm, derived from progeny of the 4d micromere. The same is true for the dissepiments and peritoneum of the coelom and probably for the gonads as well. The second and third quartets of micromeres give rise to a considerable part of the epidermis. The $2d^{111}$ blastomere is a source of two ectodermal streaks; these are the precursors of a part of the epidermis, the bristle sacs, ring muscles (from rows M_1 and M_2), and the entire nervous system (from N rows) including the ventral chain and the pharyngeal ring.

The description of stages presented below (p. 55–63) has been compiled for the first time by the author. To distinguish the stages up to four blastomeres we have used the classification of Lehmann and Hadorn [33]; the later stages have been distinguished and described according to Penners [43, 44] and Meyer [40, 41] and partially to Lehmann and Hadorn [33]. A detailed description of the embryo structure as determined by microscopy may be found in the aforementioned papers of Penners, as well as in the publications of Hirao [20], Matsumoto [37-39], and Shimizu [51]. The submicroscopic characterization of the early stages has been given in a number of papers [16, 31, 49, 40, 52-58, 69, 70].

It should be pointed out that Oligochaeta in general show a marked variability of cleavage [67, 68]. Our observations point to *Tubifex* being no exception in this respect, particularly beginning from stages 10-12 and when culturing outside the cocoon. Therefore, people studying *Tubifex* development may find deviations from the description of stages as regards the order of appearance of individual blastomeres and their size and relative arrangement.

There appears to be no information in the above-mentioned studies about the time of onset of maternal and paternal genome function in the development of *Tubifex* or about the timing of macromolecular syntheses.

The papers cited contain controversial information on the completion time of endoderm epiboly by the epidermis and embryonic streaks [40, 44]. Here we take it that epiboly is completed by stage 28.

The stages described are shown in Figs. 4.5 to 4.8; the number of each drawing corresponds to a stage. The table of normal development (Table 4.1) lists, in a summary form, the diagnostic features of different stages as well as their timing. The data of Lehmann and Hadorn [33] are used for temperatures of 13, 18, and 23°C; these authors observed the development of *Tubifex* embryos in salt solutions (the temperatures were maintained at a constant ±1°C). Unfortunately, the authors did not specify the intervals and methods of observation. Their data regarding the early stages raise some doubts.

For stages 1-2 the Henzen data [16] at 20 ± 1°C has been used. It refers to embryos developing in salt solutions and the dating refers to the beginning of a stage for the most advanced embryos in the observed series.

We have observed the development of *Tubifex* embryos inside the cocoons at stages 3-22 at 21 ± 0.5°C. The observation interval at stages 6-21 was 30 min; the beginning of a given stage (appearance of cleavage furrows) has been recorded in the most advanced embryos. At the early developmental stages asynchrony in the development of different embryos appears to be more prominent owing to the shorter duration of each stage. At stages 24-35 the external diagnostic features of stages seem to be rather ambiguous. The emergence of juvenile worms from the cocoon (stage 35) could take as long as several days. Table 4.1 shows the mean time of the onset of stages 6-22, averaged over 3-4 cocoons; the data for one cocoon is given for stages 3-5.

Lehmann and Hadorn [33] have established that the duration of different stages in *Tubifex* development changes proportionally at different temperatures. Therefore relative values may be used to characterize the temporal aspects of *Tubifex* devel-

---

Fig. 4.5–4.8. Normal developmental stages of *Tubifex tubifex.* Sketch numbers correspond to stage numbers with abbreviations referring to the direction of the view. an, animal view; veg, vegetal view; ps, view from behind; lef, view from the left; r, view from the right; ven, ventral view. Other abbreviations: an, animal pole; veg, vegetal pole; 1(2) pb, first (second) polar body; protub., protuberances; pr, pronuclei; vm, vitelline membrane (shown only at 1/1); fc, fertilization cone; app and vpp, animal and vegetal pole plasm; M_l and M_r, left and right mesoteloblasts; T_l and T_r, left and right primary ectoteloblasts; m_l and m_r, elements of mesodermal streaks; TN_l and TN_r, left and right neuroteloblasts; en, endoderm; ec, ectoderm outside the streaks; TM_{1+2}, TM_3, TM_1, TM_2, secondary and definitive myoteloblasts; Mes, mesoblast position; S_{ec}(l, r), ectodermal streak (left and right); m.ep., median epidermis; 1–24, segment numbers; at and pt, anterior and posterior tip of the embryo; a. bl., anterior edge of the blastopore; ph, pharynx; ph. gl., suprapharyngeal ganglion; sm, mesodermal streak; rm, ring muscle cell nuclei; T, ectoteloblasts; st, stomodeum; dv, dorsal blood vessel; b. som., border of somites closure on the dorsal side; bs, bristle sacs; nephr, nephridium; gon, gonoblasts; mo, mouth; M_1, M_2, M_3, muscle elements of ectodermal streaks; n, neural elements of the streaks (for designation of blastomeres see the description of stages). Sketches 1, 3, 6, 8, 29–33 are original; 4, 5, 7, 9–21, 23–25, 26, 27a, 28 are reproduced from Penners [43, 44] with modifications; 22, 26a, 27b are from Meyer [40] with modifications; 35 is from Chekanovskaya [11].

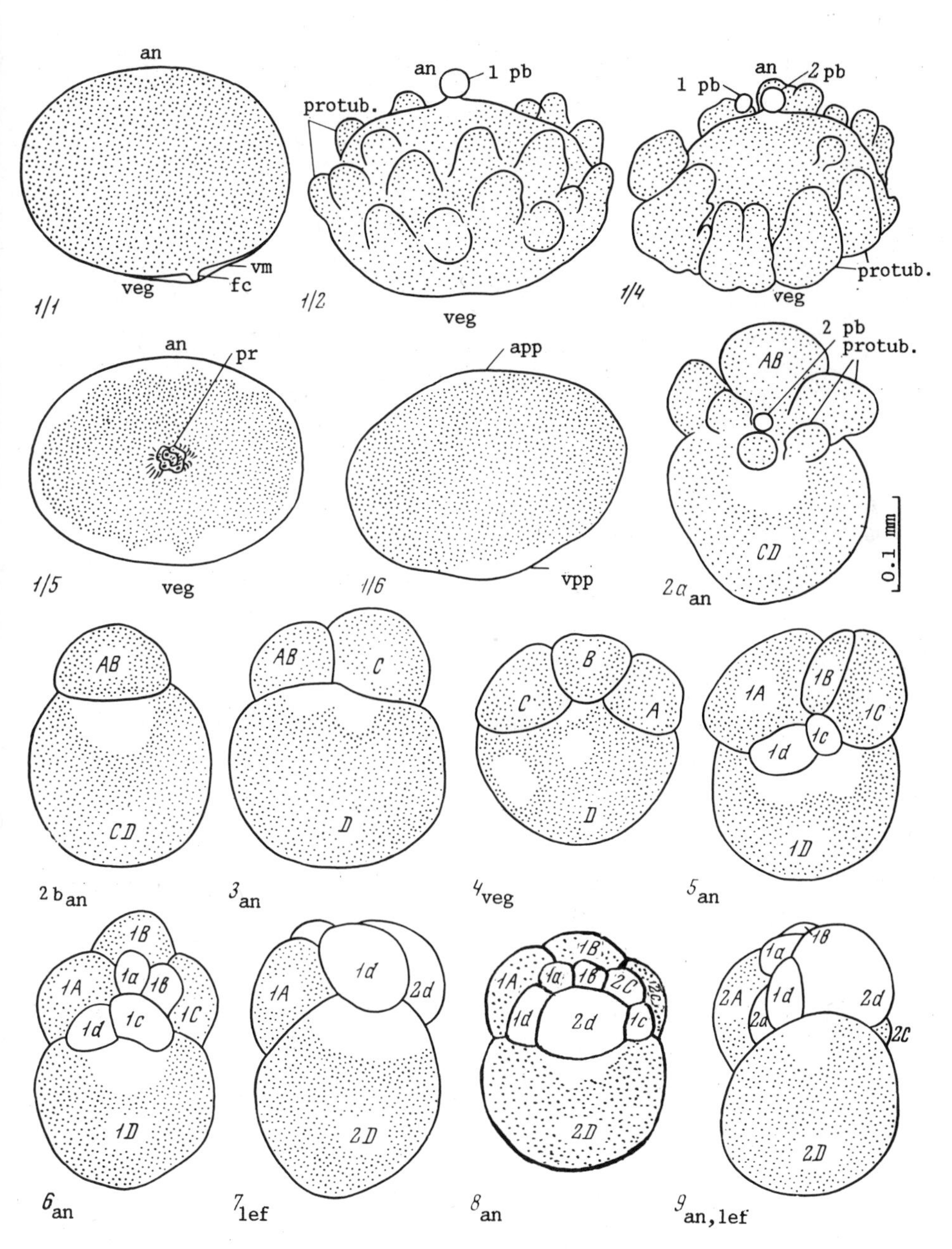
an
vm
veg
fc
1/1
an
1 pb
protub.
1/2
veg
an
2 pb
1 pb
protub.
1/4
veg
an
pr
1/5
veg
app
1/6
vpp
2 pb
protub.
AB
CD
0.1 mm
2a an
AB
CD
2b an
AB
C
D
3 an
B
C
A
D
4 veg
1A
1B
1C
1c
1d
1D
5 an
1B
1A
1a
1b
1C
1d
1c
1D
6 an
1A
1d
2d
2D
7 lef
1B
1A
1a
1b
2C
2c
1d
2d
1c
2D
8 an
1a
1B
2A
1d
2d
2a
2C
2D
9 an, lef

Fig. 4.5

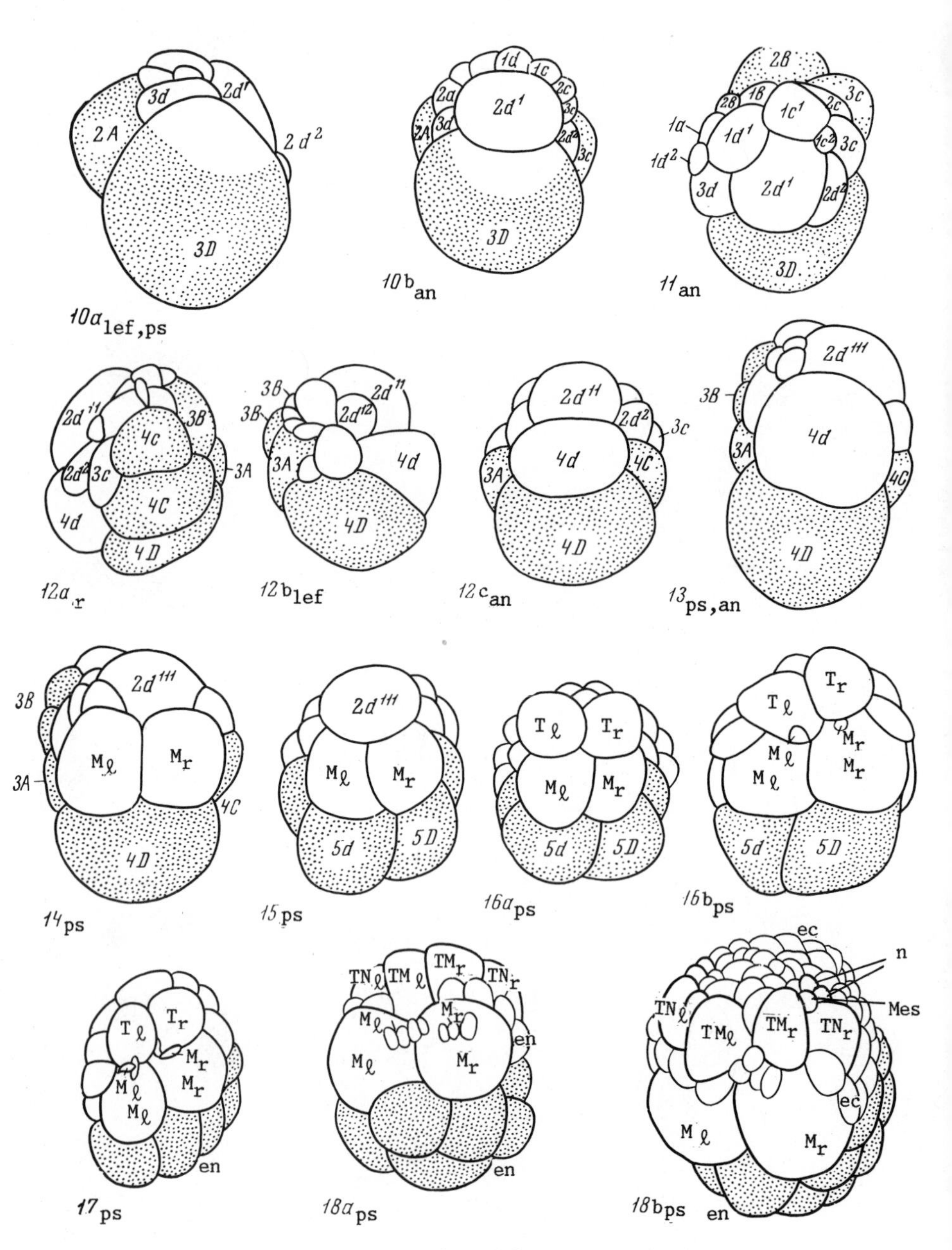
3d
2d'
2A
2d²
3D
10a lef,ps
1d
1c
2a
2d'
2c
3c
2A
3d
2d²
3c
3D
10b an
2B
3c
2B
1B
1c'
2c
1a
1d'
1c²
3c
1d²
3d
2d'
2d²
3D
11 an
2d''
3B
4c
3B
2d²
3c
3A
4C
4d
4D
12a r
3B
2d''
3B
2d¹²
3A
4d
4D
12b lef
2d''
2d²
3c
4d
3A
4C
4D
12c an
2d'''
3B
4d
3A
4C
4D
13 ps,an
2d'''
3B
Mℓ
Mr
3A
4C
4D
14 ps
2d'''
Mℓ
Mr
5d
5D
15 ps
Tℓ
Tr
Mℓ
Mr
5d
5D
16a ps
Tr
Tℓ
Mℓ
Mr
Mℓ
Mr
5d
5D
16b ps
Tℓ
Tr
Mr
Mℓ
Mr
Mℓ
en
17 ps
TNℓ
TMℓ
TMr
TNr
Mℓ
Mr
en
Mℓ
Mr
en
18a ps
ec
n
TNℓ
Mes
TMℓ
TMr
TNr
ec
Mℓ
Mr
18b ps
en

Fig. 4.6

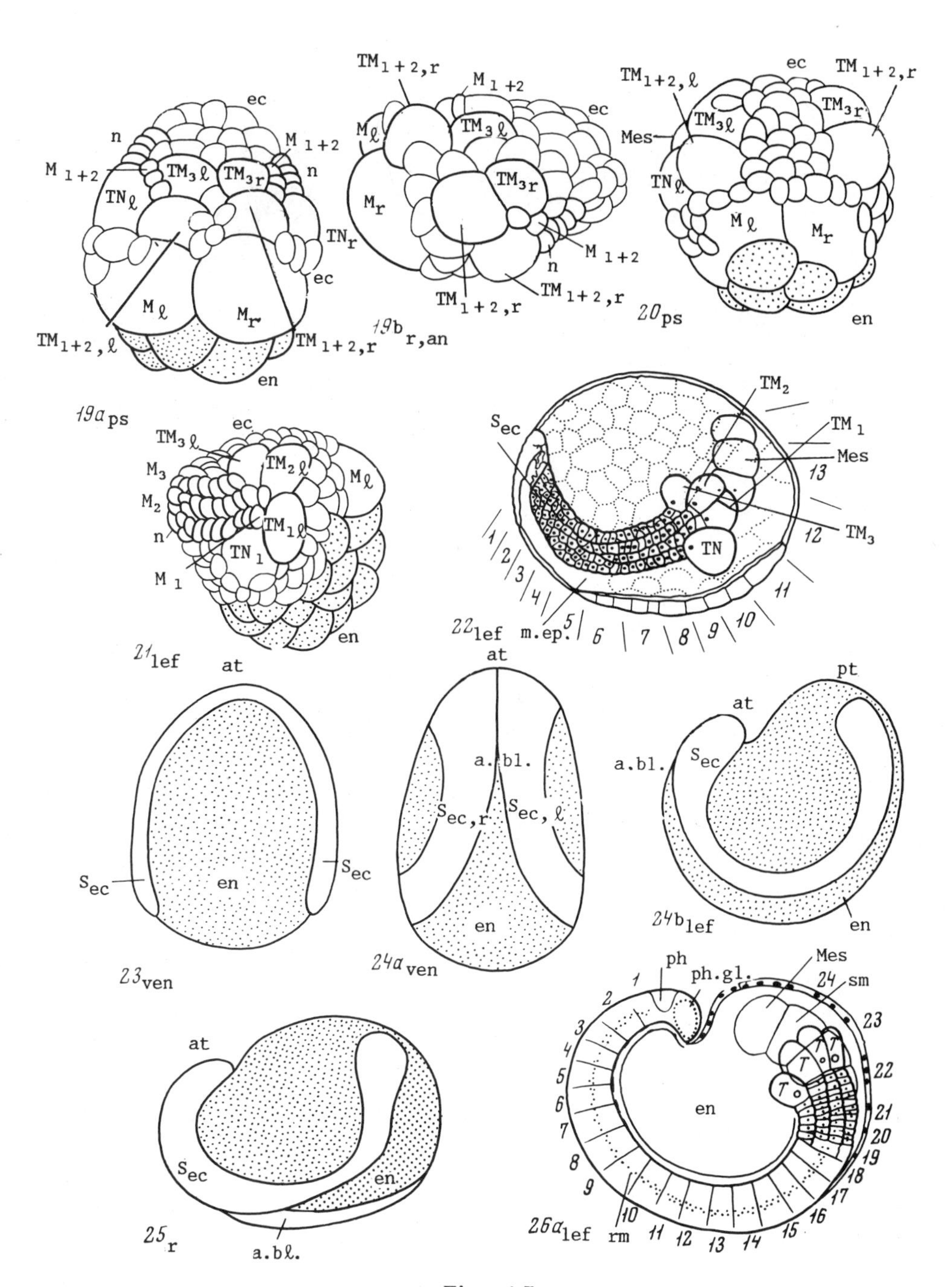

Fig. 4.7

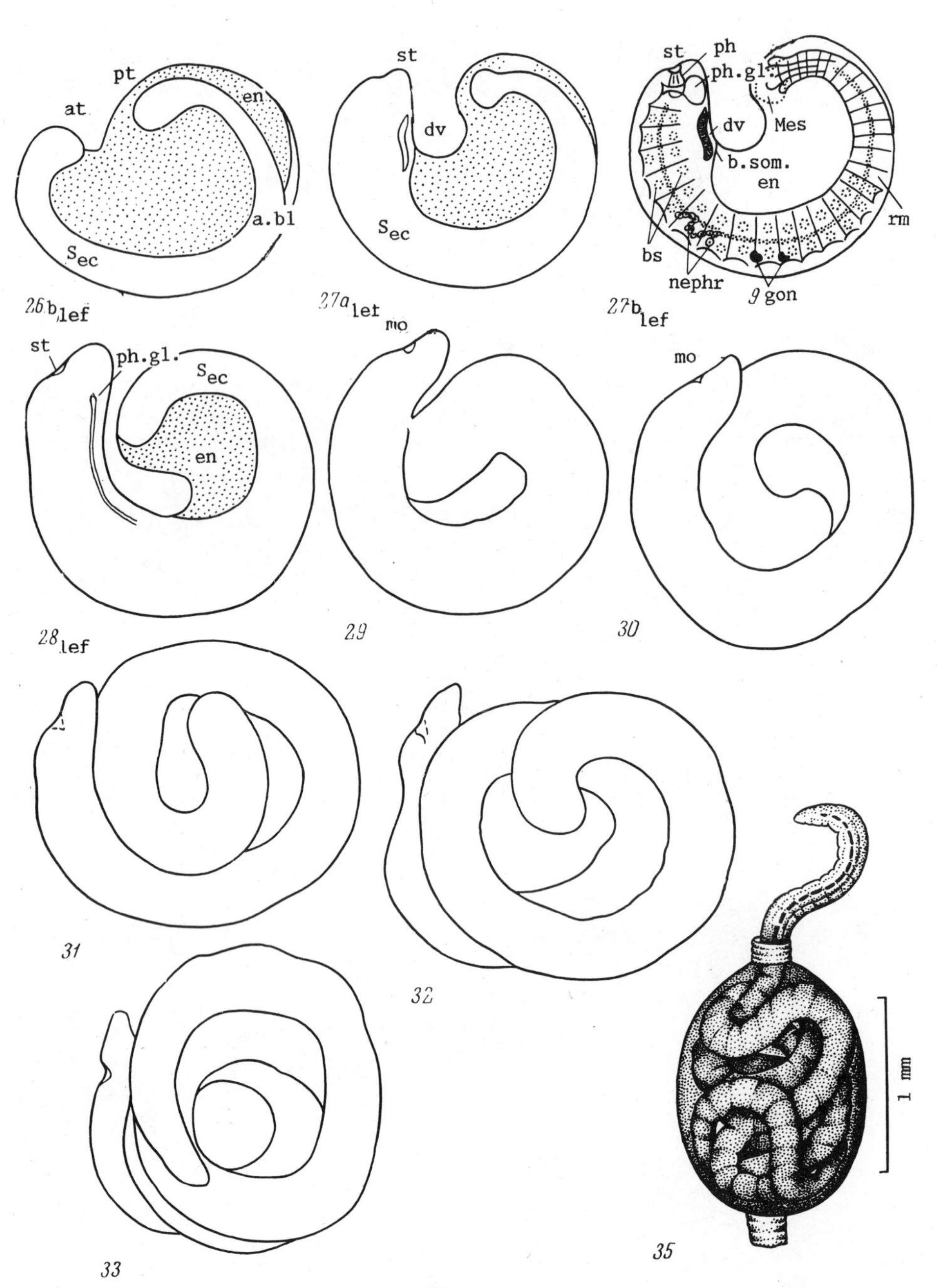
pt
at
en
a.bl
Sec
26b lef
st
dv
Sec
27a let
st
ph
ph.gl.
dv
Mes
b.som.
en
rm
bs
nephr
gon
27b lef
mo
st
ph.gl.
Sec
en
28 lef
mo
29
30
31
32
33
1 mm
35

Fig. 4.8

opment just as they are for other species. However, since synchronous cleavage divisions are lacking during *Tubifex* development, the duration of one synchronous cleavage division (τ_0) cannot be used as a criterion to characterize development (see p. 41).

4.8. DESCRIPTION OF STAGES

Stage 1: One-Cell Embryo 1/1. Laying and insemination. Cocoon formation takes place at metaphase I. Mitochondria are uniformly distributed under the cortical layer; they are also present, however, in the deeper parts of the cytoplasm.

Insemination in the Japanese species *T. hattai* takes place roughly 30 min after the cocoon is formed. The spermatozoon enters the oocyte in the region located close to the vegetal pole where the fertilization cone is formed.

1/2. Extrusion of the first polar body. During anaphase I, amoeboid protrusions or protuberances [33] appear on the egg surface. These protuberances are radially symmetrical to the animal-vegetal axis. They seem rather chaotic during the first meiotic division and appear predominantly in the animal hemisphere [36]. At this stage the vitelline membrane separates from the egg surface, yielding a well-defined perivitelline space (the membrane is not stretched but is folded at this stage). Microfilaments are only in the animal hemisphere. The meiotic spindle contains centrioles, as it also does in the next cleavage division [56, 57].

1/3. Acquisition of a stable shape. After the extrusion of the first polar body the protuberances merge and disappear. By metaphase II the egg acquires the stable shape of a rotational ellipsoid, attaining maximal diameter in the equatorial plane.

1/4. Extrusion of the second polar body. The second series of protuberances appears during the anaphase of the second meiotic division; they are more ordered than previously and are separated from the egg surface by rather deep false furrows or even narrow circular constrictions in the equatorial zone. The cortical layer has maximal thickness (0.8-0.8 μm) in the region of these furrows and contains numerous microfilaments which can also be found in other regions of the animal and vegetal hemisphere. Numerous mitochondria are accumulated near the bases of the protuberances.

The polar bodies of *Tubifex* are rather loosely attached to the egg surface and, because the vitelline membrane is separated from the egg surface, they can only rarely be found near the animal pole after the first cleavage divisions. Therefore they cannot be used to identify polarity.

1/5. Formation of the zygote. Soon after the disappearance of the second series of protuberances, the male and female pronuclei migrate toward the center of the egg where they fuse. Both pronuclei still possess a karyomeric structure at the beginning of the fusion. The spindle, possessing asters of different sizes, begins to form in the center of the nucleus during the prophase of the first cleavage division. This period corresponds to the first stage of pole plasm formation, namely to the appearance of a subcortical layer on the cytoplasmic periphery resulting from the centrifugal movement of the membranous organelles.

1/6. Formation of pole plasms. The accumulation of large quantities of mitochondria and cytoplasmic reticular elements, which resemble two hemispherical convex lenses with their flat surfaces turned toward the cytoplasm, takes place at both egg poles by metaphase of the first cleavage division. These lens-like accumulations appear semitransparent and bluish in color.

TABLE 4.1. Normal Development of *Tubifex tubifex*

Stage No.	Time after egg-laying (hours and minutes) at the following temperatures:					Diagnostic features	
	13°C	18°C	20°C	21°C	23°C	External features	Internal features
1/1	00.00	00.00	00.00	00.00	00.00	Perivitelline space is not seen under the light microscope	Metaphase of the I maturation division
1/2	–	01.00	00.30	–	–	Perivitelline space is noticeable Amoeboid protuberances on the egg surface. One polar body at the animal pole	
1/3	06.00	02.00	01.45	–	02.00	Ellipsoid shape without any irregu- larities	Metaphase of the II maturation division
1/4	–	04.00	03.55	–	–	Two polar bodies. Amoeboid pro- tuberances	
1/5	–	–	04.25	–	–	Undistorted shape. Pole plasms are not visible	Fusion of pronuclei
1/6	12.20	06.00	05.50	–	05.20	Pole plasms are well seen	Prophase and metaphase of the first cleavage division
2	–	08.00	07.30	–	–	Two blastomeres	
3	23.00	–	–	11.15	09.10	Three blastomeres	
4	–	–	–	14.25	–	Four blastomeres	
5	–	–	–	16.05	–	Appearance of 1d and 1c	
6	–	–	–	17.25	–	Appearance of 1a and 1b (segregation of the first quartet	
7	36.20	24.00	–	22.10	14.00	Appearance of 2d	
8	–	–	–	26.00	–	Appearance of 2c	

9	–	–	–	–	–	Appearance of 2a and 2b (segregation of the second quartet)	
10	–	–	–	–	–	Appearance of 3d, $2d^2$, $2d^1$ (14 cells)	
11	–	–	–	27.10	–	Division of 1d and 1c. Appearance of 3c (17 cells)	
12	–	28.00	–	28.30	–	Appearance of 4d. Division of $2d^1$, 3C, 2A, and 2B (22 cells)	
13	–	–	–	31.15	–	Formation of $2d^{111}$ (24 cells)	
14	54.00	36.20	–	33.25	23.30	Two mesoblasts. (First manifestation of bilateral cleavage)	
15	–	–	–	34.30	–	Division of 4D	
16	–	–	–	36.10	–	Division of $2d^{111}$	
17	–	–	–	42.45	–	4 entoblasts under the mesoblasts	
18	–	–	–	47.00	–	4 ectoteloblasts over the mesoblasts	
19	–	–	–	54.20	–	Continuous row consisting of 6 ectoteloblasts over mesoblasts	
20	–	–	–	61.00	–	3 ectoblasts are present on each side over mesoblasts. Embryonic streaks appeared. Beginning of epiboly	
21	–	72.00	–	62.10	–	Embryonic streaks are located equatorially	Mesoblasts migrated under the ectoderm. 4-10 somites
22	–	–	–	71.15	–	The streaks cover the endoderm over two-thirds of the embryo length along the animal–vegetal axis	13-17 somites. The rudiment of stomodeum
23	–	–	–	–	–	The streaks reach the ventral surface of the embryo	
24	168.00	108.00	–	–	74.0	The streaks bend to the dorsal side of the head tip. They come into contact on the ventral side of the head tip	Appearance of the ventral chain ganglia

TABLE 4.1 (continued)

Stage No.	Time after egg-laying (hours and minutes) at the following temperatures:					Diagnostic features	
	13°C	18°C	20°C	21°C	23°C	External features	Internal features
25	–	–	–	–	–	On the ventral side the streaks are closed over half the embryo length. Muscle movements in response to a prick	Formation of the pharynx and of the suprapharyngeal ganglia. Beginning of somite differentiation into muscles
26	180.00	120.00	–	–	88.00	The streaks are closed over two-thirds of the embryo length. Spontaneous head movements	24-27 somites. Appearance of nephridia. Gonoblasts
27	192.30	132.00	–	–	96.00	The head region is well developed. On the dorsal side the streaks are closed up to segment IV	Formation of bristle sacs. The dorsal blood vessel is functional
28	–	–	–	–	–	End of epiboly. The streaks on the ventral side fuse completely; on the dorsal side they fuse up to segment X	Nephridia are multicellular. 34-36 somites
29	–	–	–	–	–	Uniform cylindrical shape	
30	264.00	168.00	–	–	120.00	The first coil is formed	
31	–	–	–	–	–	1.5 coils	
32	348.00	216.00	–	–	156.00	2 coils	
33	–	–	–	–	–	3 coils	
34	–	–	–	–	–	Emergence from the vitelline membrane	
35	552.00	360.00	–	–	264.00	Hatching. Long hairlike bristles are present	

TABLE 4.2. Duration of Stage 1 Periods for *T. hattai* at 20°C

Period	Time, min
Oviposition	0
Beginning of the first meiotic deformation	40-60
End of the first meiotic deformation. Extrusion of the first polar body	120
Beginning of the second meiotic deformation	170
End of the second meiotic deformation Extrusion of the second polar body	250-270
End of phase I of pole plasm formation (until the prophase of the first cleavage division)	300
End of phase II of pole plasm formation (until the metaphase of the first cleavage division)	330
Completion of pole plasm formation (metaphase of the first cleavage division)	360
Beginning of the first cleavage division	570-600

The egg shape is rather stable in this state. Lipid droplets are concentrated on the boundary separating the pole plasms from the yolk. According to Inase [24], the animal pole plasm later goes into the 2d blastomere and the vegetal pole plasm into the 4d blastomere. This contradicts statements from earlier authors.

The second stage of the pole plasm formation (when the subcortical cytoplasm "flows" to the poles) is divided by Shimizu [58] into two phases. During these phases the volume of the subcortical layer does not expand, but its distribution does become extremely nonuniform. The first phase corresponds to a transition from prophase to metaphase of the first cleavage division. The lens-shaped pole plasms are actually formed during the second phase. Protuberances and microvilli appear again on the animal pole; blebbing appears to occur in this place.

Table 4.2 supplements Table 4.1. It shows the duration of stage 1 periods for *T. hattai* [49, 55] at 20°C.

Stage 2: Two Blastomeres. During anaphase a large heteropolar spindle is displaced by a smaller aster to the lateral egg surface. Surface activity is observed (see Fig. 4.5, 2a), which may accompany the division of large blastomeres containing the pole plasm up to the formation of the 4d mesoblast. This results in the formation of unequal AB and CD blastomeres; the bulk of pole plasm material remains in the CD blastomere (see Fig. 4.5, 2b). A small cavity may appear between the two blastomeres.

Stage 3: Three Blastomeres. During metaphase the CD blastomere begins to undergo an asymmetrical deformation. Eventually blastomere C segregates on the right side of AB (when viewed from the animal pole). C is equal in size to, or slightly larger than, AB. The bulk of the pole plasm material remains in D.

Stage 4: Four Blastomeres. AB divides to yield A (located on the left) and B (sometimes somewhat smaller in size). All three first cleavage divisions are thus meridional.

Stage 5: Six Blastomeres. Two micromeres, 1d and 1c (the former is usually somewhat larger) are segregated synchronously in the dexiotropic direction. They contain little yolk.

Stage 6: Eight Blastomeres. The formation of the first micromere quartet, later to yield the main bulk of the ectodermal part of the head (see Fig. 4.4), is completed by the segregation of 1a and 1b.

Stage 7: Nine Blastomeres. A large 2d blastomere containing abundant pole plasm is segregated at this stage. This is the first somatoblast, the descendants of which later yield ectodermal teloblastic streaks. Despite laeotropic segregation, 2d is located virtually in the center of the animal hemisphere.

Stage 8: Ten Blastomeres. After the formation of the 2d blastomere, the 1C blastomere laeotropically segregates a small 2c micromere.

Stage 9: Twelve Blastomeres. The formation of the second micromere quartet is completed by a laeotropic division of 1A and 1B macromeres. This stage may be recognized by the position of the 2a micromere.

Stage 10: Fourteen Blastomeres. The following blastomeres undergo cleavage: 2D segregates the rather large 3d micromere to the left of 2d; and 2d micromere buds the $2d^2$ micromere toward the C quadrant, the size of the latter micromere being slightly less than that of its sister $2d^1$ (Fig. 4.6, 10a, b).

Stage 11: Seventeen Blastomeres. The micromeres of the first quartet divide. 1d yields a smaller $1d^2$ and a larger $1d^1$ and 1c yields a smaller $1c^1$ and a larger $1c^2$. Directions of these divisions are more or less symmetrical to the B–D axis, the future anteroposterior axis of the embryo. The micromere 3c undergoes dexiotropic segregation.

Stage 12: Twenty-two Blastomeres. The second 4d somatoblast is formed at this stage. This is a gigantic blastomere segregating under $2d^{11}$ which, in turn, forms at this stage as a result of the $2d^1$ division, which buds a small $2d^{12}$ micromere to the left (see Fig. 4.6, 12a, b). 4d contains virtually all the remaining pole plasm but differs from $2d^{11}$ in that it has far more yolk. Therefore its consistency may sometimes be confused with yellow-rose macromeres. In addition, 3C divides into the two yolk-rich endodermal cells 4C and 4c (Fig. 4.6, 12a) with cells 2A and 2B completing the formation of the third quartet budding the small 3a and 3b blastomeres (Fig. 4.6, 12b).

Stage 13: Twenty-four Blastomeres. The $2d^{111}$ blastomere, which is an immediate precursor of ectodermal teloblasts, forms at this stage. It retains the median position of the $2d^{11}$ blastomere. One of the macromeres, either 1A of 1B, divides.

Stage 14: Two Mesoblasts. The second 4d somatoblast divides to yield the two identical mesoblasts M_l and M_r (left and right). Cleavage becomes distinctly bilateral.

Stage 15: Division of the 4D. The 4D macromere divides to yield two identical blastomeres. This division may sometimes come after stage 16.

Stage 16: Two Primary Ectoteloblasts, Two Mesoblasts. The $2d^{111}$ cell divides to yield two identical ectoteloblasts (T_l and T_r) (see Fig. 4.6, 16a). Teloblastic activity of the mesoblasts begins somewhat later (see Fig. 4.6, 16b), the mesoblasts budding small cells from their anterior edges. The beginning of this activity may, however, be delayed.

Stage 17: Second Division of 5d and 5D Macromeres. The 5d and 5D macromeres divide once more, forming four endoblasts under the mesoblasts; during this period each endoblast buds yet another small cell (macromeres after the segregation of 4d are called endoblasts). Each of the primary ectoteloblasts at this stage yields one small cell in the posterior direction that is toward the animal pole.

Stage 18: Four Ectoteloblasts. Each ectoteloblast divides in the direction parallel to the B–D axis to yield a neuroteloblast located externally and a more internal myoteloblast (Fig. 4.6, 18a). According to our observations, the left and right teloblasts frequently divide asynchronously at an interval of up to 1 h. By this time the mesoblasts may segregate up to 4-5 small cells by budding. Each myoteloblast can yield two cells of an epidermal nature by budding in the posterior direction. Cells of quartets, 1, 2, and 3, which by this time have multiplied, start creeping up on the endomeres and mesoblasts (Fig. 4.6, 18b). Ectodermal streaks formed from cells budded by the ectoteloblasts in the forward direction (Fig. 4.6, 18b; n, m) begin between stages 18 and 19.

The next stage begins after five or six cells have been segregated by each myoteloblast.

Stage 19: Six Ectoteloblasts. The myoteloblasts divide normally with respect to the B–D axis (again frequently asynchronously) to yield myoteloblasts, 3 (TM_3) and 1 + 2 (TM_{1+2}) (Fig. 4.7, 19a, b). Initially the teloblast triplets are packed together (Fig. 4.7, 19a). This is not an abnormality as Penners suggested. TM_{1+2} buds a row of cells which continue the TM row, while TM_3 begins a new row.

Stage 20: Beginning of Epibolic Gastrulation. Owing to intensive proliferation, ectodermal cells wedge in between the ectoteloblast triplets, displacing the ectoteloblasts to the lateral sides of the embryo, and then creep up on the mesoblasts. From this time onward the whitish embryonic streaks can be clearly seen among the semitransparent epidermal cells. Cells comprising the streaks, however, cannot be seen, and it is difficult to discern the ectoteloblasts.

Stage 21: Eight Ectoteloblasts; Equatorial Position of the Ectodermal Streaks. After segregating five or six cells, the TM_{1+2} myoteloblasts divide in a direction perpendicular to the animal-vegetal axis; therefore on each side of the embryo four ectoteloblasts, TM, TM_1, TM_2, and TM_3, can be found above the other; TM_1 begins to bud a new row. Cell proliferation in the streaks begins closer to the head region where the streaks touch one another. The mesoblasts are almost completely covered by the ectoderm. The segmentation of mesoderm begins at this stage (between 4 and 10 somites are present). Each somite on one side of the body originates from a single cell budded by the mesoblast. Mesoblasts at this stage are located close to the X segment.

Stage 22: Thirteen to Seventeen Somites. The embryonic streaks obviously bent in the vegetal direction as well as the epidermis cover the endoderm by roughly two-thirds along the animal-vegetal axis. The ectoteloblasts can no longer be seen in vital observation. In contrast to Penners [44] and Ivanov [25], Meyer [40] believes that, at this stage, the embryo possesses the first three mesodermal segments which later fuse together. The rudiment of the stomodeum becomes discernible close to the site where the streaks touch each other on the head tip.

Stage 23: Emergence of the Embryonic Streaks onto the Ventral Surface. The embryonic streaks emerge onto the ventral surface of the embryo but are not yet fused together. The embryo is still egg-shaped.

Stage 24: Beginning of Streak Fusion and Elongation of the Embryo. The embryonic streaks on the head region start bending toward the dorsal side of the embryo and, at the same time, come into contact on the ventral side of the head region (Fig. 4.7, 24a, b). The embryo begins to acquire an elongated and bent shape (the dorsal side is concave). From this stage onward, during the course of streak fusion, differentiation proceeds gradually in the anteroposterior direction. The differentiation of the mesodermal streaks proceeds alongside. At stage 24 the neural part of the ectodermal streaks is multicellular and partially covered by epidermis from the outside. Pairs of rudiments corresponding to the ventral chain ganglia have already appeared.

Stage 25: Streaks on the Ventral Side Are Closed Approximately Halfway. The rudiment of the pharynx appears for the first time as a dense ectodermal mass with the suprapharyngeal ganglion appearing alongside. The muscle part of the ectodermal streak is now multilayered, the external one yielding part of the epidermis while the internal one (predominantly rows m_1 and m_2) yields the ring muscle system. The external parts of the somites begin to differentiate into muscle elements of the longitudinal muscle system.

Stage 26: Streaks on the Ventral Side Are Closed Together for Roughly Two-Thirds of Their Length. Streak fusion approximately reaches segment XVI. The head region of the embryo makes small spontaneous movements. Pairs of rudiments corresponding to the dorsal blood vessels, and protruding up to segment X, appear in the upper part of the mesodermal streaks. Between 24 and 27 somites are present at this stage. Pairs of nephridia rudiments appear in segments VII and VIII; they are probably all of mesodermal origin. One or two gonoblasts appear in the region of segments X and XI; these gonad anlage exist in a two-cell state until hatching.

Stage 27. Segregation of the Head Region. Movements are obvious (Fig. 4.8, 27a, b). The streaks are closed on the dorsal surface in the anterior part of the body approximately up to segment IV. On the dorsoventral surface they are separated over a very short distance. The stomodeal invagination is quite distinct. The ventral blood vessel is formed in the anterior part of the body. Ectodermal bristle sacs consisting of a follicle and 5-7 large cells of the bristle gland have formed. The ventral ganglia are clearly delimited. The dorsal blood vessel begins to function. Somites have a small lumen. Precursors of the future epithelial cells of the midgut appear inside the endoderm.

Stage 28: Thirty-four to Thirty-six Somites, the End of Epiboly. The streaks on the posterior end are completely closed; in the dorsal region their closing is advanced up to segment X. According to Meyer [40, 41] the mesoderm of the three first segments loses its segmented character. Nephridia are multicellular. Bristle anlage are connected with muscles. Ectoteloblasts are present but are small. Ganglia and longitudinal connectives have differentiated within the central chain approximately to segment XXII. The length of the embryo in its extended state is about 0.7 mm.

Stage 29: Acquisition of a Wormlike Shape. The embryo acquires a uniformly cylindrical shape and the characteristic posterior thickening disappears. From now until hatching the number of segments stays at 36 (in rare cases, 38).

Stages 30-33: Coiling of the Body. When the embryo is still inside the vitelline membrane, body elongation and a decrease in cross section result in helical coiling of the body. Different stages are characterized by a different number of turns. At *Stage 30*: the first coil of the body is complete. The length of the embryo is about 2.2 mm. *Stage 31*: about 1.5 coils. *Stage 32*: about 2 coils. *Stage 33*: about 3 coils.

Stage 34: The juvenile worm emerges from the vitelline membrane, unfolds, and creeps around inside the cocoon.

Stage 35: Hatching. Juvenile Stage. The juvenile worm emerges from the cocoon through an orifice, head forward. Meyer [40] points out that the worm still does not have long hairlike bristles, but we managed to observe some. By the time of hatching, nephridia also appear in other segments (beginning with segment XIII). The contractile perivisceral vessel is prominent and the midgut possesses a lumen. The oral cavity establishes a connection with the pharynx only after hatching. The length of the hatched juvenile worms is between 3.00 and 5.50 mm (mean length 4.50 mm) [5].

REFERENCES

1. A. Arillo, "Scambio di molecole organiche fra ambiente e *Tubifex tubifex*. I. Assimilazione di acido lattico e relativi effetti sugli acidi del ciclo di Krebs," *Boll. Mus. Ist. Biol. Univ. Genova* **42**, 133–144 (1974).
2. A. Arillo, "Scambio di molecole organiche fra ambiente e *Tubifex tubifex*. II. Espulsione di acidi bi-tricarbossilici da parte di esemplari sotto-posti a diverse condizioni ambientali," *Boll. Mus. Ist. Biol. Univ. Genova* **43**, 105–144 (1975).
3. P. Braidotti and M. Ferraguti, "Two sperm types in the spermatozeugmata of *Tubifex tubifex* (Annelida, Oligochaeta)," *J. Morphol.* **171**, 123–136 (1982).
4. J. Brachet, *The Biochemistry of Development*, Vol. 2, Pergamon Press, New York (1960).
5. R. Brandsch and P. Jitariu, "Der Einfluss eines pulsierenden elektromagnetischen Feldes auf die Entwicklung und die ersten Furchungsteilen der Embryonen von *Tubifex tubifex*," *Rev. Roum. Biol., Ser. Zool.* **15**, 431–436 (1970).
6. R. Brandsch and P. Jitariu, "Influence of colchicine and theophilline combined with an electromagnetic field on the first divisions of *Tubifex* eggs," *Rev. Roum. Biol., Ser. Zool.* **16**, 215–220 (1971).
7. R. Brandsch, "Cleavage of *Tubifex* eggs under various conditions of magnetic field applied at different periods in the cell cycle," *Rev. Roum. Biol., Ser. Zool.* **17**, 121–130 (1972).
8. R. Brandsch, "Embryogenetic and genetic consequences of the treatment of *Tubifex* eggs with magnetic fields," *Rev. Roum. Biol., Ser. Zool.* **18**, 221–225 (1973).
9. F. Carrano, "L'azione del γ-dinitrofenolo sulle nuova di *Tubifex rivulorum*," *Atti Accad. Naz. Lincei, Cl. Sci. Fis., Mat. Nat. Rend.* **19**, 85–88 (1955).
10. J. M. Cather, "Cellular interaction in the regulation of development in annelids and molluscs," *Adv. Morphogenesis* **9**, 67–125 (1971).
11. O. V. Chekanovskaya, *Aquatic Oligochaeta of the USSR Fauna. Keys to the USSR Fauna* [in Russian], Nauka, Moscow–Leningrad (1962).
12. J. R. Collier, "Morphogenetic significance of biochemical patterns in mosaic embryos," in: *Biochemistry of Animal Development*, Vol. I, R. Weber (ed.), Academic Press, New York–Amsterdam (1965).
13. H. Degn and B. Kristensen, "Low sensitivity of *Tubifex* sp. respiration to hydrogen sulfide and other inhibitors," *Comp. Biochem. Physiol.* **B69**, 809–817 (1981).

14. E. Fischer and I. Horváth, "Cytological and cytochemical studies on the chloragocytes of *Tubifex tubifex* Müll. with special regard to their role in hemi-metabolism," *Zool. Anz.* **201**, 31–43 (1978).
15. T. P. Fleming, "The ultrastructure and histochemistry of the spermathecae of *Tubifex tubifex* (Annelida: Oligochaeta), *J. Zool.* **193**, 129–145 (1981).
16. M. Henzen, "Cytologische und mikrocytologische Studien über die ooplasmatische Segregation während der Meiose des *Tubifex*-Eies," *Z. Zellforsch.* **71**, 415–440 (1966).
17. O. Hess, "Entwicklungsphysiologie der Anneliden," *Fortschr. Zool.* **16**, 347–349 (1963).
18. Y. Hirao, "Reproductive system and oogenesis in the freshwater oligochaete, *Tubifex hattai*," *J. Fac. Sci. Hokkaido Univ., Ser. VI Zool.* **15**, 439–448 (1964).
19. Y. Hirao, "Cocoon formation in *Tubifex* with its relation to activity of the clitellar epithelium," *J. Fac. Sci. Hokkaido Univ., Ser. VI Zool.* **15**, 625–632 (1965).
20. Y. Hirao, "Cytological study of fertilization in *Tubifex* egg," *Dobutsugaku Zasshi* (*Zool. Mag.*) **77**, 340–346 (1968).
21. K. H. Hoffman, T. Mustafa, and J. B. Jørgensen, "Role of pyruvate kinase, phosphoenolpyruvate carboxykinase, malic enzyme, and lactate dehydrogenase in anaerobic energy metabolism of *Tubifex* sp.," *J. Comp. Physiol.* **B130**, 337–345 (1979).
22. M. Inase, "On the double embryo of the aquatic worm *Tubifex hattai*," *Sci. Rep. Tohoku Univ., Ser. 4* **26**, 59–64 (1960).
23. M. Inase, "The culture solution of the eggs of *Tubifex*," *Sci. Rep. Tohoku Univ., Ser. 4* **26**, 65–67 (1960).
24. M. Inase, "Behavior of the pole plasm in the early development of the aquatic worm, *Tubifex hattai*," *Sci. Rep. Tohoku Univ., Ser. 4* **33**, 223–231 (1967).
25. P. P. Ivanov, *Textbook of General and Comparative Embryology* [in Russian], Uchpedgiz, Leningrad (1945).
26. H. Jaana, T. Shimizu, C. Suzutani-Shiota, and R. Yasumara, "Fine structure and chemical properties of the wall of the *Tubifex* cocoon," *Dobutsugaku Zasshi (Zool. Mag.)* **89**, 130–137 (1980).
27. J. R. Kaster, "Morphological development and adaptive significance of autotomy and regeneration in *Tubifex tubifex* Müller," *Trans. Am. Microsc. Soc.* **98**,473–477 (1979).
28. D. Kosiorek, "Development cycle of *Tubifex tubifex* Müll. in experimental culture," *Pol. Arch. Hydrobiol.* **21**, 411–422 (1974).
29. F. E. Lehmann, "Die Zucht von *Tubifex* für Laboratoriumzwecke," *Rev. Suisse Zool.* **48**, 559–561 (1941).
30. F. E. Lehmann, "Zur Entwicklungsphysiologie der Polplasmen des Eies von *Tubifex*," *Rev. Suisse Zool.* **55**, 1–43 (1948).
31. F. E. Lehmann, "Plasmatische Eiorganization und Entwicklungsleitung bei Keim vom *Tubifex* (Spiralia)," *Naturwissenschaften* **43**, 289–296 (1956).
32. F. E. Lehmann, "Synergie et antagonisme des substances antimitotiques et morphostatiques et leur influence sur l'activité morphogénétique de l'hyaloplasme embryonnaire," in: *L'action antimitotique et caryoclasique de substances chimiques, Colloque Intern.* **B88**, 125 (1960).
33. F. E. Lehmann and H. Hadorn, "Vergleichende Wirkungsanalyse von zwei antimitotischen Stoffen, Colchicin und Benzochinon, am *Tubifex*-Ei," *Helv. Physiol. Pharmacol. Acta* **4**, 11–42 (1946).

34. F. Lehmann, M. Henzen, and F. Geiger, "Cytological and electron microscopic study of *Tubifex* eggs and *Amoeba proteus* cytoplasmic components in the living and fixed state," in: *Cell Ultrastructure and Function* [Russian translation], Mir, Moscow (1965).
35. I. I. Malevich, "Oligochaeta," in: *Life of Animals*, Vol. 1 [in Russian], Prosveshchenye, Moscow (1968).
36. M. Matsumoto and M. Kusa, "Time-lapse cinematographic recording of *Tubifex* eggs during maturation and early cleavage," *Dobutsugaku Zasshi (Zool. Mag.)* **75**, 270–275 (1966).
37. M. Matsumoto, "A cytological study of ash distribution in the embryo of an oligochaete, *Tubifex hattai*," *Dobutsugaku Zasshi (Zool. Mag.)* **77**, 12–18 (1968).
38. M. Matsumoto, "Concerning mucoid substances in the region of cleavage furrows of the developing embryos of aquatic Oligochaetes," *Dobutsugaku Zasshi (Zool. Mag.)* **77**, 44–51 (1968).
39. M. Matsumoto, "Histochemical characteristics of a mucoid substance present in the region of the cleavage furrow of *Tubifex* eggs," *Dobutsugaku Zasshi (Zool. Mag.)* **77**, 81–86 (1968).
40. A. Meyer, "Die Entwicklung der Nephridien und Gonoblasten bei *Tubifex rivulorum* Lam. nebst Bemerkungen zum natürlichen System der Oligochäten," *Z. Wiss. Zool.* **133**, 517–562 (1929).
41. A. Meyer, "Ein atavistischer *Tubifex*-Embryo mit überzähligem Gonoblastenpaar und seine Bedeutung für die Theorie der Segmentstauchung bei den Oligochäten," *Zool. Anz.* **85**, 321–329 (1929).
42. F. Palazzo, "Effetti dell'azide sodico sull'uovo di *Tubifex rivulorum*," *Ric. Sci.* **25**, 2873–2876 (1955).
43. A. Penners, "Die Furchung von *Tubifex rivulorum* Lam," *Zool. Jahrb. Abt. 2* **43**, 323–368 (1922).
44. A. Penners, "Die Entwicklung des Kemstreifs und Organbildung bei *Tubifex rivulorum* Lam," *Zool. Jahrb. Abt. 2* **45**, 251–308 (1923).
45. A. Penners, "Experimentelle Untersuchungen zum Determinationsproblem am Keim von *Tubifex rivulorum* Lam. I. Die Duplicitas cruciata und organbildende Keimbezirke," *Arch. Mikrosk. Anat.* **102**, 51–100 (1924).
46. A. Penners, "Experimentelle Untersuchungen zum Determinationsproblem am Keim von *Tubifex rivulorum* Lam. II. Die Entwicklung teilweise abgetöteter Keime," *Z. Wiss. Zool.* **127**, 1–140 (1926).
47. T. L. Poddubnaya, "About parthenogenesis in Tubificidae (Oligochaeta)," *Tr. Inst. Biol. Vnutr. Vod. AN SSSR* **44/47**, 3–13 (1980).
47a. U. Schöttler and G. Schroff, "Untersuchungen zum anaeroben Glykogenabbau bei *Tubifex tubifex* M.," *J. Comp. Physiol.* **B108**, 243–254 (1976).
48. J. Seuss, T. Mustafa, and K. H. Hoffmann, "Anpassungen der Citratsynthetase von *Tubifex* spec. an eine fakultativ anaerobe Lebensweise," in: *Verh. Dtsch. Zool. Ges. 73 Jahres versamml., Berlin*, Stuttgart–New York (1980).
49. T. Shimizu, "Occurrence of microfilaments in the *Tubifex* egg undergoing the deformation movement," *J. Fac. Sci. Hokkaido Univ., Ser. VI* **20**, 1 (1975).
50. T. Shimizu, "The fine structure of the *Tubifex* egg before and after fertilization," *J. Fac. Sci. Hokkaido Univ., Ser. VI* **20**, 253–263 (1976).
51. T. Shimizu, "The staining property of cortical cytoplasm and the appearance of pole plasm in *Tubifex* egg," *Dobutsugaku Zasshi (Zool. Mag.)* **85**, 32–39 (1976).

52. T. Shimizu, "Mode of microfilament arrangement in normal and cytochalasin-treated eggs of *Tubifex* (Annelida, Oligochaeta)," *Acta Embryol. Exp.* **1**, 59–74 (1978).
53. T. Shimizu, "Deformation movement induced by divalent ionophore A 23187 in the *Tubifex* egg," *Dev. Growth Differ.* **20**, 27–33 (1978).
54. T. Shimizu, "Surface contractile activity of the *Tubifex* egg: its relationship to the meiotic apparatus functions," *J. Exp. Zool.* **208**, 361–377 (1979).
55. T. Shimizu, "Cyclic changes in shape of a non-nucleate egg fragment of *Tubifex* (Annelida, Oligochaeta)," *Dev. Growth Differ.* **23**, 101–109 (1981).
56. T. Shimizu, "Cortical differentiation of the animal pole during maturation division in fertilized eggs of *Tubifex* (Annelida, Oligochaeta). I. Meiotic apparatus formation," *Dev. Biol.* **85**, 65–76 (1981).
57. T. Shimizu, "Cortical differentiation of the animal pole during maturation division in fertilized eggs of *Tubifex* (Annelida, Oligochaeta). II. Polar body formation," *Dev. Biol.* **85**, 77–88 (1981).
58. T. Shimizu, "Ooplasmic segregation in the *Tubifex* egg: mode of pole plasm accumulation and possible involvement of microfilaments," *Wilhelm Roux's Arch. Dev. Biol.* **191**, 246–256 (1982).
59. T. Shimizu, "Organization of actin filaments during polar body formation in eggs of *Tubifex* (Annelida, Oligochaeta)," *Eur. J. Cell Biol.* **30**, 74–82 (1983).
60. T. Shimizu, "Dynamics of the actin microfilament system in the *Tubifex* egg during ooplasmic segregation," *Dev. Biol.* **106**, 414–426 (1984).
61. T. Shimizu, "Movements of mitochondria associated with isolated egg cortex," *Dev. Growth Differ.* **27**, 149–154 (1985).
62. T. Shimizu, "Bipolar segregation of mitochondria, actin network, and surface in the *Tubifex* egg: role of cortical polarity," *Dev. Biol.* **116**, 241–251 (1986).
63. F. Stéphan-Dubois and G. Biver, "Action de l'actinomycine D sur la régénération caudale de *Tubifex tubifex* (Annélide, Oligochète)," *C. R. Soc. Biol.* **168**, 1068–1071 (1974).
64. F. Stéphan-Dubois and S. Schwartz, "Cycle annuel de l'activité neurosécrétrice et de l'activité sexuelle chez l'Annélide *Tubifex tubifex*," *C. R. Soc. Biol.* **172**, 697–700 (1978).
65. P. Stöckel, A. Mayer, and R. Keller, "X-ray small-angle-scattering investigation of a giant respiratory protein: haemoglobin of *Tubifex tubifex*," *Eur. J. Biochem.* **37**, 193–200 (1973).
66. C. Suzutani-Shiota, "Ultrastructural study on cocoon formation in the freshwater oligochaete, *Tubifex hattai*," *J. Morphol.* **164**, 25–38 (1980).
67. P. G. Svetlov, "Early developmental stages of *Rhynchelmis limosella* Hoffm.," *Izv. Biol. Nauchn.-Issled. Inst. Permsk. Univ.* **2**, 141–152 (1923).
68. P. G. Svetlov, "Embryonic development of Naididae fam.," *Izv. Biol. Nauchn.-Issled. Inst. Permsk. Univ.* **4**, 359–372 (1926).
69. R. Weber, "Zur Verteilung der Mitochondrien in frühen Entwicklunsstadien von *Tubifex*," *Rev. Suisse Zool.* **63**, 277 (1956).
70. R. Weber, "Uber die submikroskopische Organization und die biochemische Kennzeichnung embryonaler Entwicklungsstadien von *Tubifex*," *Wilhelm Roux's Arch. Entwicklungsmech. Org.* **150**, 542–580 (1958).

71. R. Weber, "The electronmicroscopic study of embryonic differentiation," in: *The Interpretation of Ultrastructure*, R. J. C. Harris (ed.), Academic Press, New York–London (1965).

Chapter 5

THE COMMON POND SNAIL
Lymnaea stagnalis

V. N. Meshcheryakov

5.1. INTRODUCTION

The common pond snail is a freshwater species widely used in embryological studies. Oogenesis and egg maturation of the pond snail involve complex events of ooplasmic segregation, including polar as well as mosaic differentiation of the cortical layer [141, 144, 174]. The role of this layer as a morphogenetic system has been clarified to a large extent in studies employing the pond snail [140, 141]. Typical spiral cleavage enables the fate of individual cell descendants to be traced [175, 176]. Cytoplasmic segregation increases in the course of cleavage and individual cells undergo early differentiation. The divisions rapidly become asynchronous, while the regular rhythm of cleavage in different cell lineages becomes an important prerequisite for normal morphogenesis [23, 31]. These events can be conveniently studied using the pond snail because of the changing structure of the mitotic cycles [18, 19, 24]. Early blastulation in the pond snail correlates with the early appearance of specialized cell junctions at the stage of two blastomeres; junctions play an important part in the physiology of the early embryo [72, 171] and during the critical periods of embryogenesis such as the emergence of the body plan and the dorsoventral organization [25, 137, 141, 143]. The structure of these junctions has been studied intensively at the submicroscopic level using various procedures, including freeze-fracturing techniques [56]. Methods have been developed which enable a three-dimensional reconstruction of the whole embryo from its thin sections to be obtained; these techniques allow a specific determination of the types and number of contacts in individual blastomeres [181]. The cell surface of early embryos has been studied using a scanning electron microscope [126].

The results of embryological and cytological studies conducted over the last 20 years suggest the existence of several different types of embryonic determination in the course of pond snail development [25, 48, 144]. These studies demonstrate convincingly the central place of cellular interactions in the mollusc embryogenesis, which is generally considered a classic example of mosaic development. Mosaic features are manifested as the early differentiation of certain larval rudiments or as the early irreversible pinpointing of regions competent to respond to induction stimuli. The most important processes of determination are associated with the contact interactions of blastomeres in the course of blastulation or of whole cellular sheets during gastrulation. It is noteworthy that early determination is associated with changes in embryo symmetry, followed by the appearance of regular differences in the rate and orientation of cell divisions [21, 22, 115]. In contrast, biochemical specialization may precede the histological regionalization of the rudiments in the course of late development [172]. Blastomere isolation in the common pond snail and other species of pulmonate molluscs, achieved by deletions or by separation using hair loops [40, 74, 123], enables certain regulatory capabilities of the early embryo to be revealed. It has been demonstrated, for example, that pulmonate molluscs belong to the small group of animals which regulate cleavage according to the whole egg pattern, and only in this case does subsequent embryogenesis proceed normally.

Detailed studies of the cleavage and cellular interactions responsible for the embryonic organizational pattern have served as the basis for the mathematical simulation of early pond snail development [16, 144].

During cleavage the pond snail embryo is a suitable model for studying cytokinesis, mitosis, macro- and micropinocytosis, as well as cellular interactions [26, 55, 72, 75, 112, 178]. Removing the vitelline membrane (using proteases) enables the cell surface to be marked by carbon particles and its movements in the course of divisions to be traced. This method enabled the detection of the spiral deformation of the cleaving zygotes which have opposite directions in dextral and sinistral molluscs [118]. A procedure for isolating the egg cortical layer has been developed; this allows an examination of the organization of contractile structures responsible for cytokinesis on large surface areas [113]. By adding fragments of skeletal muscle myosin to cortices containing intact microfilaments, we have developed a contractile model simulating cytokinesis upon the addition of ATP [114, 119]. The possibility of testing various macromolecules makes this model a useful tool, not only when studying contractile mechanisms in the nonmuscle systems, but also for elucidating mechanisms underlying dextral and sinistral cleavage. Zygotes of the pond snail can be used for studying the mechanisms of receptor-mediated endocytosis which may be easily evoked by various multivalent ligands (Meshcheryakov, unpublished).

The factors responsible for the orientation of cleavage spindles in the pond snail may be studied by changing the shape and mutual orientation of the blastomeres; this can be achieved by the centrifugation of embryos and by changing the proportion of nuclear and cytoplasmic volumes in zygotes [59, 108, 110, 111]. Simple methods for isolating mitotic structures from the eggs and embryos of the pond snail allow the characteristics of multipolar mitoses to be examined, for example in the binucleate blastomeres which are produced by cytochalasin treatment and which have different initial distances between the nuclei [109, 116]. Cytokinesis and mitosis may also be studied using smaller zygotes with different cytoplasm composition [108].

The capsular and vitelline membrane can be permeated by a wide range of substances. This enables experimental studies of cell cycles to be performed and autoradiographic and radiometric experiments to be conducted [18-20, 63, 102, 141]. Studies of protein synthesis with the aid of electron microscopic autoradiography are also relevant [142, 179, 180] because they throw new light on the function of yolk granules in development.

Pond snail embryos are being used to study genetics and the molecular biology of development. A large number of eggs present in one egg mass, which is undergoing almost synchronous development, enables nucleic acids at different stages of embryogenesis to be isolated [36, 39, 169]. Changes in the spectra of synthesized proteins, activities of different enzymes, and the morphogenetic function of nuclei have been studied [125, 130, 131]. Self-fertilization in the pond snail has been used to convert genes into a homozygous state; this enables their periods of active functioning in the course of embryogenesis to be detected [177].

From the period 1923-1930 [35, 168] much attention has been paid to the problem of genetic control underlying dextrality–sinistrality (dissymmetry) in pulmonate molluscs; this trait is inherited maternally and appears for the first time in the course of spiral cleavage cytokinesis [117]. The latest studies performed with *Lymnaea peregra* enabled numerous genetic aspects of the problem to be defined. They also allowed us to demonstrate that dissymmetry of the dextral cleavage type is the result of a dominant gene product located in the egg cytoplasm [61].

Later stages of embryogenesis, cytodifferentiation in connection with polyploidization, postembryonic growth, and histogenesis in the pond snail are also the subject of numerous studies [29, 162]. Lastly, embryogenesis in the pond snail may be used as an inexpensive and readily available model for teratological experiments. These studies are aimed, for example, at testing reservoirs for industrial pollution [100].

The pond snail is quite common in natural freshwater habitats and may be easily cultured in the laboratory [79, 86, 107, 163]. The adult animals are used, for example, for studying the properties of enzymes, hormones, and calcium metabolism [54, 160, 184], water metabolism, the function and composition of the hemolymph [158, 183], defense reactions and phagocytosis [97], neurophysiological processes [96, 129, 173], the physiology of nutrition [156], behavior [50, 81, 166], and the dynamics of natural populations [27]. Considerable advance has been made in the study of the reproduction biology of the pond snail which shows endogenous rhythms under a complex hormonal control [52, 66, 78, 84, 85, 95, 134, 153, 164, 165].

Obviously, the common pond snail is one of the popular laboratory animals and is widely used in experimental biology.

We now list some of the drawbacks of using the pond snail as a laboratory species. These include:

a. Difficulties in culturing embryos outside capsules, which is only possible up to gastrulation.
b. Few opportunities for performing microinjections and conducting electrophysiological studies because the plasma membrane of the egg or blastomeres is highly sensitive to mechanical lesions.
c. Rather high mortality rate among young snails, thereby making it difficult to maintain constant laboratory cultures.
d. High degree of infestation among natural populations by fluke larvae. This means that a considerable number of specimens collected in the field have to be discarded.

Some of these drawbacks can be overcome by using other species of the genus *Lymnaea*, also used in embryological studies: *L. acuminata, L. auricularia, L. exilis, L. elodes, L. ovata, L. palustris, L. peregra*, and others. The embryos of *L.*

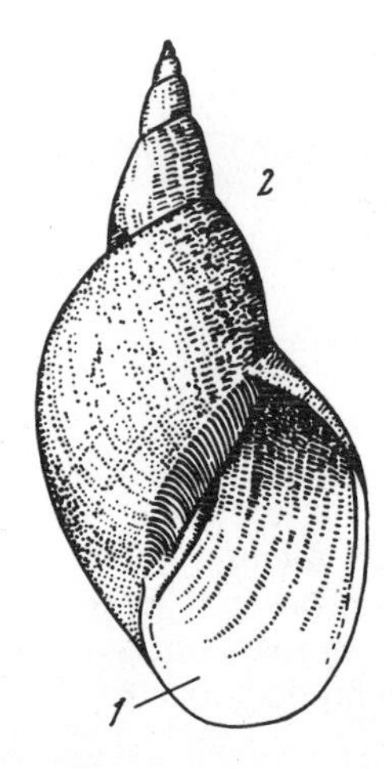

Fig. 5.1. Shell of *L. stagnalis* (natural size) [187]. 1) Aperture; 2) helix.

peregra, for example, allow us to perform microinjections amounting to about 2-4% of the egg volume. This has been used in experiments to determine the temporal interval in which the genes responsible for coiling are active during development [61].

Morrill [127] has recently published a comprehensive survey, concerning the normal development and experimental embryology of *L. palustris* as well as other species of Lymnaeidae.

5.2. TAXONOMY AND DISTRIBUTION

The genus *Lymnaea* (Gastropoda, Pulmonata, Basommatophora, Lymnaeidae) contains numerous species of the pond snail. About 20 species can be found in the USSR [161]. They include *L. stagnalis* L., which can be easily distinguished by its light-colored shell (Fig. 5.1), the helix of which has height no less than the diameter of the mouth, and whose last turn is greatly inflated. The largest shells attain 70 mm in height while the average height is 40-47 mm [187]. The shape of shells may vary considerably. Normally the shell has right-handed coiling, exceptions to this being extremely rare. The location of the inner organs is shown in Fig. 5.2 [187]. Further details on the anatomical and microscopic structure of different organs can be found in specialized studies [2, 12, 28, 47, 70, 91, 133, 134, 159, 185, 186].

L. stagnalis is common in Europe, northern Asia, and North America. In the USSR it can be found throughout the entire European part as well as in the Trans-Caucasian region, central Asia, eastern and western Siberia, and in the Far East.

5.3. ECOLOGY AND CULTURE

The pond snail lives in the littoral zone of stagnant reservoirs or slow rivers (e.g., the Moskva). It prefers reservoirs where the pH of water is 6.0-7.0 [135].

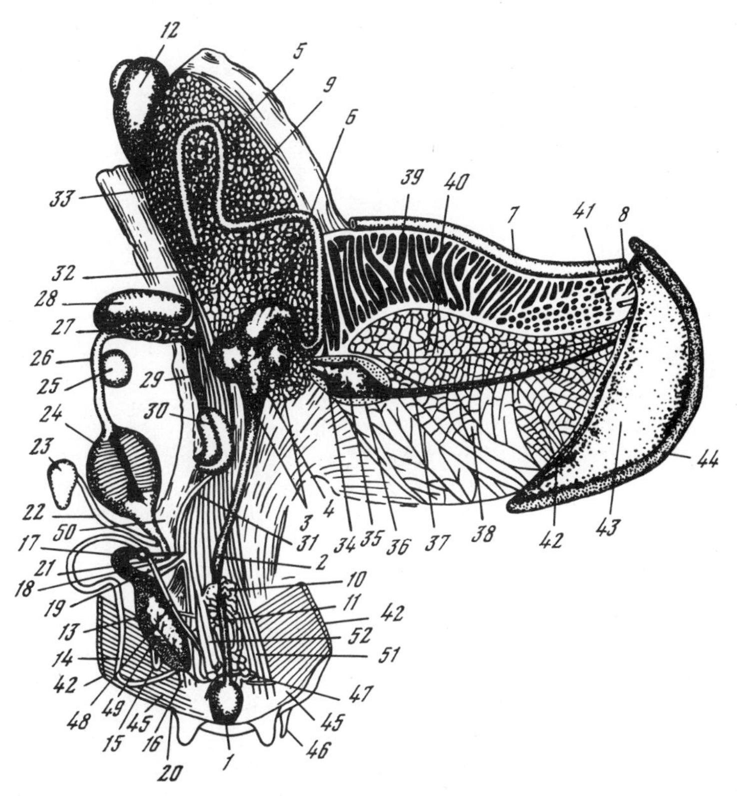

Fig. 5.2. Sketch illustrating internal organs of *L. stagnalis* [187]. 1) Oral sac; 2) esophagus; 3) muscular stomach; 4) crop; 5) intestine; 6) pyloric part of stomach; 7) rectum; 8) anus; 9) "liver" or digestive gland; 10) salivary glands; 11) salivary gland ducts; 12) spiral part of body containing one lobe of digestive gland; 13) penial sac; 14) vas deferens; 15) preputium protractor muscle; 16) male genital pore; 17) penis; 18, 19) penial sac retractors; 20) body wall; 21) female genital pore; 22) seminal receptacle duct; 23) seminal receptacle; 24) first accessory albumen gland; 25) second accessory albumen gland; 26) oviduct; 27) uterine part of oviduct; 28) albumen gland; 29) posterior part of prostate; 30) anterior part of prostate; 31) prostatic region of vas deferens; 32) hermaphroditic duct; 33) hermaphroditic gland; 34) ventricle; 35) atrium; 36) pericardial cavity; 37) pulmonary sac; 38) plexus of blood vessels on left kidney lobe; 39) pulmonary sac; 40) glandular part of kidney; 41) urethra; 42) cut layer of muscles; 43) velum; 44) muscles of velum; 45) cut tissue of the head; 46) tentacles; 47) suprapharyngeal ganglion; 48) nerve to head; 49) nerve to penial sac; 50) nerve to female genitals; 51) foot muscles; 52) right retractor of pharyngeal sac.

Because of its capacity to withstand water pollution it is classed as a mesosaprobe.

In a moderate climatic zone the pond snail feeds from May to October. For food it used bacteria, algae, detritus, higher plants (frequently in the early stages of stagnation), and in summer it may eat ticks, snails, and other small animals, as well as insect eggs [157]. It is generally quite voracious.

During the warm season the pond snail uses atmospheric air, breathing with its lung; this, however, does not exclude the possibility of filling the pulmonary cavity with water and using the oxygen dissolved in it. For wintering the animal digs into the ground and the pulmonary respiration is replaced by skin respiration. *L. stag-*

nalis lives at least a year on average, sometimes longer, from 2 to 5 years [15, 104]. It has been reported that the maximum age of the pond snail is 7 years [27].

In nature, pond snails often serve as the intermediate hosts of various flukes, including *Fasciola hepatica*; the larval forms of these flukes frequently infest the internal organs, such as the liver, of a considerable proportion (sometimes up to 40%) of molluscs in a given water reservoir. Since larval infestation affects fertility [87], the possibility of infestation should be checked immediately when animals collected in the field are destined for the laboratory. For this purpose the collected molluscs are placed individually into glass beakers containing 100-150 ml of clean tap water and kept without food. If they are infested with trematodes, the fluke larvae appear in the water as small white mobile points after 1-3 days. Such molluscs are then discarded. Warming the water by several degrees increases the reliability of the method.

Although heavier body weight implies greater weight of gonads [103], moderately sized snails should be selected for the laboratory. This is because the largest specimens are usually infested by trematodes. If a constant laboratory population is needed, molluscs can be grown from eggs in the laboratory. This procedure, however, is rather cumbersome because the mortality rate of juvenile snails is high and correspondingly many batches should be cultured. However, pond snail progeny reared in the laboratory are always free from fluke larvae.

Sometimes reproduction is impeded because of infestation by semiparasitic oligochaetes such as *Chaetogaster lymnaei*, which resemble large rotifers [155].

Conditions for keeping pond snails can vary depending on the purpose of the future experiments. Generally a constant laboratory population with minimal mortality and moderate reproduction activity is desirable. In the laboratory each animal should be provided with at least 0.5 liter of water. Increased population density not only adversely affects the feeding intensity and fertility, but may even result in cannibalism [45, 81, 107, 121]. The animals should preferably be kept in large aquariums with constant aeration and running water [122, 132, 163]. Placing aquatic plants into aquariums is not a necessitv. but may be desirable as a source of small animals for food. If the aquarium is stationary and there is constant aeration, the water need only be changed one every 2 weeks. Without aeration at 17-20°C the water should be changed weekly, and if plenty of food is supplied it should be done even more frequently. The following may be used to feed the molluscs: carrots, radish leaves, cucumbers, cabbages, various aquatic plants. Pond snails prefer lettuce leaves and *Hydrocharis*. Dead pond snails may be left in the aquarium because they are usually eaten by the remaining molluscs. Fraser [60] recommends feeding with boiled wheat grains. According to some observations, food should contain mineral particles; in other words, some sand or river mud should be put into the aquariums [167]. Plenty of food generally has a favorable effect on the growth and reproduction of the pond snail [165].

In winter, aquariums should be equipped with additional illumination facilities. An illumination period lasting 12 h is sufficient to maintain moderate intensity of metabolism; a lower temperature and a shorter illumination period may stimulate the snail's preparation for hibernation [89]. Growth and gametogenesis proceed at temperatures of no less than 11-12°C [157].

Genetic and embryological experiments may require self-fertilization in order to obtain homozygous embryos. For this purpose snails hatched from one egg mass are kept together until they reach 5 mm in length. Thereafter each animal is kept in an individual vessel [177].

Mass production of normal fertilized eggs for embryological studies requires the pond snails to be placed periodically under conditions which stimulate oviposition. Such conditions include a higher temperature, low content of metabolic products such as ammonia and urea in the water, no food, and the presence of green plants. Food, however, should be plentiful during the preceding period, and the illumination period should be sufficiently long; the latter generally maintains reproduction at a high rate (for details see Section 5.4). According to the author's experience, this can be achieved by keeping two or three pond snails together in special vessels with 2-3 liters of water but without artificial aeration. Each group of snails should be stimulated to lay eggs no more than twice a week. A smaller volume of water in the vessels leads to rapid water pollution, thereby preventing spontaneous egg-laying. Care should be taken to keep this contamination at an acceptably low level, avoiding water acidification and the appearance of bacterial growth on the surface of the water.

5.4. REPRODUCTION AND STIMULATION OF OVIPOSITION

Pond snails are hermaphrodites but, as a rule, mutual insemination takes place during mating. Self-fertilization is less frequent. Furthermore, when the snails are cultured individually, egg-laying begins later than when several animals are placed together. The common pond snail reaches maturity approximately 1.5 months after hatching from the cocoon. When embryogenesis has been completed, the gonads pass through the juvenile stage, then the male maturity stage, and then achieve complete hermaphroditism [13]. Ovulation first begins when the shell height attains about 25 mm. It can, however, be stimulated even in younger animals by injecting an extract prepared from the caudal-dorsal cells of the cerebral ganglia [53]. The onset of reproduction may also be affected by the duration of the photoperiod. If the photoperiod is long (16 h), at 20°C the snails begin to lay eggs as early as 7 weeks after hatching. If its duration is 12 h, they start oviposition after 9 weeks [30].

In a moderate climatic zone adult pond snails begin to lay eggs late in May. Oviposition reaches its maximum in June and ends in August [15, 27]. The temperature at which pond snails lay their eggs is generally 15-18°C [103]. In nature the reproduction of pond snails begins during their second summer as they only grow to 8-10 mm in size during the first summer. Under laboratory conditions (20-23°C) the molluscs lay eggs from March to November [49]. Although the author managed to obtain eggs from pond snails in December and January under the conditions of a 12-h photoperiod, it is not usually recommended that embryological experiments be conducted during the winter months. As previously mentioned, reproduction is stimulated by a long photoperiod, abundant nutrition, optimal temperature, clean water, and moderate population density of pond snails. The function of gonads is quickly affected by starvation [87, 165]. According to Levina [104], the pond snail may lay about 8700 eggs during its lifetime, although other data put the number at above 10,000 [103].

Reproduction in the pond snail is controlled by several regions of the nervous system [85]. Two groups of caudal-dorsal cells located in the cerebral ganglia (see Fig. 5.19, stage 25 K) secrete the ovulation hormone, usually no later than 2 h prior to egg-laying [95, 153]. The dorsal bodies produce another hormone acceler-

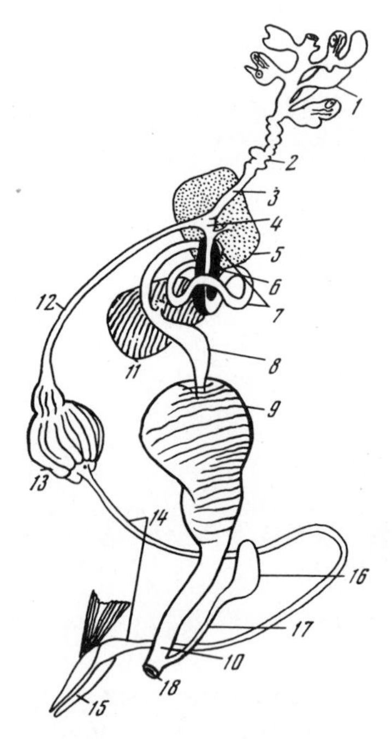

Fig. 5.3. Genital tract of *L. stagnalis* [136]; penial sac and muscles are not shown. 1) Ovotestis; 2) vesicular part of hermaphroditic duct with seminal vesicles; 3) straight part of hermaphroditic duct; 4) bifurcation of hermaphroditic duct; 5) albumen gland; 6) ventricular part of oviduct; 7) convoluted part of oviduct (uterus); 8) straight part of oviduct; 9) oothecal part of oviduct; 10) vaginal part of oviduct; 11) accessory mucous gland; 12) pars afferens of seminal duct; 13) prostate; 14) vas deferens; 15) penis; 16) seminal receptacle; 17) seminal receptacle duct; 18) female genital pore.

ating the differentiation of accessory organs of the female part of the gonad [66]. The lateral lobes of the cerebral ganglia stimulate the secretion of both these hormones and inhibit the neurosecretory light-green cells which produce the growth hormone [152]. A balance between growth and reproduction can be partly controlled by the duration of the photoperiod. Hormones of vertebrate animals have been used to control reproduction of the pond snail [69].

Gametogenesis in the pond snail has been analyzed extensively [4-10, 34, 83, 94, 139, 151, 174]. The duration of oogenesis from the completion of oogonial mitoses (see Fig. 5.12) to oviposition is about a month [42].

During the course of mating, sperms enter the spermatheca through the female genital pore (see Figs. 5.2, 5.3) and some time thereafter small packets of spermatozoa are delivered into the vagina. They pass through the oviduct and fertilize the eggs in the straight part of the hermaphroditic duct [41, 134, 136].

Fertilized eggs from the hermaphroditic duct individually enter the ventricular part of the oviduct, which simultaneously receives a portion of the secretion produced by the albumen gland.

Thereafter the egg, along with some of the liquid, enters the pouches of the convoluted part of the oviduct (the uterus), where a two-layered capsular membrane and the mucous capsule pellicle are being formed. In the straight region of the oviduct the capsules are glued together by the secretion of the accessory mucous gland and, finally, the envelope of the cocoon is formed in the nidamental or oothecal part. The egg masses are generally deposited onto green plants such as *Nymphaea alba, Nuphar luteum, Potamogeton natans,* and *Hydrocharis morsus ranae.*

The envelope of the cocoon immediately after laying is opaque, but 1.5-2 h later it swells and becomes transparent. According to Raven [136] the whole process at 17°C, from the entry of the first egg into the oviduct to laying, takes 2 h from the time the molluscs were stimulated.

When the cocoon is laid, it emerges blunt end first from the vagina. The eggs located at this end begin to develop earlier (the difference in the developmental stages of anterior eggs and those located in the posterior part of the cocoon may be 45-50 min at 22°C). The first snails to hatch from the capsules, however, are those located in the external region of the central cocoon part. They are followed by those at the anterior tip and finally by the capsules of the posterior cocoon end [49]. With regular feeding and aerated water the reproduction may be stimulated by several factors, such as lower water temperature (by several degrees) [77] or the replacement of rougher feed by lettuce leaves and a longer photoperiod (our observations). Other methods for stimulating reproduction are as follows:

a. Molluscs are kept in the aquarium without plants, and fed regularly; thereafter *Hydrocharis* leaves are placed into the aquarium and intense illumination is supplied [136].
b. Molluscs are kept at 17-18°C, fed regularly, and then the temperature is gradually raised to 25-26°C [146].

We use the following technique. Molluscs are kept in groups of 2-3 animals in aquariums or finger bowls without aquatic plants but with a small amount of river sand. The water is not changed for a week while the animals are regularly fed with lettuce. Two to three days prior to stimulation, cucumber pulp or cabbage leaves are added. This rapidly pollutes the water. The illumination is kept intensive at room temperature. On the day of the experiment the polluted water is changed for fresh water, an intense light is switched on, no food is given, and the water is gradually heated to 25-26°C. *Hydrocharis* leaves or transparent plastic dishes with a flat bottom are placed onto the water surface; the snails use them for oviposition, which takes place several hours after the water has been changed. The snail does not usually change its position for some time prior to egg-laying.

5.5. THE GAMETES

5.5.1. Eggs

Pond snails lay eggs in elongated cocoons or syncapsules [14], which at the beginning of the reproductive period contain several, and later up to 300, eggs each of which is located in an individual egg capsule (Fig. 5.4).

The mean capsule length is 1.3 mm with a diameter of 0.8 mm [49]. The presence of several eggs in a single capsule (polyvitelliny) is abnormal. Fertilized eggs are laid at metaphase I [138]. The egg is somewhat irregular in shape immediately after oviposition. However, after about 5 min it acquires the shape of

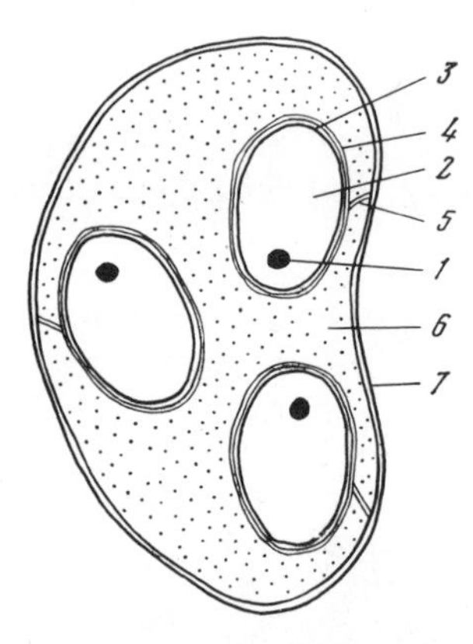

Fig. 5.4. Sketch illustrating structure of *L. stagnalis* cocoon. 1) Egg covered by vitelline membrane; 2) albumen fluid of egg capsule; 3) membrane of egg capsule; 4) outer membrane of egg capsule; 5) pedicle of egg capsule; 6) inner mucus of cocoon (tunica interna); 7) outer envelope of cocoon (tunica capsulae).

a rotation ellipsoid; that is, it becomes flattened along the animal–vegetal axis. Its maximum diameter at this stage is 122 μm (the average is 120 μm). The egg volume varies from 1.1 to 1.6 nl [171]. However, during the period between laying and the onset of cleavage, the egg volume slowly increases by about 45% owing to the uptake of water [136, 141].

A small part of the egg in the animal pole region is free of yolk and contains the first maturation division spindle. Usually the egg has a yellowish-white pigmentation. In addition to a plasma membrane, the egg is covered by a thin (no thicker than 1 μm) vitelline or yolk membrane. The perivitelline space is virtually absent (0.5 μm or less) [124]. The vitelline membrane, which consists of loosely interwoven fibrils and small round bodies, is formed from the peripheral layer of the oocyte zona radiata [174].

The egg may be regarded as telolecithal because of the distribution of yolk; the animal half is less rich in yolk and contains the nucleus.

The plasma membrane forms microvilli, which contain bundles of microfilaments. Microvilli are first formed at the free oocyte pole corresponding to the future animal pole of the egg. The oocyte region attached to the gonad basal membrane corresponds to the vegetal pole [151].

Under the plasma membrane of the mature egg, actin microfilaments form several successive layers, the microfilaments of adjacent layers being inclined to one another at twist angles of 30-60° or 90°. Each layer is composed of many small domains containing 2-20 parallel microfilaments of identical polarity as revealed by decoration with myosin subfragment I [114, 119, 120]. It cannot be excluded that such domain-helicoid organization of the microfilament network is directly related to the spiral nature of cleavage cytokinesis (see below). The pond snail egg does not contain cortical granules. The growing oocyte vigorously takes up exogenous ferritin, along with other substances, and ferritin taken by endocytosis involving coated pits and vesicles forms paracrystalline inclusions in the yolk granules [34].

According to our observations, these coated structures are also present in the mature egg cortex and may be intimately associated with microfilaments [119].

Prior to cleavage microtubules are not numerous [105] in the cortical layer, whose thickness does not exceed 0.1 μm; by contrast, during the interphase of cleavage divisions they may accumulate under the outer blastomere surface as parallel and oblique rows [124]. Mitochondria and ribosomes are distributed nonuniformly along the polar egg axis. During the first cleavage division a well-developed Golgi system (with bulky trans-reticular elements and many coated pits) can be observed in the subcortical cytoplasm (Meshcheryakov, unpublished data).

The pond snail egg has yolk granules of different types containing hydrolases, complex proteins, phospholipids, basic proteins, and RNA [142, 174, 179, 180]; in addition large lipid droplets are present.

Yolk β-granules with a diameter of 1-3 μm and coated with a membrane are formed earlier during oogenesis than are the γ-granules. The diameter of the γ-granules exceeds 4 μm after oviposition. The γ-granules are surrounded by a vacuole which takes in water from outside. It is assumed that they play the role of lysosomes during development. β-Granules, which are consumed at a slower rate in the course of embryogenesis than the γ-granules, also appear to participate in the intracellular nutrition [3, 26]. There is evidence that yolk granules participate in the formation of cell membranes [179, 180].

Complex ooplasmic segregation may be traced cytochemically in the pond snail zygote [141, 143, 144, 174]. While the egg passes through the oviduct, a vegetal pole plasm containing abundant yolk β-granules is being formed. After oviposition this plasm moves to the animal pole under the membrane and is transformed into the subcortical plasm. Animal pole plasm rich in mitochondria appears soon after the extrusion of the second polar body. By the third cleavage division both plasms merge together. In the course of subsequent cleavage division the plasms undergo a nonuniform distribution between blastomeres in the form of ectoplasm surrounding the nuclei. At the sites where the oocyte contacts inner follicular cells (see Fig. 5.12), 6-9 subcortical accumulations (SCA) are formed under the equatorial streak. They are very rich in RNA, which does not, however, incorporate any labeled precursors after oviposition [22, 141]. The SCA partially fuse together in the course of synchronous cleavage divisions and can be located under the vegetal surface of the macromeres at the 8-blastomere stage. Later, concomitantly with the determination of the 3D macromere (see below) the SCA undergo compactization. They are transformed into ectosomes migrating to the center of the embryo along the inner contact zones of the blastomeres. It may well be that they participate in inductive interactions [22]. Submicroscopically the SCA correspond to large accumulations of polysomes, endoplasmic reticulum, and small vesicles; the aggregation of these components appears to increase during the course of cleavage [55]. Raven [141, 143] assumed that, prior to cleavage, the SCA system is related to the determination of the embryo symmetry plane, which always passes through the pair of blastomeres B and D.

The egg is suspended in the capsular fluid which is the main source of nutrition for the embryo. It contains galactogen (3-6% dry weight), proteins (6-8%), as well as salts and mucopolysaccharides [3, 26, 57, 80, 170]. The content of individual components depends on the composition and quantity of food [77]. At the 40-49 blastomer stage the capsular fluid undergoes pinocytosis in the animal zone, while at the 120-blastomere stage (corresponding to a late blastula) pinocytosis spreads all over the surface of the embryo [26]. The extracellular digestion of the capsular fluid entering the gut lumen begins at the trochophore stage [3].

The pond snail egg is protected by the cocoon envelope, the jelly, the capsules, the capsular membrane, the protein-rich fluid present in the capsule, and finally the vitelline membrane.

The content of the freshly laid egg is isotonic to a 0.093 M nonelectrolyte solution. The salt content of the egg corresponds to 0.042 M NaCl or 0.05 M LiCl solution. The egg membrane as well as the vitelline membrane is permeable to numerous nonelectrolytes, cytidine, thymidine, uridine, uracyl, amino acids [18-20, 39, 82, 136, 141], puromycin [82], colcemide (Meshcheryakov; our data), and inorganic ions such as alkali metal cations with the possible exception of lithium [58].

It has been shown that the intensity of sodium exchange between the embryo and the capsular liquid increases 50-100 times soon after the onset of cleavage, while water transport increases by a factor of 5 [171].

The cocoon envelope is permeable to inorganic ions, water, and various gases [11, 141, 185], but appears to be only slightly permeable to bile acid salts [90]. The mucus present inside the cocoon appears to exert an absorptive action on alkali metal cations [62]. The capsular membrane is permeable to water, inorganic ions, amino acids, cytochalasin, bile salts, sucrose, raffinose [11, 65, 128], ATP, EDTA [37], sodium azide [46], ethyl alcohol, and urea [138]. The capsule membrane, however, is not permeable to polyethylene glycol (molecular weight 3000-3300) [11]. Actinomycin D has poor penetration [64] and needs to be used in concentrations up to 200 μg/ml for a long period [38]. Neyfakh [131] has observed a rapid effect of actinomycin after puncturing the egg capsules. Puromycin penetrates the egg easily [31, 82, 102]. The colloid osmotic pressure of the capsular fluid is 4-5 mOsm [11].

5.5.2. Spermatozoa

Characteristically, the pond snail spermatozoon possesses a very long intermediate region. The length ratio of the head, the intermediate region, and the tail is approximately 1:12:11. The intermediate region consists of an axonema and a strongly modified mitochondrial layer containing paracrystalline structures. The caudal part is rich in glycogen [1, 67, 93]. The head and middle part of the sperm have been found to contain bundles of filaments with a diameter of 10-12 nm, which appear to be similar to the intermediate filaments [92].

5.6. CULTURE OF EMBRYOS

Observations of development and certain other types of experiments may be performed using the cocoons. It is much more convenient, however, to work with individual capsules. To obtain these, the egg mass is placed on a filter paper and cut into several parts with a surgical knife or a razor blade; then the capsules are pressed out with forceps onto filter paper. The capsules are rolled on paper to remove the jelly. They must be cleansed of the jelly particularly thoroughly if there are no plans to extract embryos, because the jelly may possess either protective or teratogenic properties after interaction with various agents [62, 102].

The capsules may be cultured in ordinary tap water, which should be changed daily and, if possible, additionally purified of copper traces. The best results are obtained if the capsules are placed onto a 2% agar surface saturated with water vapor (cracks in the agar may be filled with water [175]). In the latter case the optimal temperature is 25°C [49]. Under these conditions the spontaneous mortality of embryos does not exceed 2% [177].

In order to facilitate permeability to different substances, the capsules may be punctured as described below. Although the capsular fluid undergoes gelatinization around the hole sometime after puncturing, the punctured capsule nevertheless becomes more prone to bacterial and fungal infection. Nevertheless, a certain proportion of the embryos in such capsules develop successfully up to hatching and almost all of them pass through the gastrula stage. Care should be taken to prevent too much of the fluid from leaking.

Finally, fixation and certain experimental manipulations require the embryos to be removed from the capsules. The best procedure for performing this is as follows: a batch of capsules is transferred by a pipette into a dry plastic dish and excess water is removed. The capsules are aligned by a glass rod along the wall under a dissecting microscope and then quickly punctured with sharp needles. During this procedure some of the fluid leaks out, thereby reducing the intracapsular pressure, and the majority of embryos remain inside. Some water is added and the fluid remaining in the capsules along with the eggs is carefully pressed through the hole in the torn capsular membrane. Decapsulated embryos are washed several times with water to remove the last traces of the viscous capsular fluid which may interfere with the fixatives. Such embryos, as well as the blastomere plasma membrane, are only covered by the vitelline membrane which is shed in the course of gastrulation. Up till now all attempts at a more or less long-term cultivation of embryos outside the capsules have been unsuccessful. This is probably because the embryo normally begins to absorb the capsular fluid at relatively early developmental stages, and this process continues almost up to hatching [141]. It has not been possible to cultivate *Lymnaea* embryos in the capsular fluid [74] although this has been possible in the case of *Limax* [71]. According to Morrill [127], post-tro-

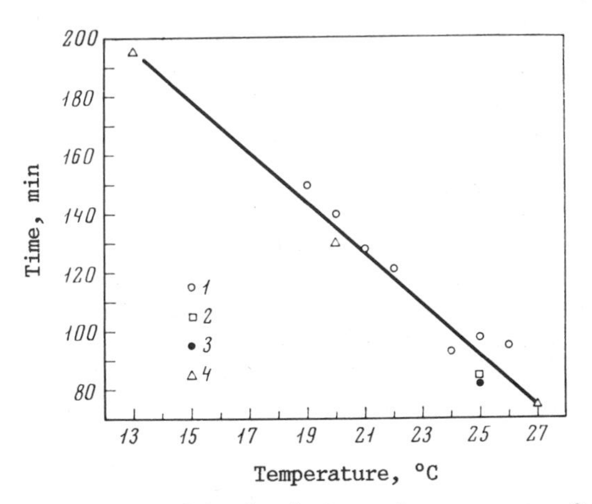

Fig. 5.5. Temperature dependence of the time between the appearance of the first and second cleavage division furrows or between the first and second division metaphases in the pond snail (mean values are given). 1) Our data; 2) Verdonk [175]; 3) Labordus [102]; 4) Jockusch [82].

chophore *Lymnaea palustris* embryos isolated from the capsules can develop normally in pond water.

We believe that the following approach is promising. After experimental manipulations with decapsulated embryos they are washed in a sterile solution. Each embryo is then transferred using a pipette with the tip of a 15-20 μm diameter into a punctured capsule where the egg was inactivated by a beam of UV irradiation. The compositions of the solutions used for the short-term culturing of early embryos are as follows:

Jockusch [82]

	g/liter
NaCl	0.60
KCl	0.42
$CaCl_2$	0.24
0.02 M tris-HCl	
(pH 6.7)	

Horstmann [80]

	g/liter
Na_2HPO_4	1.246
$NaH_2PO_4 \cdot 5H_2O$	0.468
NaCl	1.265
KCl	0.185
$CaCl_2$	0.115
$MgSO_4 \cdot 7H_2O$	0.185

Taylor [171]

	mmole/liter
$NaHCO_3$	2.00
$KHCO_3$	0.40
$CaCl_2$	1.72
$MgSO_4$	1.00
(pH 7.4)	

Jockusch has pointed out that eggs left inside the cocoon, eggs in capsules placed in water, and eggs decapsulated and incubated in the salt solution which she described develop almost synchronously during the early phase.

According to Raven [141], early embryos develop normally in tap water or in 0.4 mM $CaCl_2$ solution. Our own observations confirm this. Sometimes development in water proceeds to late blastula, but more frequently to the stage of 24-29 blastomeres. From the 16 blastomere stage onward, a slight developmental delay as compared with the control may be observed. Thus, when studying early cleavage one can successfully culture embryos in tap water provided that the decapsulation was careful and accurate. Performing experiments and observations at 22-25°C is very suitable.

More refined experiments may require the eggs to be pooled at synchronous developmental stages. It has already been mentioned that eggs in a batch do not show quite synchronous development. More crude "synchronization" involves cutting the anterior and posterior fifth from the cocoon [82] and using the middle part. According to another technique [102], the cocoon, immediately after laying but prior to the first cleavage division, is cut into 10-12 transverse parts. In each part 2-4 eggs are used as a control and the rest for experimental purposes. If the experiment begins after the first cleavage division, then groups of capsules are selected for which the appearance of the first cleavage furrow occurs within 1 min.

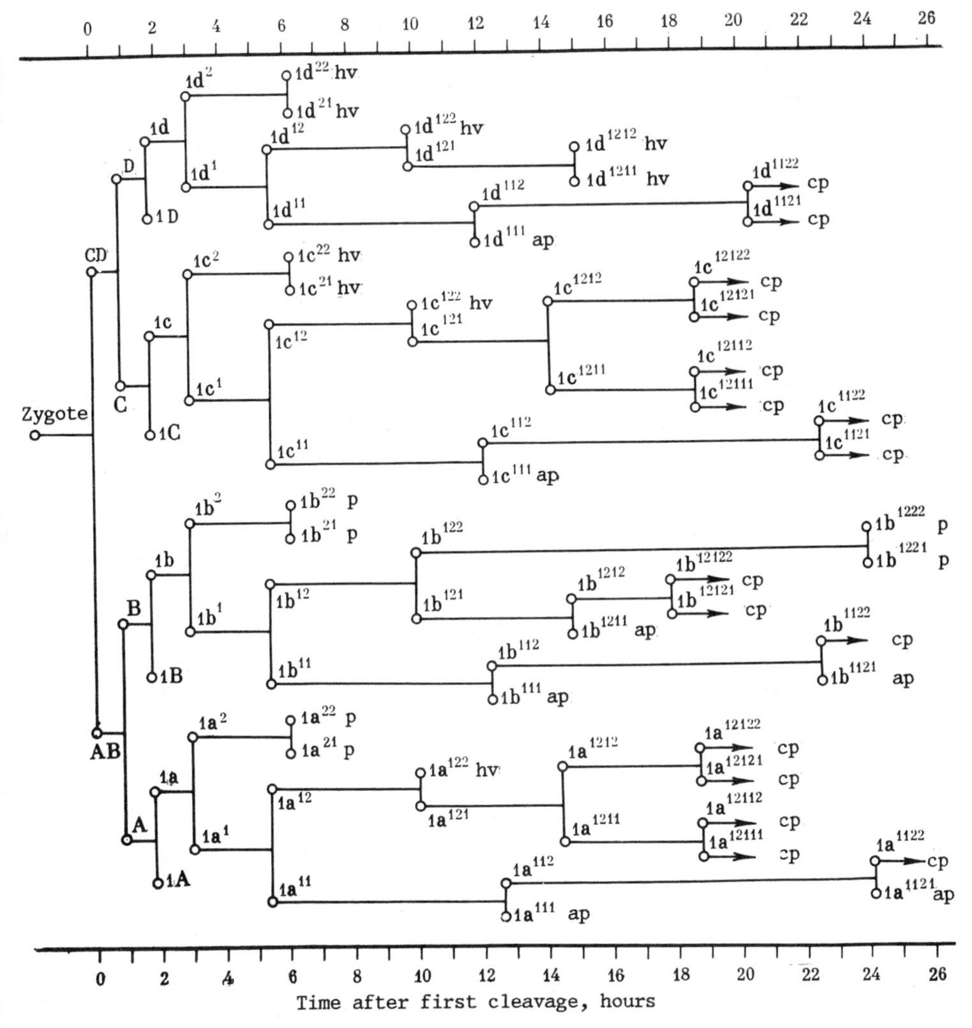

Fig. 5.6. Lineage of the first micromere quartet up to early gastrula in *L. stagnalis* [175] at 25°C. ap, apical plate; cp, cephalic plate; hv, head vesicle; p, prototroch.

The first and last of such groups serve as controls [63]. If the embryos are decapsulated, a similar method may be used even prior to cleavage. In this case the extrusion of polar bodies may be used as a time marker; this can be observed using trans-illumination.

If it is not planned to follow up late morphogenetic events, which may be impaired after treating the early embryo at 37-38°C [32], one may successfully synchronize the embryos at the first cleavage division by heating and subsequent accumulation in the cold. For this purpose one requires a heating table with a temperature of about 35°C and an ice bath (preferably a plate of ice wrapped in a polyethylene film with a surface temperature of 4-6°C). The embryos with a furrow on the animal pole (Fig. 5.9) are immediately transferred to the cold (all manipulations are made in water droplets in plastic dishes), while the others are placed for 2-5 min at 35°C. Thereafter the embryos with furrows are again selected and transferred to the cold. The procedure is repeated until all the embryos are in the cold. They can then be used for experiments. It should be pointed out that long-term contact with 0-6°C may prevent the appearance of the first or second polar bodies, and this whole process results in the formation of tetra- or tripolar mitoses during the first cleavage division [108]. Therefore, temperature treatments for the sake of synchronization should be employed when the maturation divisions have been completed.

5.7. RATE OF DEVELOPMENT AND ITS DEPENDENCE ON TEMPERATURE

Pond snail development is possible at temperatures ranging from 10 to 30°C. In nature oviposition usually takes place at 15-18°C and ceases at temperatures exceeding 27°C [103]. However, in some pond snail species development may proceed at temperatures as low as 2-5°C [73]. Richards [150] has studied the duration of pond snail development as a function of temperature (from the outset of oviposition until 50% of embryos have been hatched); his results are presented below:

Temperature, °C	Percent of embryos hatched	Time of development (h)
5	0	–
9	1	>2000
12	90	1262 ± 41
15	94	866 ± 18
18	84	688 ± 14
20	93	588 ± 20
22	96	489 ± 15
25	97	448 ± 17
28	96	416 ± 27
32	0	–
35	0	–

If the time of development is T, then the relationship of ln $(1/T)$ to temperature is a straight line in the temperature interval 13-26°C. Thus, taking into account the percentage of successful hatchings, the optimal temperature range is 15-26°C. It should be remembered that the optimal temperature is 25°C when capsules with embryos are cultured on agar [49]. A special study dealing with the effect of different constant and varying temperatures on the duration of development has been published for *L. auricularia* [154]. These authors point out that the embryos may be frozen for short periods without losing their capacity for subsequent development.

Since the first three cleavage divisions in the pond snail are synchronous, the interval between the appearance of the first and second cleavage division furrows (τ_0) may be used, as in other animal species, as a dimensionless unit to define the relative duration of development. Figure 5.5 presents available data on τ_0 for the pond snail at different temperatures [82, 102, 175; Mesheryakov; our data]. Jockusch's data [82] refer to the mean duration of the interval between the metaphases of the first and second cleavage divisions. In these experiments the eggs present in the cocoon, inside the isolated capsules, and decapsulated eggs in the salt solution underwent synchronous development. All other data are based on vital observations of cleavage, and they refer to the average interval of time between the initial appearance of first and second cleavage division furrows in the egg batch used. Temperatures quoted in the above account are accurate to only ±0.5°C.

5.8. NORMAL DEVELOPMENT

Cleavage in the pond snail is of the spiral type and is highly deterministic. In other words, the fate of individual blastomeres can be followed accurately until the early stages of organogenesis. Figures 5.6 and 5.7 show the pattern of cleavage for the first quartet of micromeres [175], and for the whole embryo up to the stage of 49 blastomeres [18-20].

We shall now consider certain peculiarities in the cell cycle during early cleavage. According to Jockusch [82], the relative duration of the mitotic phases in cell cycles II and III at 20°C are as follows (figures in percentages):

	Interphase	Prophase	Metaphase	Anaphase + telophase
II	23.4	26.2	21.0	29.4
III	25.0	27.3	22.7	25.0

Van den Biggelaar [18] has reported somewhat different figures for these cycles (in percentages):

	Interphase	Prophase	Metaphase	Anaphase	Telophase
II	46.5	12.8	11.6	5.8	23.3
III	44.2	16.2	11.6	4.7	23.3

The structure of the interphase has been studied in detail by van den Biggelaar [18, 19] using autoradiography and cytophotometry. In the course of the first cycles (whose duration in the course of synchronous divisions has been designated as *T*), the proportion between phases of the cell cycle is as follows: mitosis, 0.48 *T*; S-phase, 0.27 *T*; G_2 phase, 0.25 *T*. G_1 is absent during these cycles as well as during all subsequent ones. The duration of mitosis is identical for synchronous

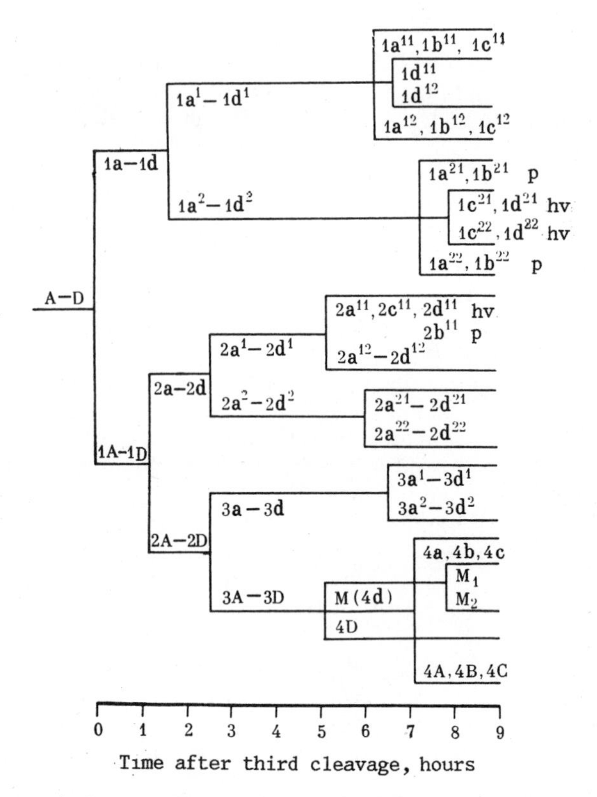

Fig. 5.7. Cell lineage in *L. stagnalis* embryos between the stages of 8 and 49 blastomeres at 25°C [19]. Cells belonging to the prototroch (p) and the head vesicle (hv) do not divide any further.

and asynchronous divisions. The karyomeric nuclei during cleavage cycles I-IV are irregular in shape and polymorphic even by the end of the G_2 phase (see Fig. 5.8). They only become rounded at the beginning of the prophase [64]. The beginning of prophase in these cycles can be easily identified on the embryo sections by the concentration of chromatin on one side of the nucleus, close to the outer surface of the blastomere [140, 141].

Until the 49-blastomere stage DNA synthesis begins at the telophase; at the 49-blastomere stage DNA is also synthesized in the cells which no longer divide. The duration of the S period begins to undergo significant changes only during cycles $1a^1$-$1d^1$ and $1a^2$-$1d^2$, that is, after the 16-blastomere stage. Any differences in the duration of cycles between individual animal blastomeres prior to this stage are due to changes of G_2; for the vegetal blastomeres the same holds true until the stage of 49 cells. In the same $1a^1$-$1d^1$ and $1a^2$-$1d^2$ blastomeres the nucleoli appear for the first time at the stage of 16 cells. At the stage of 24 cells they appear in other blastomeres, too. Although the first divisions of cells belonging to quartets 1, 2, and 3 are not synchronous, DNA synthesis proceeds in them synchronously.

Three periods, each possessing certain morphological peculiarities, may be distinguished in the course of early cleavage by the criterion of cytokinesis syn-

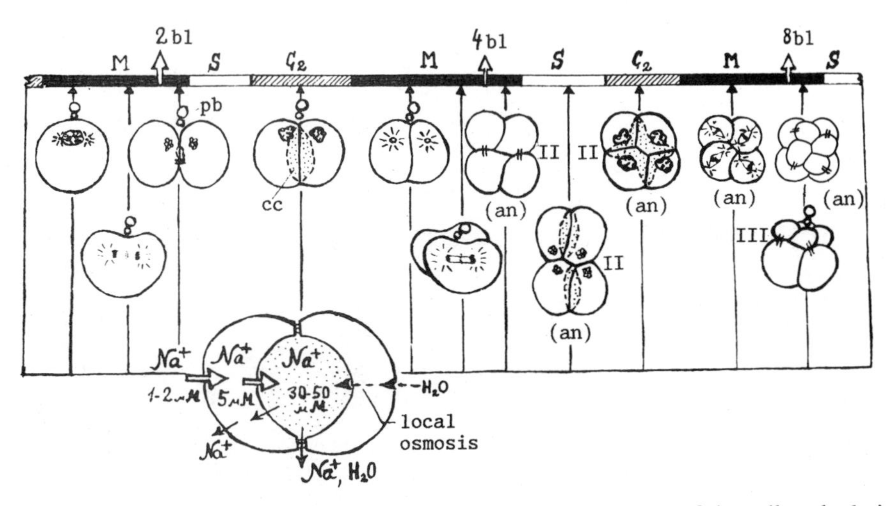

Fig. 5.8. Correspondence between the shape of the embryo and phases of the cell cycle during synchronous cleavage divisions in the pond snail (from [18]). Open arrows on the cell-cycle scale indicate end points of cytokinesis. Overlapping of the S-phase and telophase is not shown for the sake of simplicity. Roman numerals refer to the number of the division where the corresponding contact zones are formed. pb, polar bodies (over the animal pole); cc, cleavage cavity (side view and animal view). Shown below is a sketch illustrating the active (open arrows) and passive (small solid arrows) flows of sodium ions and water accompanying the cleavage cavity formation [171]. Concentration of exchangeable sodium is shown in capsular fluid, blastomere cytoplasm, and cleavage cavity. bl, blastomeres; an, animal view.

chrony. The first includes the synchronous divisions (from zygote to the 8-blastomere stage); the second involves synchronous divisions of cells with an identical index (e.g., $1a^1$-$1b^1$-$1c^1$-$1d^1$), that is, of cells located at an identical level relative to the poles of the embryo (8-24 blastomeres); the third is a period of general asynchrony involving differentiation of the quadrants A–D in the timing and direction of divisions (24-29 blastomeres). At first we will consider certain phenomena common to all three periods and associated with the contact interactions of newly appearing sister blastomeres.

The vitelline membrane surrounding the whole embryo and providing anchorage for part of the egg-surface microvilli does not take part in forming cleavage furrows. The opposing membranes of new blastomeres begin to establish contact only when cytokinesis has been completed. The growth of contact zones at the place of earlier furrows ends at the G_2 period; it is achieved not by the adhesion of blastomere outer surfaces but through the intercalation of a new surface from the inside [109-112]. Therefore there is a continuity between the surface of the zygote and the outer surface of the embryo. This continuity is partly broken only during the third period, when the new contact zones of the blastomeres emerge, not only in the course of cytokinesis but also as a consequence of interphase cell movements.

The so-called recurrent cleavage cavity appears at the 2-blastomere stage. It is formed as a result of the secretion of fluid from the vacuolized inner walls of the blastomeres. The cavity attains maximum size during the G_2 or prophase when its volume is 0.2-0.4 nl, while the volume of the embryo minus that of the cavity is about 1.1 nl [171]. The cavity is emptied through apertures located on the poles of

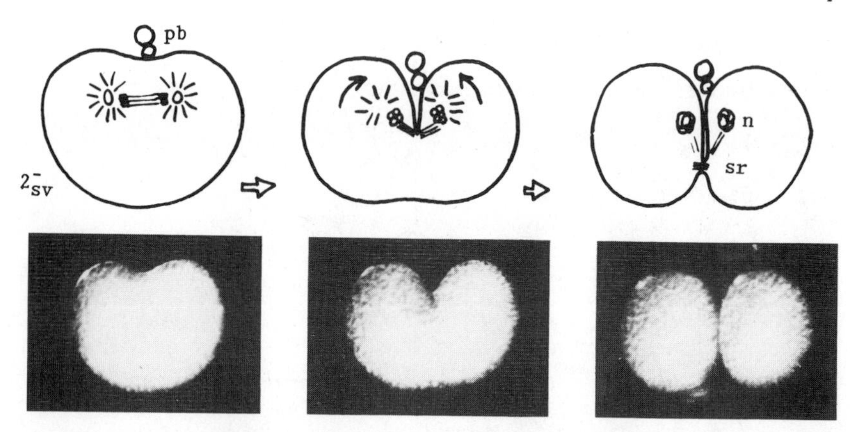

Fig. 5.9. Beginning, middle phase, and end of cytokinesis of the first cleavage division in the pond snail embryo. Side view. Sketches of sections and photographs of live material. Arrows show the movement of spindle halves. pb, polar bodies; sr, spindle remnants; n, nucleus.

the embryo during metaphase, anaphase, or early cytokinesis. After each division, the recurrent cleavage cavity is rapidly restored along the previous contact zones, but forms slowly in the region of the growing contact zones (see Fig. 5.8). As the cavity continues to increase, the shape of the embryo becomes more regular. With experience the shape of the embryo, as well as the location and size of the sister nuclei during the first cycles, may provide information on the cycle phase (Fig. 5.8).

As early as 1850 [182] it was assumed that the cavity plays a part in osmotic regulation, and this has recently been proved using radioactive isotopes. Figure 5.8 shows the fluxes of exchangeable sodium ions which serve to actively pump water into the cavity [171]. It has also been assumed that the cavity serves to remove unnecessary metabolites and maintains a particular inner medium which is necessary for the functional interaction between blastomeres [140, 141]. A permanent central cavity forms during the third period of late cleavage, the local cavities, which are more stable in the animal hemisphere, coexisting alongside the central one.

The growth of the blastomere contact zones is accompanied by the appearance of specialized cell junctions as early as at the two-cell stage. Interlocking and unusually long microvilli have been seen in the contact zones [126]. According to our observations, junctions of the unusual folded shape of the simple apposition type can often be found at the peripheral part of the contact zone at the four blastomere stage . Septate desmosomes and intermediate contacts of the zonula adhaerens type are the first of the more specialized contacts to be formed [26, 56]; it appears that they separate the growing cavity from the outer medium. The gap junctions appear about 60 min after the beginning of the second division. Contacts of this type are found ubiquitously at the 24-blastomere stage [55, 56]. We shall now describe more characteristic morphological features during each period of early cleavage.

Period I. Cleavage furrows are initially unilateral (Fig. 5.9) and displace the middle part of the anaphase spindle, which contains the stem bodies, so that the midbody and spindle remnants appear below the level of the sister nuclei. Such a "kink" in the spindle is no longer present during period III. The midbody may be seen until the prophase. This should be taken into account during experiments on

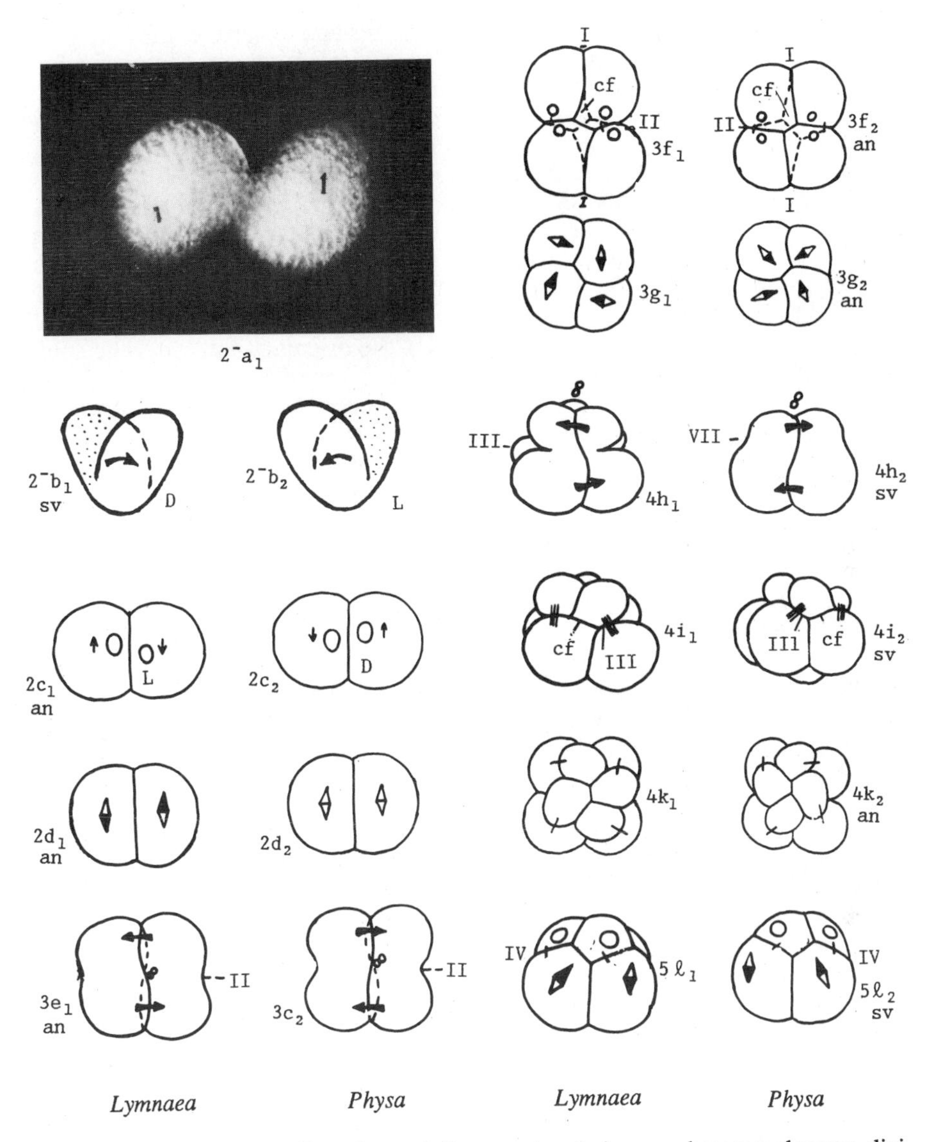

Fig. 5.10. Morphological manifestations of dissymmetry during synchronous cleavage divisions in the dextral species *L. stagnalis* and the sinistral species *Physa acuta*. 2a) Dexiotropic turn of the emerging blastomeres in the pond snail after the vitelline membrane has been removed; animal view. 2b) Sketch showing direction of the turn when viewing the embryo along the spindle axis; shaded halves of the cleavage spindles are closer to the observer. Arrows in 3e and 4h show direction of spiral deformations at the beginning of cytokinesis. I–IV, division number; cf, cross furrow; an, animal view; sv, side view; D, dexiotropic; L, laeotropic rotation around the spindle axis.

blastomere isolation. Because of the telophase movements the sister nuclei are located close to the new contact zone of blastomeres and, therefore, they "mark" this zone (Figs. 5.9, 5.10). At the end of G_2 the nuclei are moved somewhat further away from the contact zone, while the distance between them and the outer surface of blastomeres is a few microns. With the exception of cycles I and II, when the location of the nuclei may depend on the rotations of blastomeres (Fig. 5.10, 2c, 3f), the centers of the nuclei are located opposite each other.

In the course of normal development cleavage figures may sometimes be classified as left or right from the two-cell stage onward and always from the stage of 4-blastomeres. Furthermore, identical stages in dextral and sinistral species have opposite dissymmetry signs. This dissymmetry may be manifested as a helical deformation of the surface and contact zone (Fig. 5.10, 2a, b, 3e, g, 4h) as well as the position of sister nuclei (Fig. 5.10, 2c, 3f), the orientation of cleavage spindles (Fig. 5.10, 2d, 3g, 51), and the location of sister pairs of blastomeres with respect to the polar axis (Fig. 5.10, 3f, 4i, k, 51). The sequence of figures with alternating signs of dissymmetry is determined during the first period. There is evidence that suggests that all manifestations of dissymmetry depend on the spiral field of stresses which appear during the metaphase of each cycle owing to the activation of cortical microfilaments [108, 114, 115, 118]. This field determines the specific inclined position of the second- and third-division spindles as well as the helical deformations of the surface, which appear when blastomeres acquire an extended shape (anaphase of divisions I and II, metaphase of division III). Rotations of blastomeres in the course of cytokinesis at the first division can usually be visualized when the restricting vitelline membrane is removed (Fig. 5.10, 2a). These dissymmetric movements, whose direction with respect to the spindle axis is identical in each division in a given mollusc species, are only possible during the first period, since the planes of cleavage furrows more or less coincide during this time. With the beginning of period II these rotations are naturally suppressed because the blastomeres are tightly packed together.

During metaphase and anaphase the cleavage spindles are located tangentially to the nearest outer surface of the zygote or blastomeres. They possess rather large asters. During unequal divisions the nonidentity of aster diameters appears no earlier than the anaphase.

Period II. Spiral surface deformations are no longer visible. Dexiotropic and laeotropic orientations of spindles and contact zones of sister blastomeres alternate during the consecutive divisions. The orientations of spindles with respect to the polar axis is completely determined by the shape of the external blastomere surfaces [115].

Despite the presence of general division synchrony for cells with identical indices at the 16-blastomere stage, Raven [143] found a slightly earlier onset of divisions in 2B, 2D, and 2d, 2b as compared with the 2A, 2C and 2a, 2c blastomere pairs. On the one hand this correlated with the different number of adjacent blastomeres (six for 2B and 2D and five for 2A and 2C); on the other hand it correlates with the slightly greater volume for B and D as compared to A and C, as early as at the 4-blastomere stage [17]. The quadrant A cannot be distinguished from C, nor B from D, during period II. Therefore, these designations should be regarded as arbitrary. The nucleoli appear for the first time during this period.

Period III. The cell cycles become markedly longer. This is accompanied by the most important event of the early embryogenesis of the pond snail: the determination of the dorsoventral axis and the quadrant D as a source of the mesoderm

[144]. At the 24-blastomere stage, the most vegetal macromeres which are in contact with each other along the cross-furrow begin to establish inner junctions with the blastomeres of the animal hemisphere. Eventually one of them disappears almost completely from the outer surface, filling the cleavage cavity and, in addition, increasing the number of its neighbors from 6 to 20. From this moment onward this macromere is determined as 3D [21-23, 143]. The RNA-containing ectosomes (former SCA) move along the inner walls of all four macromeres in the course of their interaction with animal pole micromeres involving gap junctions. The ectosomes undergo progressive compactization in 3A-3C, but become dispersed in 3D [22, 55]. Each of the two vegetative macromeres may enter the cavity. This relates to Raven's idea that the selection of 3D is accomplished by an equally balanced competition or "conflict" of the vegetal macromeres [23, 25, 143]. Studies of the cell surface during *L. palustris* cleavage [126] revealed some interesting changes in the distribution of microvilli correlated with the migration of SCA. A special zone of microvilli has been found at the blastomere stage in the region of SCA localization at the vegetal pole. These microvilli disappear when 3D sinks in at the 24-blastomere stage. With the onset of the bilateral stage of cleavage only the surface of the 4d and $2d^{22}$ cells carries a large number of microvilli. The initial polarity of the zygote seems to be reflected in the segregation of the bulk of microvilli into micromeres of the first three quartets.

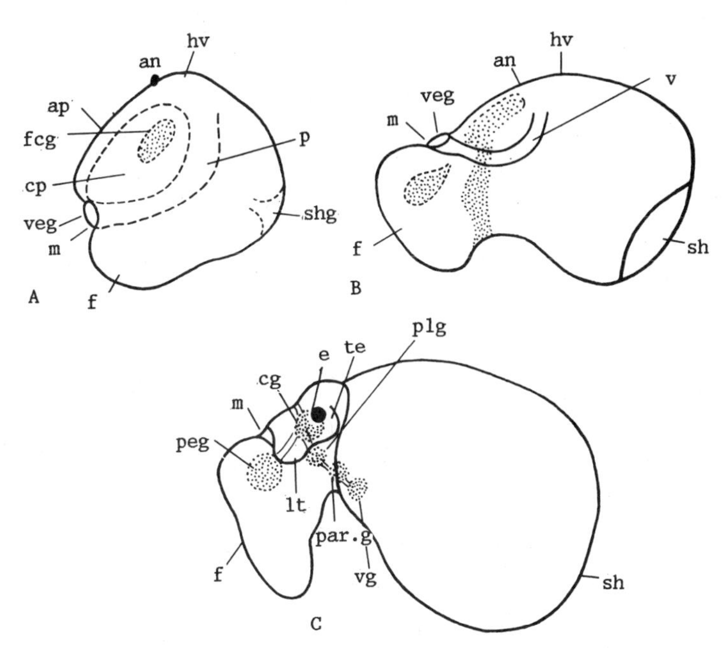

Fig. 5.11. Sketch illustrating embryogenesis in *L. stagnalis* (modified from [140]), side view. A) Late trochopore (stage 21); B) mid-veliger (stage 23); C) late veliger (stage 24, "Hippo"). an, initial animal pole; veg, initial vegetal pole; ap, apical plate; p, prototroch; v, velum; m, mouth; cp, cephalic plates; hv, head vesicle; shg, shell gland; f, foot; fcg, field of cerebral ganglion; sh, shell; te, tentacles; lt, labial tentacles; cg, cerebral ganglia; plg, pleural ganglia; peg, pedal ganglia; par. g, parietal ganglia; vg, visceral ganglion; e, eye. Presumptive areas of the nervous system (A, B) or ganglia (C) are shaded.

After the interaction between 3D and the micromeres the quadrants of the embryo undergo differentiation, not only in the rhythm of divisions, but also in their direction and the extent of their inequality. Some accessory contacts remain in the cleavage cavity; they are realized through lobopodia or filopodia. The orientations of spindles perpendicular to the outer surface appear; latitudinal and meridional orientation with respect to the polar axis may be seen. Spindle asters become progressively smaller and eventually continue to appear only in divisions of the 4d mesentoblast descendants [140]. As cytoplasmic segregation increases from the outer surface to the center of the embryo, the interphase nuclei are found deeper than during periods I and II; they are located roughly at the boundary between the ectoplasm and endoplasm [26].

Thus, cleavage during period III acquires a bilateral nature and individual blastomeres begin to differentiate into larval structures.

We shall now briefly consider the cellular origin of the main organ systems in the pond snail (see also Figs. 5.7 and 5.11).

Temporary larval head structures consist of the following elements: the apical plate – $1a^{111}$-$1d^{111}$, $1a^{1121}$, $1b^{1121}$, $1b^{1211}$; prototroch – $1b^{1221}$, $1b^{1222}$, $1a^{21}$, $1a^{22}$, $1b^{21}$, $1b^{22}$, $2b^{111}$, $2b^{112}$; head vesicle – $1c^{21}$, $1c^{22}$, $1d^{1211}$, $1d^{1212}$, $1d^{122}$, $1d^{21}$, $1d^{22}$, $2d^{11}$, $2a^{11}$, $2c^{11}$, $1a^{122}$, $1c^{122}$. The right cephalic plate appears from the cells $1a^{12111}$, $1a^{12112}$, $1a^{12121}$, $1a^{12122}$, $1a^{1122}$, $1b^{12122}$, $1d^{1211}$, $1d^{1122}$. The left cephalic plate originates from the cells $2c^{12111}$, $1c^{12112}$, $1c^{12121}$, $1c^{12122}$, $1c^{1121}$, $1c^{1122}$, $1b^{12121}$, $1b^{1122}$. The cephalic plates give rise to the tentacle epithelium, the dorsal part of the labial tentacle epithelium, eyes, and cerebral ganglia. The stomodeum or foregut is formed from the micromeres of predominantly the third and second quartets, and subsequently forms the oral cavity epithelium, pharynx, radular sac, tongue epithelium, salivary glands, esophagus, and buccal ganglia. The derivatives of the second and third quartet, 2d in particular, give rise to the foot, body wall ectoderm, and mantle. The foot epithelium produces statocysts, pedal glands, and pedal ganglia. The body wall ectoderm yields posterior head ganglia, copulatory organs with part of the outer region of the sperm duct, and a small region of the hindgut near the anus. The mantle forms the shell, the primary urethra, and the genital tract (with the exception of the vas efferens of the sperm duct). The development of the shell field and growth of the mantle have been studied in great detail by Timmermans [172] and Kniprath [98, 99]. The larval muscles of the foot, head, and prototroch form from the mesenchyme and probably have an ectodermal origin.

The following organs are formed from the 4d offspring corresponding to mesodermal embryonic streaks: heart, pericardium, vessels, definitive kidney, gonad, connective tissue integuments of the gastrointestinal tract, muscles of the radular sac and tongue, sublingual "cartilage," muscles of the copulative organ, the columellar muscle, the connective tissue layer of the skin (cutis), and, probably, in part, the body wall muscles.

The so-called nuchal cells, which (like the protonephridium) are absent in the adult animal, also have a mesodermal origin.

Finally, the endodermal derivatives 4A-4D, 4a, 4b, 4c, and enteroblasts (early derivatives of the 4d) result in the larval liver, hindgut, and stomach, whose walls later form the definitive liver; muscles covering the gastrointestinal tract also form from these derivatives.

The lung, as well as the osphradium, appears independently on the mantle cavity from the body-wall ectoderm. Therefore the lung cavity of the pulmonates can-

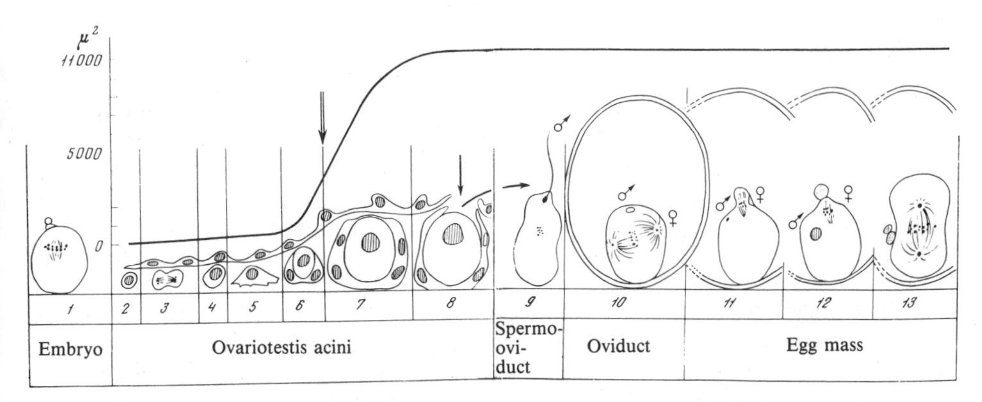

Fig. 5.12. Diagram of life cycle phases of a *Lymnaea* egg (from [174]) with modifications. 1) Embryonic development; 2) formation of primary germ cells; 3) multiplication of primary germ cells; 4) oogonia; 5) amoeboid phase (migration within the gonad); 6) formation of follicle; 7) growth phase; 8) rest phase; 9) ovulation and insemination (stages 0 and 1/1); 10) phase of pre-maturation (1/2); 11) first maturation division (1/3); 12) second maturation division (1/4); 13) first cleavage division and beginning of the embryonic development (2). Arrow shows time of complete formation of inner follicle.

not be considered a homolog of the mantle cavity of Prosobranchia. This outline of organogenesis is based on the studies of Fraser [60], Raven [137, 138], Regondaud [147, 148], Verdonk [177], and Cumin [49] (see also [43, 75, 172]).

To conclude this section, studies dealing with the molecular biology and developmental genetics of the pond snail are discussed briefly. Although RNA synthesis can be detected using biochemical techniques even in the zygote (see [36]), the first major increase is noted during the second cleavage period; this corresponds to the formation of nucleoli ([20]; see also [33]). According to Neyfakh [130, 131], morphogenetic nuclear function in the course of pond-snail development begins at the 12-blastomere stage and supports embryogenesis when period II has been completed. A somewhat different conclusion has been reached by Verdonk [177], who studied the expression of pond snail lethal genes which are made homozygous by self-fertilization. Gastrulation is the first period depending on the embryonic genome, but a particularly high number of genes undergoes expression at the later stage of the intensive differentiation of rudiments in the early trochophore. The expression of a few genes essential to development may be observed after the early veliger stage. Morrill and associates [125], on the basis of the results with *L. palustris*, believe that the bulk of proteins prior to the early trochophore stage is synthesized on long-living mRNA molecules; gastrulation involves molecular events which are critical to the differentiation of the definitive organs.

In the last few years some progress has been made toward understanding the genetic control of dextrality–sinistrality. Traditionally, such studies are performed using *L. peregra*, for which dextral and sinistral races have been described. It has been assumed for a long time that the dissymmetry sign for cleavage and adult animal organization depends on the dominant dextral allele and the recessive sinistral allele of one gene; the offspring phenotype is determined by the maternal genotype, that is, before maturation divisions [35, 168]. The problem was complicated by the permanent formation of dextral snails in the sinistral "pure" strains. This has been explained by reverse mutations and the presence of the complex system of modifying genes [51]. A thorough genetic analysis recently carried out by Freeman and Lundelius [61] enabled a new mechanism to be proposed for the inheritance of dextrality–sinistrality and the formation of abnormal dextral specimens in the sinistral strains of *L. peregra*. The gene determining dextrality consists of two regions which have a dominant effect only when present in one chromosome in the "cis" position (s^{1+} s^{2+} or a tandem duplication of s^{1+} s^{1+} type). In sinistral snails this structure is formed spontaneously by crossing-over. This is possible in three variants: mitotic crossing-over at the oogonial stage; meiotic equal crossing-over during prophase I of meiosis; and meiotic unequal crossing-over at the same stage. Genetic analysis demonstrates the existence of all three types with a marked prevalence of the equal meiotic crossing-over.

Injections of the cytoplasm from phenotypically dextral eggs into sinistral ones (but not vice versa) result in the inversion of the dissymmetry sign of cleavage if these injections are performed prior to the extrusion of the second polar body [61]. It may be noted that the animal–vegetal polarity of the zygote appears to undergo irreversible determination at the same stage [71]. Microinjection experiments have also enabled the relationship between the dose of dextrality gene and its effect to be established, as well as dependence on the developmental stage of the cytoplasm donor. Methods are now available for purifying the products of chirality genes and for studying their mechanism of action, which appears to be due to the effect on the organization of the cortex actin cytoskeleton [114].

5.9. DESCRIPTION OF STAGES OF NORMAL DEVELOPMENT

A table of normal development is presented below for the first time. Stages 0-1/5 are based on Raven [138, 141] and Ubbels [174]; this period is described by us in less detail. Here the description of stages is based on morphological peculiarities. Stages 2-18 are derived from Verdonk's study [175]. Stages 19-21 correspond to the description of embryogenesis taken from Raven [137] but in far less detail. Finally, stages 22-29 are taken from the Normal Table of pond-snail development published by Cumin [49]. They correspond to his stages E_4-E_{11}, recorded at daily intervals during development at 25°C. In Raven's and Verdonk's studies the stages have been identified by studying the structure of the embryo at certain standard intervals, with the exception of early cleavage where the stages are distinguished on the basis of the number of cells. During early developmental stages the time of onset of a given stage has been determined in our observations by the first appearance of its characteristic features. After early blastula the interval between observations was 3 h. In Table 5.2, we show the time up to a corresponding stage at 22 ± 0.5°C (our data) and at 25°C [175, 49] in dimensionless units, and the size of the embryo (our data to stage 21 and further information according to [49]); a short diagnosis of the stage is attached. Drawing numbers in Figs. 5.13 to 5.21 correspond to those of stages; the drawings of stages according to Cumin are marked with the letter C (for example 22C). Our original drawings referring to stages presented according to Verdonk and Cumin are marked with the letter M.

It should be pointed out that our data are valid for the development of embryos in capsules incubated in tap water, while Verdonk and Cumin have observed the development in capsules incubated in moist air upon agar surfaces. Therefore the data regarding the time of onset of a given stage at 22°C and 25°C should be compared cautiously. The time of onset of stages at 25°C, up to and including stage 24, is given according to Verdonk, and from stage 25 onward according to Cumin. At 22°C the value of τ_0 is 120 min, and 85 min at 25°C.

Stage 0: The Ovulated Egg.

Ovulation proceeds with autolysis of the follicle and gonad epithelium. The oocytes pass into the gonad lumen, then into the hermaphrodite duct where they are accumulated in the straight part. The germinal vesicle is retained for some time after ovulation (Fig. 5.12).

Stage 1: One-Cell Embryo.

1/1. Insemination takes place in the straight region of the hermaphroditic duct. At the moment of insemination the germinal vesicle disintegrates (Fig. 5.12). The inseminated eggs may remain in this part of the oviduct for up to 1.5 h.

1/2. Metaphase of the first maturation division. The deposition of eggs in the cocoon proceeds at this stage. Immediately after laying, the egg is somewhat irregular in shape but soon acquires that of a rotation ellipsoid. A small part of the egg is free of yolk near the animal pole where the first maturation spindle is located. This region is semitransparent when seen with transmitted light.

1/3. Extrusion of the first polar body. At this stage the egg may display some amoeboid deformation. The first polar body, which has the appearance of a small transparent sphere, is extruded at the animal pole and is coated with its own membrane as well as by part of the vitelline membrane; it is located outside the perivitelline space over the animal pole.

1/4. Extrusion of the second polar body. The second polar body is extruded directly under the first; it is somewhat smaller and becomes located between the vitelline membrane and the egg plasma membrane. Some amoeboid activity of the egg can be observed during extrusion.

1/5. Fusion of pronuclei. This stage is preceded by the fusion of the female karyomeres into the pronucleus. True fusion of the male and female pronuclei does not take place. On the sagittal section it can be seen that the pronuclei contacting each other are located almost directly under the animal pole.

The axes of the two pronuclei may have different orientation. Sometimes the pronuclei, as well as the spindle of the first cleavage division, may be seen in the

TABLE 5.1. Intervals between Developmental Stages of the One-Celled Embryo (22°C – our data; 25°C – data of Labordus [102] or Van den Biggelaar [18]; in min)

Stages	22°C	25°C
1/2-1/3	Time varies; usually it is 45-60	53
1/3-1/4	60	38
1/4-1/5 (up to the beginning of the prophase of the 1st cleavage division)	–	53*
1/4-2	102	76, 90*

*Data of Van den Biggelaar [18].

TABLE 5.2. Normal Development of *Lymnaea stagnalis* L.

Stage No.	Stage, size in mm	Time after first cleavage furrow: 22 ± 0.5°C* h, min	Time after first cleavage furrow: 22 ± 0.5°C* τ_n/τ_0	Time after first cleavage furrow: 25 ± 0.1°C† h, min	Time after first cleavage furrow: 25 ± 0.1°C† τ_n/τ_0	Diagnostic features: External features	Diagnostic features: Internal features
2	2 blastomeres. Embryo diameter about 0.13	00.00	0.0	00.00	0.0		
3	4 blastomeres	02.00	1.0	01.25	1.0		
4	8 blastomeres	05.00	2.5	02.45	1.9	Last synchronous cleavage division	
5	12 blastomeres	07.00	3.5	04.00	2.8		
6	16 blastomeres	07.35	3.8	05.00	3.5		RNA synthesis. Appearance of nucleoli
7	24 blastomeres	08.55	4.5	06.00	4.2	First appearance of bilateral symmetry at the vegetal pole; rounded 3D blastomere	Disappearance of the cleavage cavity. Nucleoli in all blastomeres
8	25 blastomeres	12.05	6.0	09.00	6.2	Segregation of the 4D primary mesoblast	
9	33 blastomers	14.45	7.4	10.0	7.1		
10	Early blastula. 49 blastomeres	16.15	8.2	11.30	8.2		
11	Resting stage of the early molluscan cross			13.00	9.2	Animal $1a^{11}$-$1d^{11}$ cells are very narrow	
12	Formation of the outer median cells of the cross	21.30	10.8	15.00	10.6		
13	Formation of apical rosette cells	23.35	11.8	17.00	12.1		Beginning of capsular fluid pinocytosis by blastula cells

14	Middle blastula			19.00	13.4	Appearance of bilateral symmetry in the animal hemisphere: $1d^{121}$ division	
15	Late blastula						
15/1	Spherical blastula	28.30	14.3	21.00	14.8		
15/2	Flattened blastula			24.00	16.9	The embryo is flattened on poles	
16	Early gastrula	32.40	16.3	26.00	18.2	Blastopore rounded and wide	Mesoblasts inside the blastocoel
17	Middle gastrula	41.30	2.8	28.00	19.8	Fissure-like blastopore. Emergence from the vitelline membrane	
18	Late gastrula. Diameter 0.15-0.2	49.30	24.8	36.00	25.4	Small blastopore located opposite the shell gland rudiment. Rotation 1 rev/105 sec. Prototroch does not rise over body surface	Protonephridia consist of four cells
19	Early trochophore. Diameter 0.17	52.30	26.3	43.00	30.3	Prototroch rises over body surface. The embryo weakly transparent. Shell gland with depressed middle part	Beginning of albumen cell secretion. Separation of stomodeum upon oral cavity and esophagus
20	Middle trochophore. Length × height × width, 0.25 × 0.25 × 0.2	67.00	33.5			Rotation 1 rev/5 sec. Embryo more transparent. The shell gland is deeply depressed, thin shell is located over it. Lateral prototroch parts are well seen	Protonephridia formed. Two groups of nuchal cells. Formation of the stomach and colon
21	Late trochophore. Diameter 0.30	90.00	45.00	72.00	50.8	Shell gland moves almost to the original position, somewhat displaced to the left. Foot region appears. Embryo is still rounded	Rudiments of cerebral ganglia and eyes

TABLE 5.2 (continued)

		Time after first cleavage furrow				Diagnostic features	
		22 ± 0.5°C*		25 ± 0.1°C†			
Stage No.	Stage, size in mm	h, min	τ_n/τ_0	h, min	τ_n/τ_0	External features	Internal features
22	Early veliger. Embryo 0.6 × 0.3 × 0.3. Shell 0.2	95	47.5			Rotation 1 rev/10 sec. Foot definitely segregated. Prototroch bulges from the side. Shell gland almost flat, mantle folds are seen	Formation of pedal ganglia. Radular sac is not connected with mouth cavity. Beginning of torsion
23	Middle veliger						
23/1	Embryo 0.7 × 0.4 × 0.3. Shell 0.3 × 0.4 × 0.3	108.00	54.00			Rotation 1 rev/15 sec. Spontaneous foot movement. Mantle fold envelops one-third of visceral sac. Heart contractions. Eyes reddish in color	Formation of the pulmonary cavity and osphradium. Appear- of buccal ganglia. Aprance of radular dental plate
23/2		115.0	57.5	85.00	60.1	Mantle fold envelops two-thirds of the visceral sac. Heart is located over the shell middle	
24	Late veliger						
24/1	Embryo 0.8 × 0.6 × 0.3. Shell 0.4 × 0.6 × 0.3	139.00	69.5			Rotation irregular. Posteriorly the foot does not reach shell, may extend horizontally. First evidence for helical shell structure. Eyes are reddish. Pulmonary cavity reaches heart located sagitally. Two albumen sacs	Beginning of lens and reno-pericardial channel formation. Appearance of connectives and osphradium ganglion. Beginning of radular teeth formation. Rudiments of definitive liver formed

24/2		145.00	72.5	110.0	77.6	Posteriorly the foot reaches the shell. Eyes are dark	
25	Veliconcha. Embryo 0.9 × 0.6 × 0.4. Shell 0.7 × 0.6 × 0.6			165.00	116.5	Foot is used for locomotion, sporadic rotation. Shell covers the whole visceral complex. Heart on the left body side	Formation of the mantle cavity and oviduct. Beginning of protonephridia and head vesicle resorption. Pericardium formed. Formation of salivary glands
26	Transition to foot movement. Embryo 1.0 × 0.65 × 0.6. Shell 1.0 × 0.6 × 0.6			189.00	132.8	Embryo creeps using foot and may withdraw into the shell. Shell with helix	Hermaphrodite gland is laid down. Prototroch absent. Kidney in definitive position
27	Embryo 1.0 × 0.7 × 0.7. Shell 1.0 × 0.6 × 0.6			213.0	148.9	Shell size increases to double. Heart is not located horizontally	Beginning of liver diverticuli formation. Osphradial sac closed. Beginning of albumen cell division
28	Embryo 1.0 × 0.8 × 0.8. Shell 1.1 × 0.6 × 0.7			237.00	166.7	The shell helix increases three times. Marked enlargement of pulmonary cavity	Formation of penis, appearance of hermaphroditic duct of sex gland. Osphradium divided into two chambers. Three diverticuli in liver
29	Hatching stage. Embryo 1.2 × 0.8 × 0.8. Shell 1.2 × 0.8 × 0.8	272.00–284.0	136.00–142.0	261.00	185.3	Embryo occupies whole capsule. Shell with 1.5 helical turns. Heart removed from stomach	Hermaphrodite duct connected with the oviduct. Four diverticuli in liver. Absence of dorsal cilia in esophagus

*τ_0 = 120 min. †τ_0 = 85 min.

course of vital observation using strong oblique illumination or when a dark-field condenser is being used.

We shall now present data on the duration of individual periods of this stage (Table 5.1). A detailed cytological description of the one-cell stage can be found in Raven [136].

Stage 2: Two Blastomeres.

At first the furrow appears at the animal pole of the egg which has become elongated in the equatorial plane (stage 2-). Later the furrow divides the whole egg (stage 2). See Figs. 5.9 and 5.10.

Stage 3. Four Blastomeres.

The furrows of the second laeotropic cleavage division are formed synchronously in both blastomeres at a small angle to one another and to the animal–vegetal axis. Consequently, blastomeres A and C contact each other at the animal pole along the so-called polar or cross-furrow, which originates at the bending of the first furrow; at the vegetal pole the blastomeres B and D are in contact, and cannot be distinguished from each other.

All blastomeres usually appear to be identical in size. However, it has been established by Bezem et al. [17] that the volume of A or C (24.34% of the whole) is significantly lower than that of B or D (25.66%).

Stage 4: Eight Blastomeres.

The third cleavage division is dexiotropic. Four micromeres of the first quartet 1a-1d and macromeres 1A-1D are formed synchronously.

This is the last division with helical deformation in the course of which cross-furrows are formed on the lateral cell surface (Fig. 5.10). It is also the last synchronous division. The ratio between the volume of blastomeres 1q (q = quartet), 1A (or 1C), and 1B (or 1D) is 5.58:18.46:20.39 [17].

Stage 5: Twelve Blastomeres.

The 1A-1D macromeres undergo laeotropic divisions forming the second quartet of micromeres 2a-2d, which are slightly larger than 1a-1d.

Stage 6: Sixteen Blastomeres.

The micromeres of the first quartet undergo laeotropic division which results in more animal $1a^1$-$1d^1$ cells and the four primary trochoblasts $1a^2$-$1d^2$. Nucleoli appear for the first time in these eight blastomeres, their number not exceeding three per nucleus. RNA synthesis may be detected by autoradiography for the first time at this stage [20]. The ratio between the volumes of blastomeres $1q^1$, $1q^2$, 2q, 2A (2C), 2B (2D) is 3.34:2.15:7.40:11.40:12.83 [17].

Stage 7: Twenty-four Blastomeres.

The macromeres 2A-2D and the micromeres of the second quartet, that is, all blastomeres of the vegetal hemisphere, undergo synchronous and dexiotropic divisions. From the surface the size of 3a-3d is greater than that of 3A-3D. The 24-blastomere stage is remarkable because it involves a cessation of cleavage for about 3 h. Nucleoli appear in all blastomeres. The 3D blastomere differs from 3A-3C in its rounded shape (Fig. 5.13, 7 veg). It fills almost the entire cleavage cavity, where it contacts blastomeres $1a^1$-$1d^1$, $1a^2$-$1d^2$, $2a^1$-$2d^1$, 3a and 3b. At 25°C these contacts become fully established 80 min after the beginning of this stage [21]. After 110-170 min the top of 3D, located inside, is oriented toward $1b^1$ and $2b^1$; the material of the ectosomes in the 3D is concentrated along the contacts with these micromeres. Two hundred minutes later 3D enters mitosis; the spindle is formed in the center of the blastomere, which is then displaced toward the vegetal pole, providing an inequality of the 3D division. Ectosomes from 3A-3C are then found in micromeres 4a-4c [21, 22]. The volume of 3D at this stage, according to Bezem et al. [17], exceeds that of 3B by about 15%. The ratio between the volumes of $1q^1$, $1q^2$, $2q^1$, $2q^2$, 3q, 3A (3C), 3B, 3D is 3.07:2.10:3.47:3.68:6.75:5.01:6.40:7.38.

Stage 8: Twenty-five Blastomeres.

The 3D macromere is divided by a plane forming a sharp angle with the polar axis, yielding the 4D macromere and 4d mesentoblast (M) which serves as the future source of most of the mesoderm. 4d is much larger than 4D.

Stage 9: Thirty-three Blastomeres.

Contacts between 4d and 3a, 3b, and $2a^1$-$2d^1$ are partially lost while the 3A-3C cells are displaced closer to the center of the embryo. $2a^1$-$2d^1$ cells are the first to divide, forming the so-called tip cells of the future cross, $2a^{11}$-$2d^{11}$; this is followed by a laeotropic division of $2a^2$-$2d^2$ cells, $2q^{21}$ blastomeres being larger than $2q^{22}$.

Stage 10: Forty-nine Blastomeres. Early Blastula.

This stage is achieved by several sequential divisions closely following each other. Initially $1a^1$-$1c^1$ micromeres undergo dexiotropic divisions (36 blastomeres); this is followed by dexiotropic divisions of $1d^1$ and by almost radial divisions of 3a-3d (41 blastomeres); the size of $3a^2$ is greater than $3a^1$, $3b^2$ is greater than $3b^1$, but $3c^2$ is smaller than $3c^1$ and $3d^2$ is smaller than $3d^1$ [21]. The primary trochoblasts $1a^2$ and $1b^2$, taking part in forming the prototroch, divide almost in the horizontal plane; 3A-3C macromeres divide laeotropically (46 blastomeres); eventually the 4d primary mesoblast divides, yielding two identical mesoteloblasts M_1 and M_2 (Fig. 5.13, 10^+ veg) and the remaining primary trochoblasts $1c^2$ and $1d^2$ (Fig. 5.14, 10_{an}). It is noteworthy that the delayed divisions of $1d^1$, $1c^2$, and $1d^2$ correlate with their longer contact with 4d.

Verdonk [174] distinguished the 48 blastomere stage with eight trochoblasts located pairwise one on top of the other (Fig. 5.14, 10_{an}) and the-as-yet-undivided 4d. The trochoblasts do not divide for a long time after this. This stage completes the early cleavage period.

The subsequent stages 11-15 are characterized by the existence of the so-called molluscan cross, formed by the cells of the first quartet in addition to trochoblasts and the tip cells $2a^{11}$-$2d^{11}$. Beginning with these stages, the details of cell localization may be distinguished predominantly on fixed embryos.

Stage 11: Resting Stage of the Early Cross.

After the primary trochoblasts divide (preceding stage) cleavage is interrupted for about 3 h. During this period $1a^{11}$-$1d^{11}$, the larger part of the animal cells move away from the surface, while the basal $1a^{12}$-$1d^{12}$ cells are displaced toward the middle of the animal hemisphere and become wider. Their position marks the branches (or arms) of the cross: d, dorsal; b, ventral; a and c, lateral (see Fig. 5.14, 11).

Stage 12: Formation of the Outer Median Cells of the Cross.

The basal $1a^{12}$-$1d^{12}$ cells divide perpendicularly in relation to the direction of the cross arms (Fig. 5.15, 12_{an}). The cells $1a^{122}$-$1d^{122}$ are now referred to as the outer median cells, while $1a^{121}$-$1d^{121}$ cells are termed basal, as before.

Stage 13: Formation of Apical Rosette Cells.

The animal cells $1a^{11}$-$1d^{11}$ divide laeotropically and form $1a^{111}$-$1d^{111}$ apical rosette cells located at the animal pole; later these cells will become part of the apical plate of the trochophore and no longer divide; in addition they yield peripheral

Figs. 5.13–5.21. Stages of *Lymnaea stagnalis* L. normal development. Numbers correspond to stage numbers. Letters accompanying numbers designate: an, animal view; veg, vegetal view; m, view from the mouth; bl, view from the blastopore; sv, side view; al, anterolateral view; lef, view from the left; r, view from the right; d, dorsal view; df, dorsofrontal view; sag, sagittal section.

Figs. 5.13–5.15. Sketches 2–24 are from preparations impregnated with silver [175].

Figs. 5.16 and 5.17. Sketches 1/3 M–1/5 M and 15 M–24/2 M are outlines of the pond snail embryo at consecutive developmental stages. an, animal pole; veg, vegetal pole; 1(2) pb, first (second) polar body; vitmem, vitelline membrane; pr, pronuclei; bl, blastopore; p, prototroch; ap, apical plate; cp, cephalic plates; m, mouth; f, foot; shg, shell gland; sh, shell; ma, mantel; vel, velum; e, eyes; te, tentacles; lte, labial tentacles; h, heart; pcs, post-trochal ciliate strand; pco, pulmonary cavity orifice; hv, head vesicle. Some of the abbreviations here also appear in Fig. 5.15.

Figs. 5.18–5.21. Normal developmental stages of the pond snail at daily intervals during 2–11 days of embryogenesis (25°C) [49]. 1) Stomodeum (mouth); 2) gastral cavity; 3) intestinal plate consisting of small cells; 4) hind gut; 5) albumen sac; 6) left albumen sac; 7) right albumen sac; 8) stomach; 9) gastric epithelium; 10) gastric wall; 11) dorsal gastric aperture; 12) ventral gastric aperture; 13) large gastric aperture; 14) small gastric aperture; 15) large liver rudiment; 16) small liver rudiment; 17) radula; 18) esophagus; 19) crop; 20) anus; 21) prospective foot region; 22) transverse ventral groove; 23) foot; 24) cerebral ganglia; 25) cerebral commissure; 26) pedal ganglia; 27) visceral ganglion; 28) buccal ganglia; 29) parietal ganglia; 30) pleural ganglia; 31) mesoderm; 32) prototroch (velum); 33) nuchal cells; 34) protonephridium; 35) remnants of protonephridium; 36) kidney anlage; 37) kidney; 38) renopericardial channel; 39) heart; 40) pericardium; 41) atrium; 42) ventricle; 43) shell gland; 44) invagination of the shell gland; 45) shell; 46) pulmonary cavity; 47) respiratory orifice; 48) mantle cavity; 49) osphradium; 50) statocyst; 51) eyes; 52) tentacle; 53) oral lobe (labial tentacle). The digestive tract is shaded.

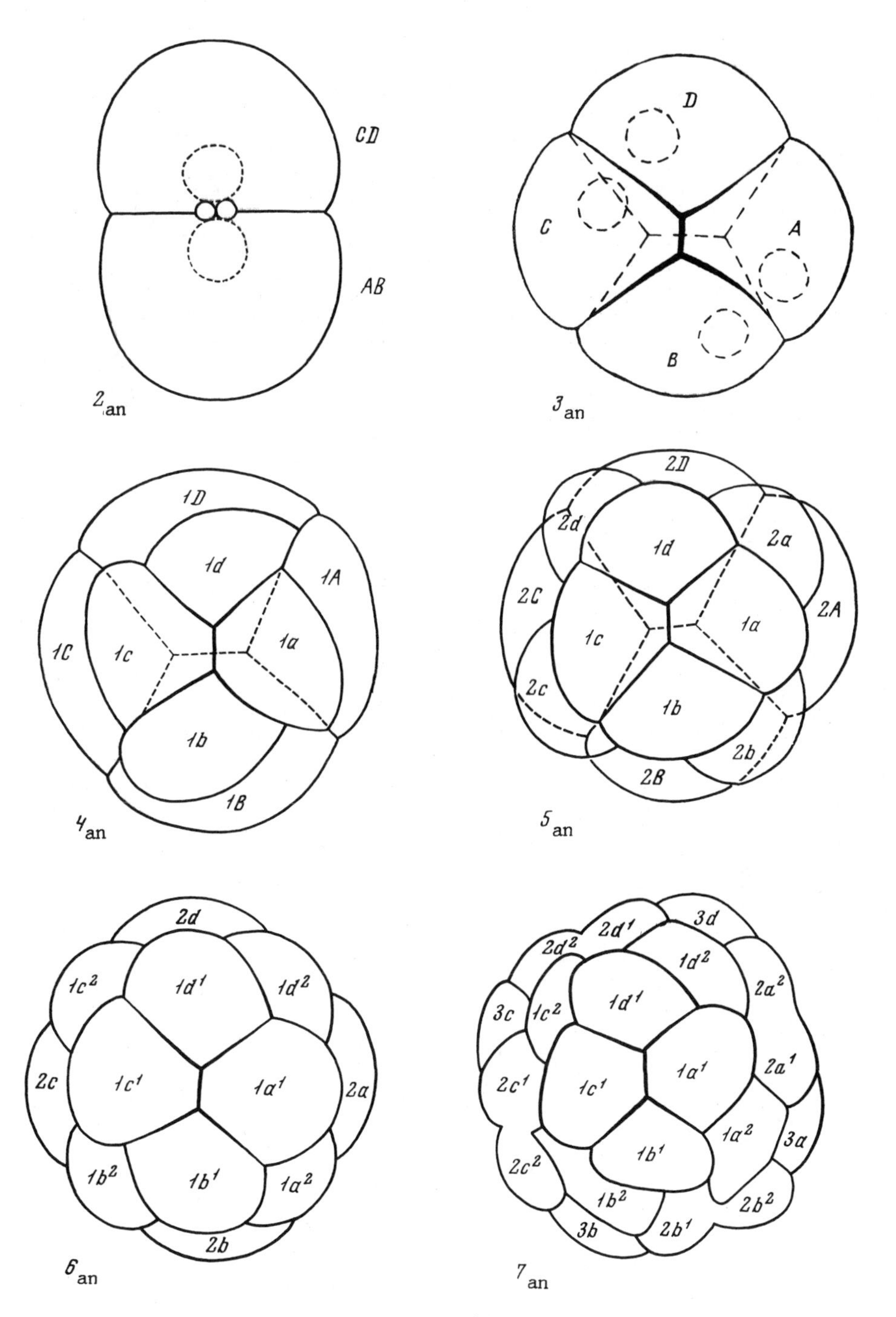

CD
AB
2_{an}
D
C
A
B
3_{an}
1D
1d
1A
1C
1c
1a
1b
1B
4_{an}
2D
2d
1d
2a
2C
1c
1a
2A
2c
1b
2b
2B
5_{an}
2d
$1c^2$
$1d^1$
$1d^2$
2c
$1c^1$
$1a^1$
2a
$1b^2$
$1b^1$
$1a^2$
2b
6_{an}
$2d^1$
3d
$2d^2$
$1d^2$
$2a^2$
3c
$1c^2$
$1d^1$
$1a^1$
$2a^1$
$2c^1$
$1c^1$
$1a^2$
3a
$2c^2$
$1b^1$
$1b^2$
$2b^2$
3b
$2b^1$
7_{an}

Fig. 5.13

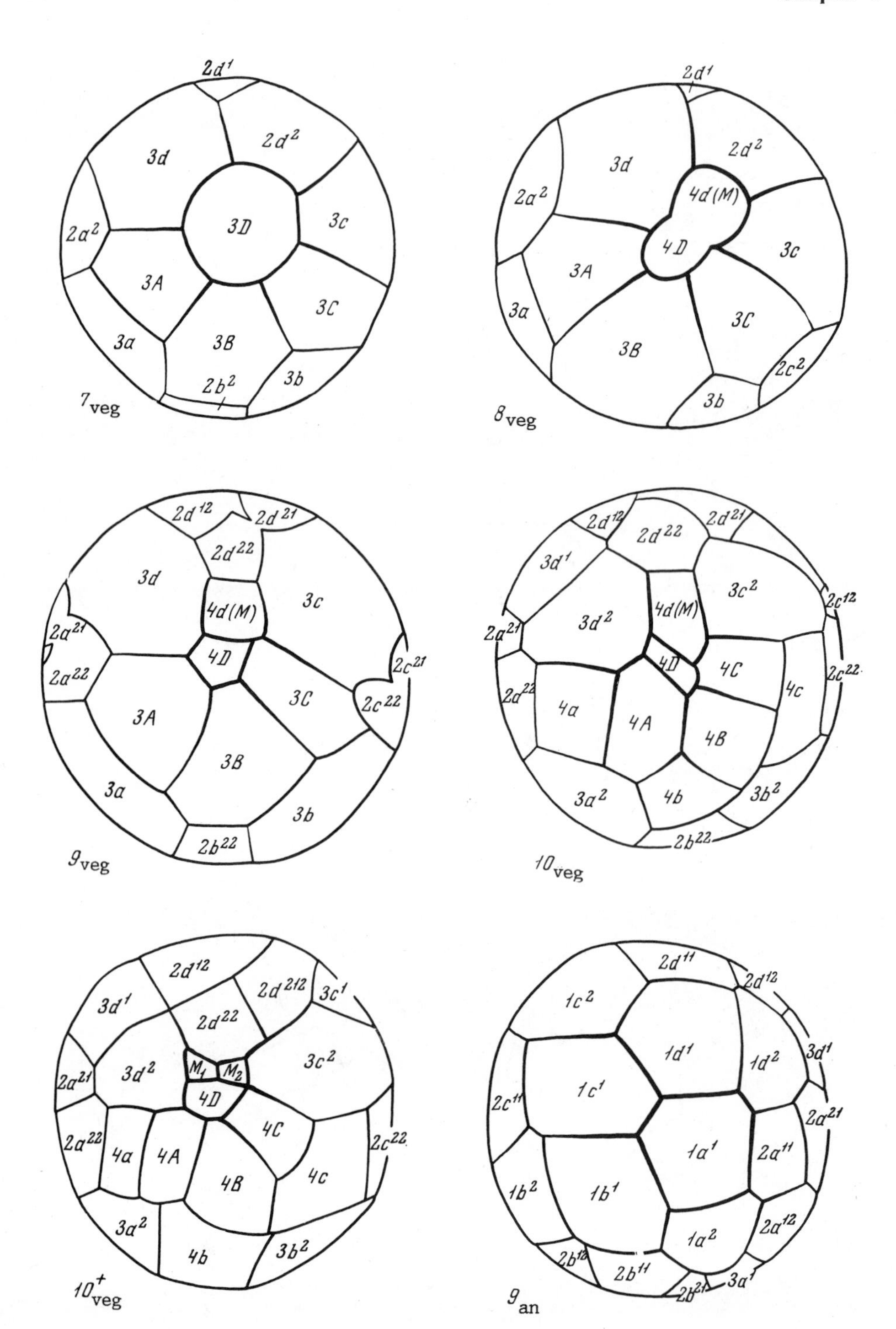

Fig. 5.13 (continued).

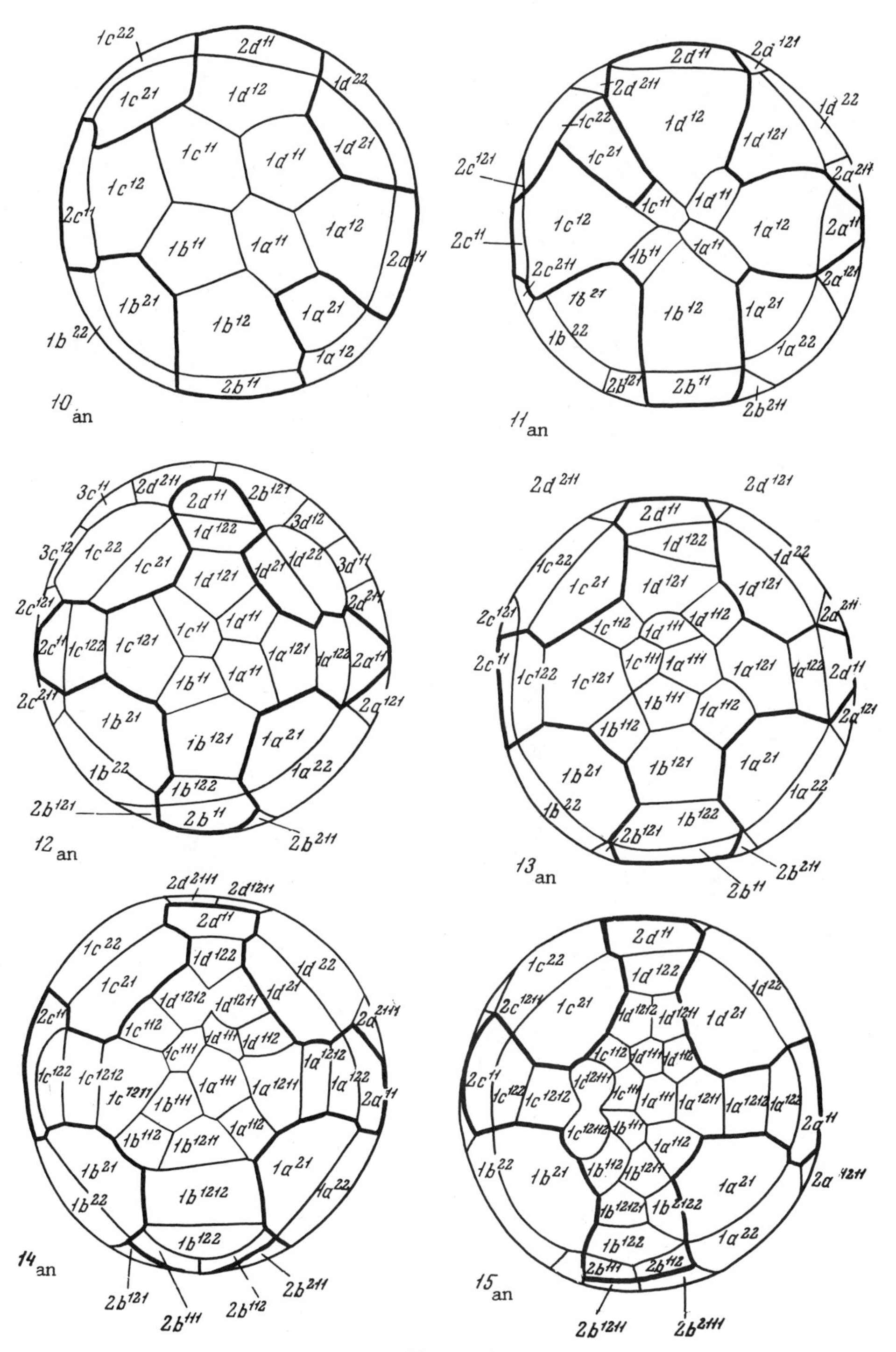
$1c^{22}$ $2d^{11}$ $1d^{22}$ $1c^{21}$ $1d^{12}$ $1c^{11}$ $1d^{11}$ $1d^{21}$ $1c^{12}$ $2c^{11}$ $1b^{11}$ $1a^{11}$ $1a^{12}$ $2a^{11}$ $1b^{21}$ $1b^{12}$ $1a^{21}$ $1b^{22}$ $1a^{12}$ $2b^{11}$
10_{an}
$2d^{11}$ $2a^{121}$ $2d^{211}$ $1d^{12}$ $1d^{22}$ $1c^{22}$ $1d^{121}$ $1c^{21}$ $2c^{121}$ $2a^{211}$ $1c^{11}$ $1d^{11}$ $1c^{12}$ $1a^{12}$ $2a^{11}$ $2c^{11}$ $1b^{11}$ $1a^{11}$ $2c^{211}$ $2a^{121}$ $1b^{21}$ $1b^{12}$ $1a^{21}$ $1b^{22}$ $1a^{22}$ $2b^{121}$ $2b^{11}$ $2b^{211}$
11_{an}
$3c^{11}$ $2d^{211}$ $2d^{11}$ $2b^{121}$ $3d^{12}$ $1d^{122}$ $3c^{12}$ $1c^{22}$ $1c^{21}$ $1d^{121}$ $1d^{21}$ $1d^{22}$ $3d^{11}$ $2c^{121}$ $2d^{211}$ $1c^{11}$ $1d^{11}$ $2c^{11}$ $1c^{122}$ $1c^{121}$ $1a^{121}$ $1a^{122}$ $2a^{11}$ $1b^{11}$ $1a^{11}$ $2c^{211}$ $2a^{121}$ $1b^{21}$ $1b^{121}$ $1a^{21}$ $1b^{22}$ $1a^{22}$ $1b^{122}$ $2b^{121}$ $2b^{11}$ $2b^{211}$
12_{an}
$2d^{211}$ $2d^{121}$ $2d^{11}$ $1d^{122}$ $1c^{22}$ $1d^{22}$ $1c^{21}$ $1d^{121}$ $1d^{121}$ $2c^{121}$ $1c^{112}$ $1d^{111}$ $1d^{112}$ $2a^{211}$ $2c^{11}$ $1c^{111}$ $1a^{111}$ $1c^{122}$ $1c^{121}$ $1a^{121}$ $1a^{122}$ $2d^{11}$ $1b^{111}$ $1a^{112}$ $2a^{121}$ $1b^{112}$ $1b^{21}$ $1b^{121}$ $1a^{21}$ $1b^{22}$ $1a^{22}$ $1b^{122}$ $2b^{121}$ $2b^{211}$ $2b^{11}$
13_{an}
$2d^{2111}$ $2d^{1211}$ $2d^{11}$ $1d^{122}$ $1c^{22}$ $1d^{22}$ $1c^{21}$ $1d^{1212}$ $1d^{1211}$ $1d^{21}$ $2c^{11}$ $2a^{2111}$ $1c^{112}$ $1d^{111}$ $1d^{112}$ $1c^{111}$ $1a^{1212}$ $1c^{122}$ $1c^{1212}$ $1a^{122}$ $1c^{1211}$ $1b^{111}$ $1a^{111}$ $1a^{1211}$ $2a^{11}$ $1a^{112}$ $1b^{112}$ $1b^{1211}$ $1b^{21}$ $1a^{21}$ $1b^{22}$ $1a^{22}$ $1b^{1212}$ $1b^{122}$ $2b^{121}$ $2b^{111}$ $2b^{112}$ $2b^{211}$
14_{an}
$2d^{11}$ $1c^{22}$ $1d^{122}$ $1d^{22}$ $2c^{1211}$ $1c^{21}$ $1d^{1212}$ $1d^{1211}$ $1d^{21}$ $1c^{112}$ $1d^{111}$ $1d^{112}$ $2c^{11}$ $1c^{12111}$ $1c^{122}$ $1c^{1212}$ $1c^{111}$ $1a^{111}$ $1a^{1211}$ $1a^{1212}$ $1a^{122}$ $2a^{11}$ $1c^{12112}$ $1b^{111}$ $1a^{112}$ $1b^{22}$ $1b^{21}$ $1b^{112}$ $1b^{1211}$ $2a^{1211}$ $1a^{21}$ $1b^{12121}$ $1b^{12122}$ $1b^{122}$ $1a^{22}$ $2b^{111}$ $2b^{112}$ $2b^{1211}$ $2b^{2111}$
15_{an}

Fig. 5.14

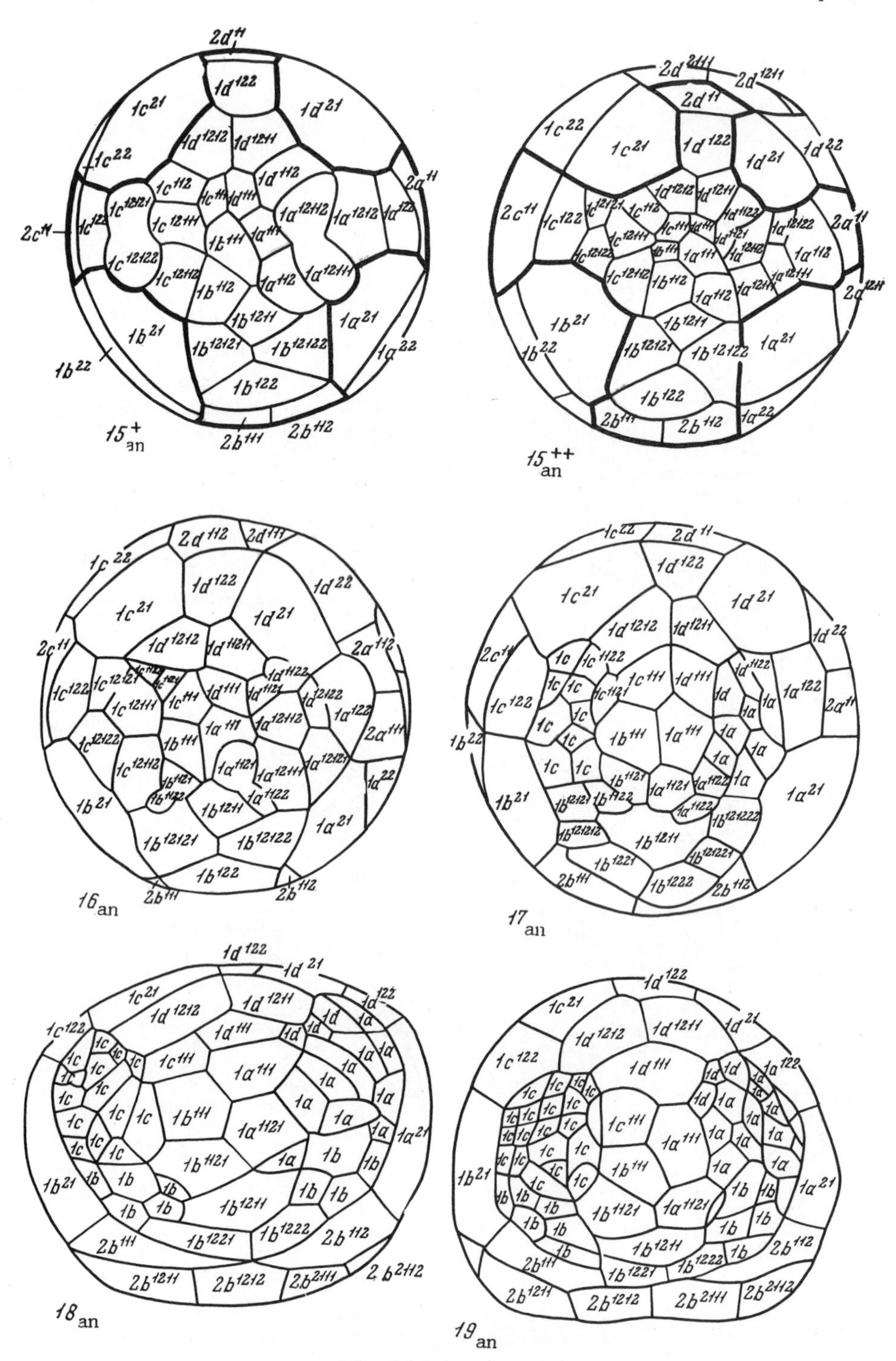

Fig. 5.14 (continued).

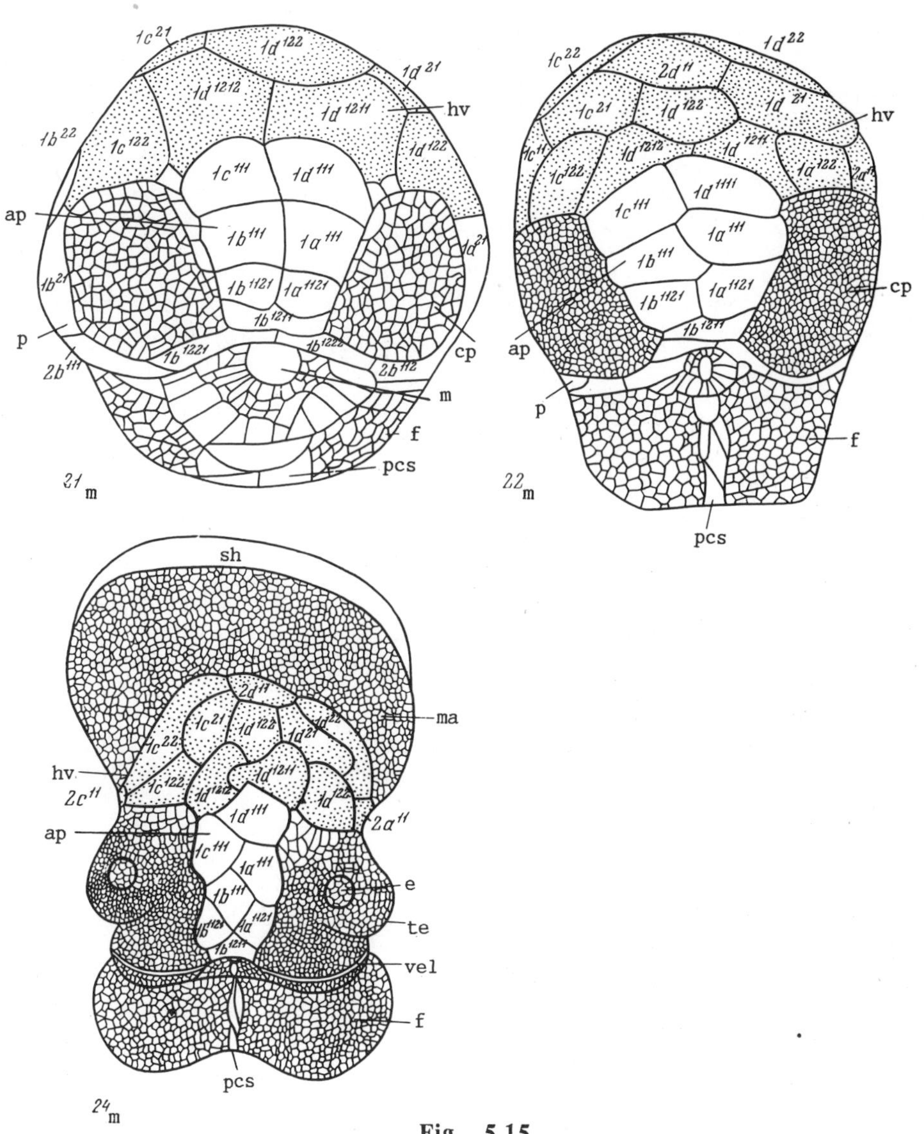

Fig. 5.15

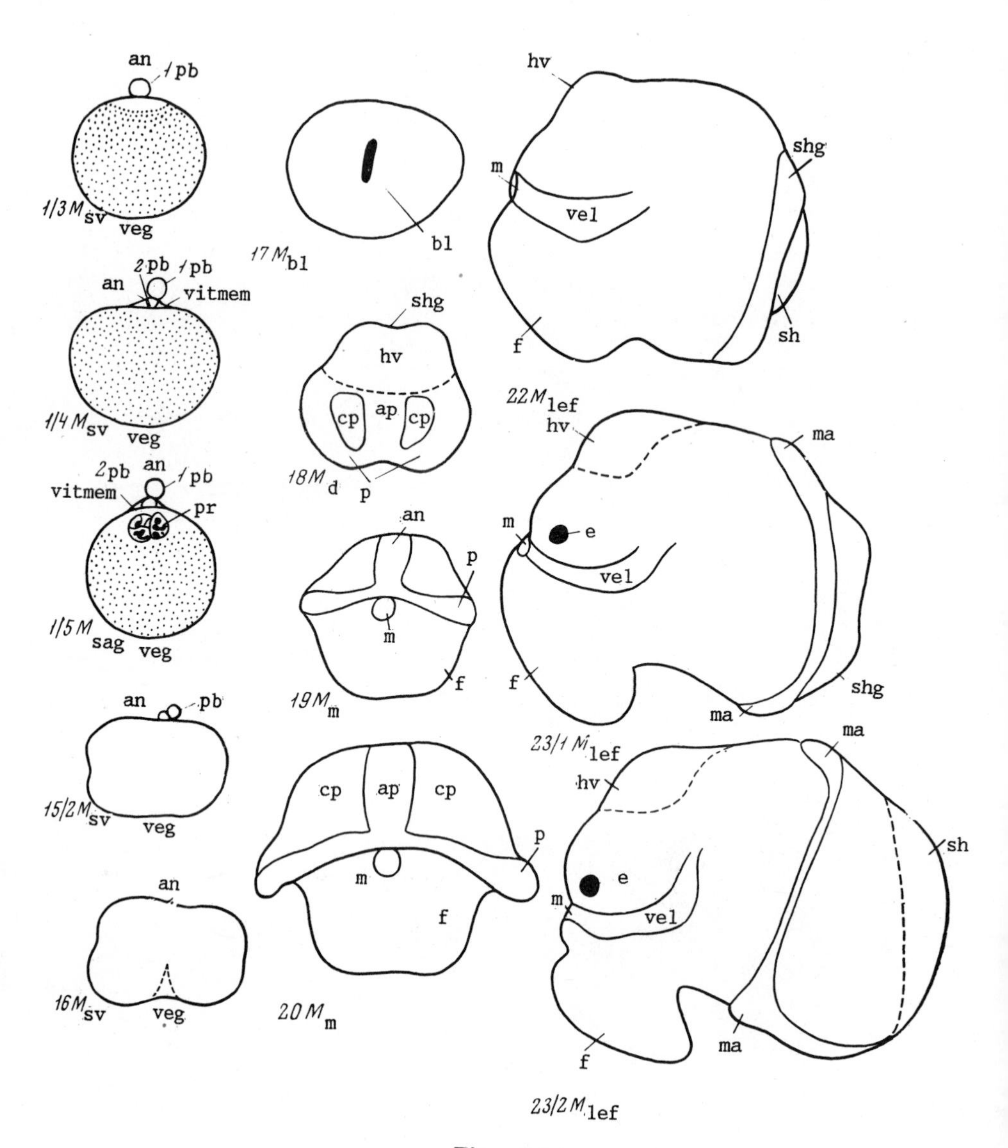
an
1pb
1/3 M sv
veg
2pb 1pb
an
vitmem
1/4 M sv
veg
2pb
an
1pb
vitmem
pr
1/5 M sag
veg
an
pb
15/2 M sv
veg
an
16 M sv
veg
bl
17 M bl
shg
hv
cp
ap
cp
18 M d
p
an
p
m
f
19 M m
cp
ap
cp
p
m
f
20 M m
hv
shg
m
vel
f
sh
22 M lef
hv
ma
m
e
vel
f
ma
shg
23/1 M lef
ma
hv
sh
e
m
vel
f
ma
23/2 M lef

Fig. 5.16

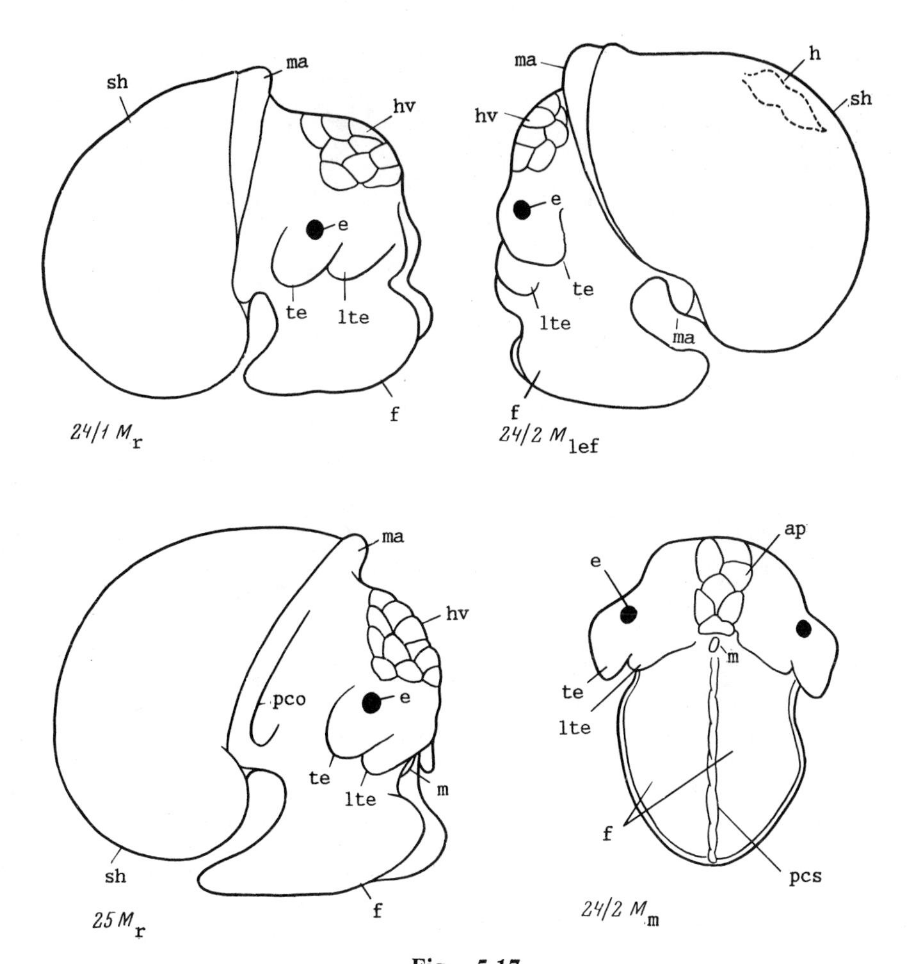
sh
ma
hv
e
te
lte
f
24/1 M_r
ma
h
sh
hv
e
te
lte
ma
f
24/2 M_{lef}
ma
hv
pco
e
te
lte
m
sh
f
25 M_r
ap
e
m
te
lte
f
pcs
24/2 M_m

Fig. 5.17

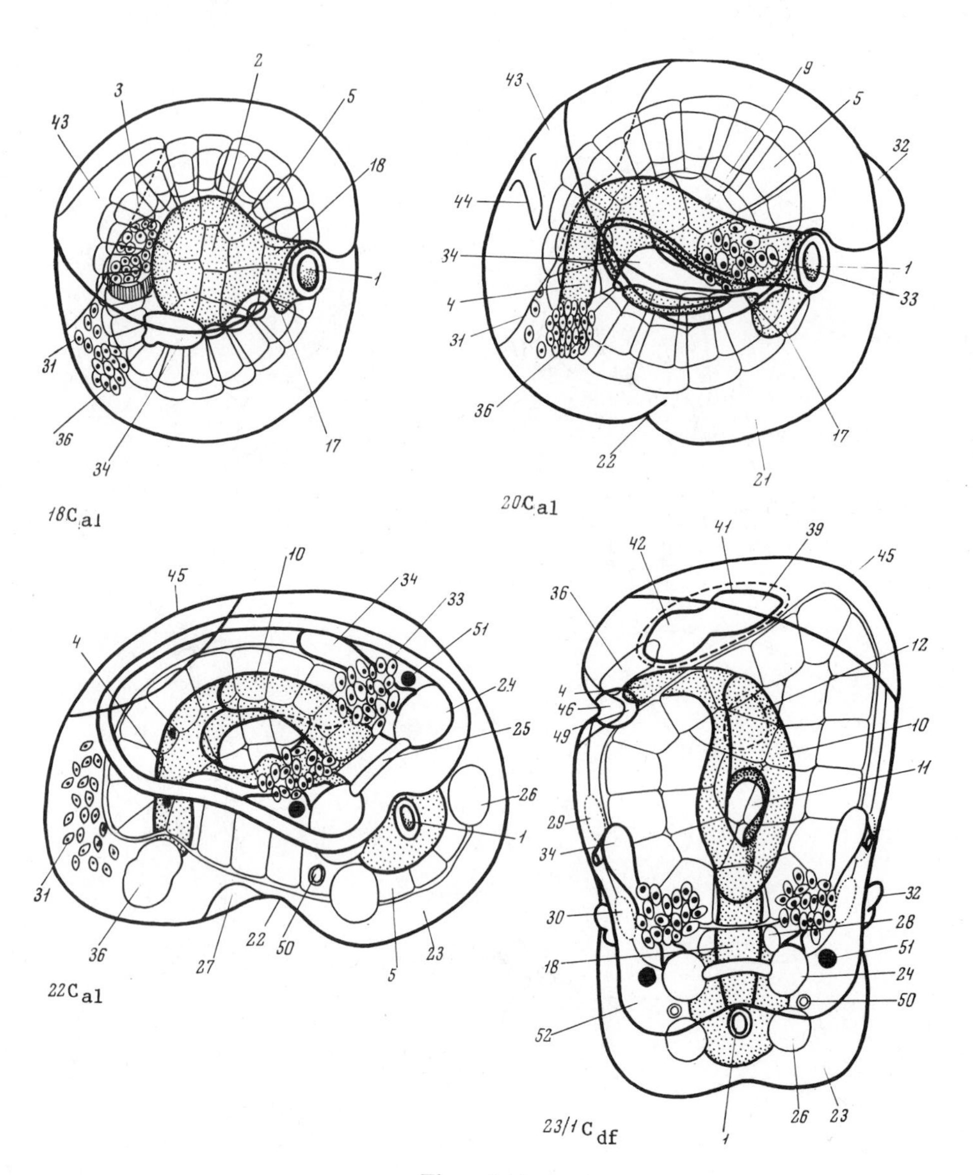

Fig. 5.18

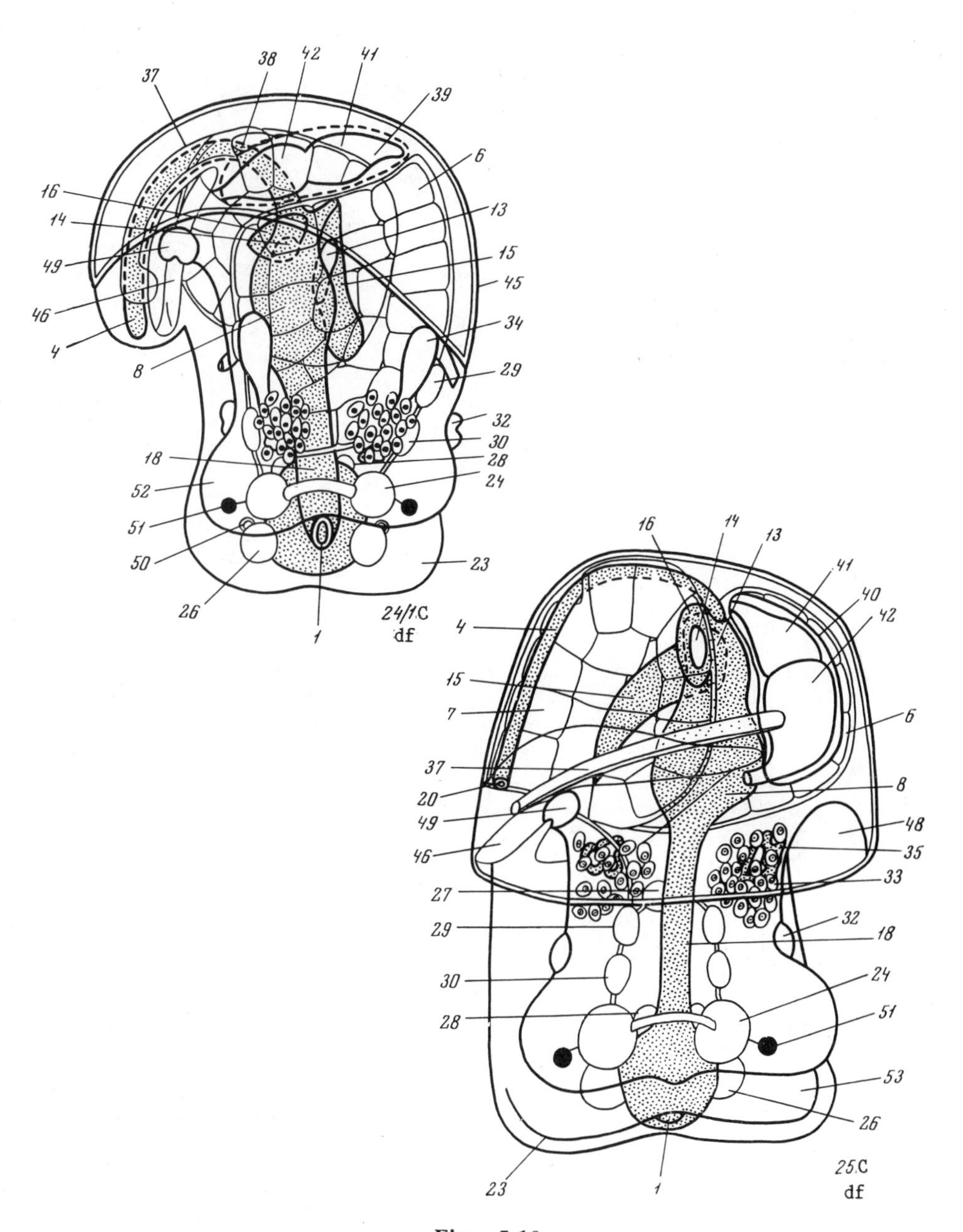

38
42
41
37
39
6
16
13
14
15
49
45
46
34
4
8
29
32
30
18
28
52
24
51
50
23
26
1
24/1C
df
16
14
13
41
40
42
4
15
7
6
37
20
8
49
48
46
35
27
33
29
32
18
30
24
28
51
53
26
23
1
25C
df

Fig. 5.19

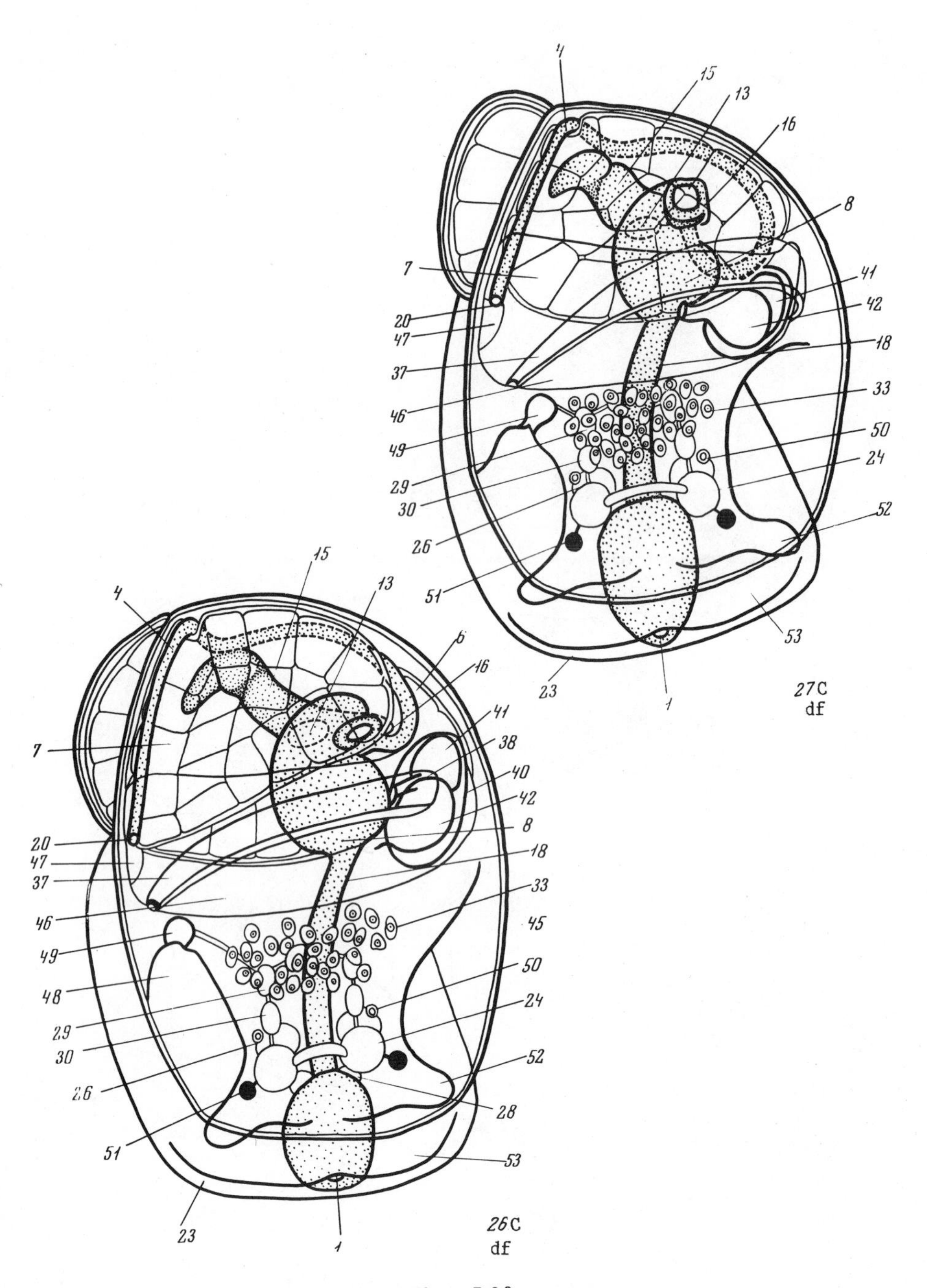
4
15
13
16
8
7
41
42
20
47
18
37
33
46
50
49
24
29
30
52
26
51
53
23
1
27C
df
4
15
13
6
16
41
7
38
40
42
8
20
47
18
37
33
46
45
49
50
48
24
29
30
52
26
28
51
53
23
1
26C
df

Fig. 5.20

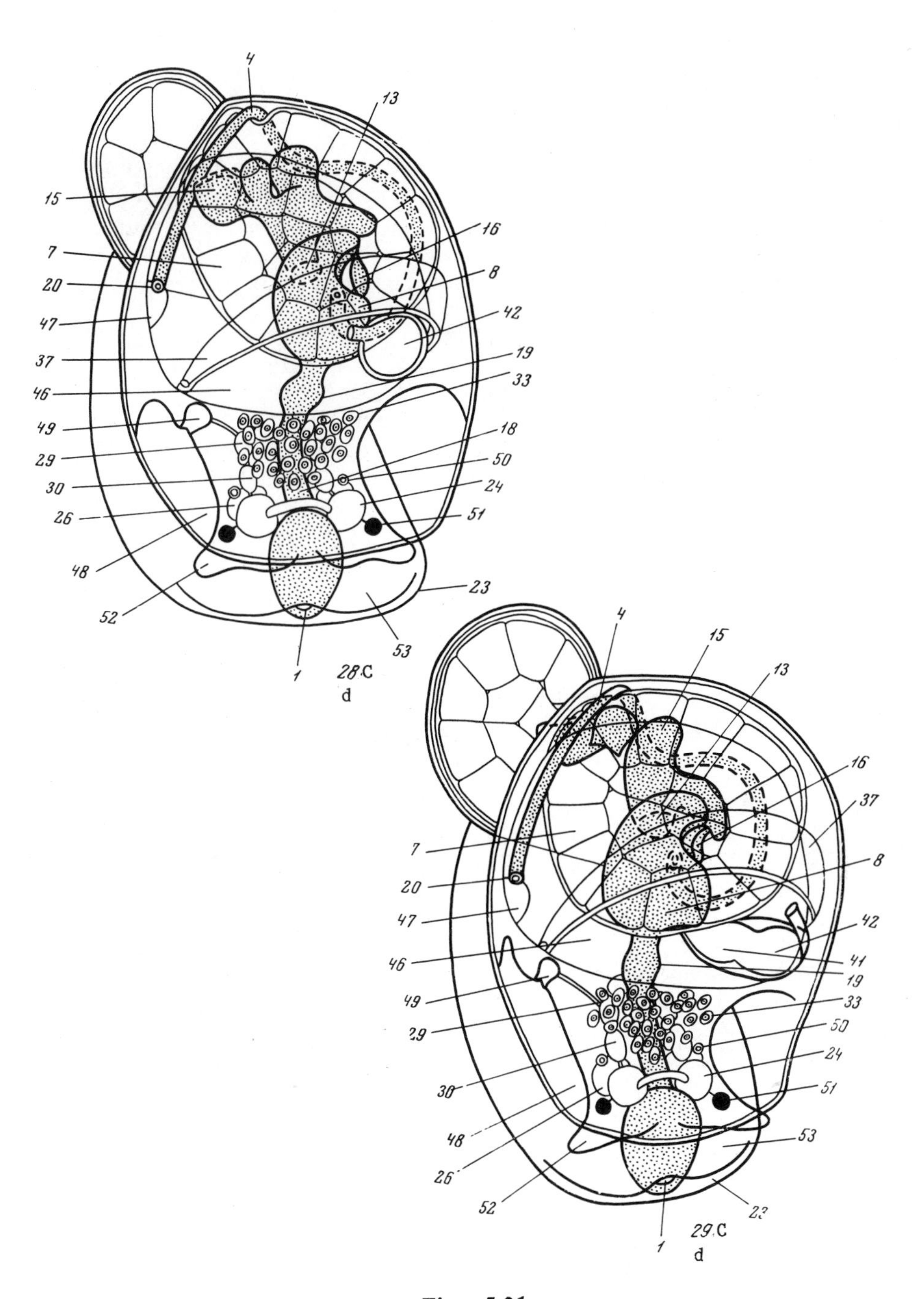
4
13
15
16
7
8
20
42
47
37
19
46
33
49
18
29
50
30
24
26
51
48
23
52
53
1
28 C
d
4
15
13
16
37
7
8
20
47
42
41
46
19
49
33
29
50
24
30
51
48
53
26
52
23
1
29 C
d

Fig. 5.21

rosette cells $1a^{112}$-$1d^{112}$ adjacent to the trochoblasts. The composition of the rosette does not change up to the disappearance of the cross (stage 16). According to Raven [141], pinocytosis of the capsular fluid by all cells of the blastula begins approximately at this or the next stage.

Stage 14: Mid-Blastula. Appearance of a Bilateral Symmetry in the Animal Hemisphere.

Each of the basal cells $1a^{121}$-$1c^{121}$ segregates a daughter cell toward the outer median cells. These cells $1a^{1212}$-$1c^{1212}$ are termed the inner median cells. The basal $1d^{121}$ cell, located in the dorsal arm of the cross, divided after some time in the plane parallel to the arm. The two sister cells $1d^{1211}$ and $1d^{1212}$ are located symmetrically to this arm. This completes the lineage of the dorsal arm.

Stage 15: Late Blastula.

The division of a number of cells comes to an end by this stage (Fig. 5.14, 15, 15^{+}, 15^{++}). The basal and inner median cells in the lateral arms of the cross undergo divisions. These divisions proceed in planes parallel to the arms. The outer median cells $1a^{122}$ and $1c^{122}$ no longer divide (Fig. 5.14, 15, 15^{+}). The tip $2b^{11}$ cell in the ventral arm of the cross undergoes meridional division. It is followed by the inner median cell $1b^{1212}$ [15]. Thereafter, all cells of the ventral arm, with the exception of the basal $1b^{1211}$ cell, begin to expand in the transversal direction. Descendants of the $1b^{1212}$ cell gradually reach the lateral arms. The cross is still distinctly discernible at the stage 15/1 (15^{++}). Prior to gastrulation (stage 15/2) the embryo, now consisting of about 200 cells, becomes markedly flattened at both poles (Fig. 5.16, 15/2M).

Stage 16: Early Gastrula.

The cross is no longer discernible. Cells located at the vegetal pole elongate in the animal direction and gradually begin to migrate inside. Mesoblasts and their descendants are located in the blastocoel by this time. At the beginning of gastrulation the blastocoel may be almost completely full of cells. The blastopore is located at the vegetal pole and has a rounded shape as seen from the surface. At the animal pole one frequently sees a flat depression formed by the apical rosette at the base of the dorsal arm of the cross. The ventral arm of the former cross is quite wide (Fig. 5.14, 16). The $1a^{112}$-$1d^{112}$ peripheral rosette cells begin to proliferate while the apical rosette cells widen considerably.

Stage 17: Mid-gastrula.

The blastopore moves to the ventral arm of the cross and acquires a fissure-like shape. Its borders appear to consist of 2d derivatives. According to Verdonk [175], the embryo at this stage emerges from the vitelline membrane but remains inside the capsule. The apical and cephalic plates are already discernible in the animal hemisphere (Fig. 5.14, 17). In other words, bilateral symmetry of the embryo is already quite distinct.

Stage 18: Late Gastrula.

The blastopore is at the definitive position opposite the shell gland rudiment; it looks like a small depression (closure of the blastopore takes place in the posteroanterior direction). The rudiment of the shell gland is composed of a wide depression of the thickened ectoderm. An open prototroch is located immediately above the blastopore; the prototroch does not as yet rise over the surface of the embryo. Cilia are active in the prototroch, foot, and several other regions of the body. The embryo rotates inside the capsule at a speed of 1 rev/105 sec. The head vesicle – a translucent structure consisting of 12 flattened cells – is located under the animal pole (see Fig. 5.6). The apical plate, extending from the head vesicle to the prototroch, divides two regions of moderately sized ectodermal cells which are known as cephalic plates. Digestion of the protein-containing fluid begins in the large "albumen" endodermal cells. Two protonephridia consist of four large cells each. The midgut plate consisting of small cells is located caudal to the albumen cells, which are covered by the common envelope, giving rise to the albumen sac and delimiting the gastral cavity. The midgut plate later yields the hindgut, stomach, and part of the definitive liver. The definitive kidney exists as a loose complex of mesodermal cells. A more detailed description of the development between stage 19 and stage 25 can be found in Raven [137, 138].

Stage 19: Early Trochophore.

The lateral parts of the prototroch begin to rise over the body surface. The first cilia appear on the surface of the apical plate cells. The rudiment of the shell gland has a somewhat depressed middle part. The stomodeum, which originates from the ectoderm, comprises the outer part of the blastopore and begins to differentiate into the oropharyngeal part and the esophagus. The albumen cells discharge secretion into the gut lumen. Two protonephridia develop. The embryo is only barely transparent at this stage.

The ultrastructure of the trochophore has been studied in detail by Arni [3].

Stage 20: Mid-trochophore.

Lateral parts of the prototroch rise above the surface of the body as two bands. A row of larger cells corresponding to the post-trochal ciliate strand is obvious in the middle of the future foot region. The embryo rotates, each rotation taking about 5 sec. It is more transparent than in the preceding stage. The shell gland is deeply invaginated toward the gut rudiment, and there is a thin translucent shell over it. Both the gland and the shell are slightly displaced to the left from the sagittal body plane. (In our description of pond-snail development we will, as usual, define as right and left the sides of the embryo which are located respectively to the right and left of the cranial–caudal axis of the animal when viewed from the dorsal side of the body.) Functional protonephridia are formed. Two groups of 15 nuchal cells are located over the protonephridia. Cell proliferation begins in the cephalic plate ectoderm and in the part of the foot which is adjacent to the mouth. The radular sac segregates from the stomodeum wall. At this stage the sac has a lumen; radular muscles form around the sac. Two epithelial strands connected with the esophagus in the frontal region detach from the gut plate on its right and left sides. Later they form the stomach walls and muscles. The newly formed stomach is open at the sides and is in contact with the cells of the albumen sac, which are no longer active in secretion. The hindgut rudiment has appeared.

Stage 21: Late Trochophore.

The shell gland depression has almost reached the outer surface of the posterior part of the body. The middle part of the gland is covered by a thin transparent shell. The foot region is well formed. The formation of cerebral ganglia begins in the cephalic plates, and eye rudiments appear as not-very-prominent fossae. The anlage of statocysts appear in the foot region. The stomodeum begins to differentiate anteriorly in the oral and pharyngeal cavities. The head vesicle occupies about half of the pretrochal region and laterally is in contact with the dorsal tips of the prototroch. The ventral part of the prototroch consists of cells $1b^{1221}$, $1b^{1222}$, $1b^{112}$, and $2b^{111}$; the dorsal part of the prototroch on each side consists of the two cells $1a^{21}$ and $1a^{22}$ on the left and $1b^{21}$ and $1b^{22}$ on the right. Generally, the embryo is still rounded.

Stage 22: Early Veliger.

The embryo makes one complete turn in 10 sec. The foot rudiment is distinctly separated from the visceral sac. The prototroch is about 0.1 mm in length; laterally it forms strongly protruding thickenings which dorsally join up with the head vesicle. The mouth is located in the middle of the prototroch. From this stage onward the prototroch begins to reduce gradually. The shell gland is almost flat; thickenings or mantle folds have formed near its borders and are in contact with the shell. The body's cranial–caudal axis is longer than the dorsal–ventral axis. The nuchal cells proliferate. Cerebral ganglia are connected by a commissure. Rudiments of the pedal ganglia are present, and the formation of pleural, parietal, and visceral ganglia is under way. The pulmonary cavity is present as a thickening of the ectoderm near the rudiment of the definitive kidney. The radular sac is not connected to the oral cavity. A cavity or lumen appears in the albumen sac on both the right and left sides of the stomach; these cavities accumulate the proteinaceous fluid which comes from the stomach. This fluid is pumped to the stomach from the mouth with the help of a ciliate thickening of the esophagus.

The process of torsion begins at this stage. It is characteristic for the development of all gastropods and involves a turn of the whole visceral complex with respect to the head region around the cranial–caudal body axis. In the pond snail this turn proceeds clockwise if the embryo is viewed in the direction from the head to the shell. This is particularly well seen in the movements of different regions of the gastrointestinal tract. Thus, parts of the albumen sac turn by 180° between stages 24 and 29; the small stomach orifice turns by 270°, and the large one by 90°. The heart and genital duct rudiment move together along the arc in the frontal plane and are displaced from their position and shifted somewhat to the right side of the sagittal plane into the left dorsal position (see Fig. 22c, d). The gut rudiment makes a loop and opens to form the anus near the right edge of the dorsal mantle fold (see Figs. 5.18 and 5.17, 23c and 29c).

Stage 23: Mid-veliger.

This stage can be divided into two substages:

23/1. The embryo makes a complete turn in 15 sec. The foot shows spontaneous movements for the first time; it is separated from the visceral complex by a deep furrow. The middle region of the shell gland is still flat. The mantle fold envelops about one-third of the visceral sac. The heart, consisting of two chambers and pulsating at a rate of 44-48 beats/min, is located somewhat to the right of the sagittal plane in the caudal direction. The ventricular and atrial walls are composed of one layer. The pericardium has not yet formed. On the right side a small open pulmonary cavity is located, and the osphradium is situated near it. The length of the prototroch is about 0.1 mm. Protonephridia are functional. The kidney does not yet show a lumen. The nuchal cells have stopped dividing. The pedal ganglia are connected by a commissure, while the buccal ganglia are being laid down. Other ganglia have not yet segregated from the ectoderm. Red-brown eyes are located at the base of the barely protruding head tentacles. A dental plate has appeared in the radular sac, which is still closed; the jaw has appeared between the oral cavity and the pharynx. Owing to torsion, the stomach apertures are now located dorsally and ventrally rather than laterally. The hindgut is connected to the stomach cavity.

23/2. The mantle and shell surround two-thirds of the visceral complex. The foot is divided longitudinally into two halves by a groove containing a strand of

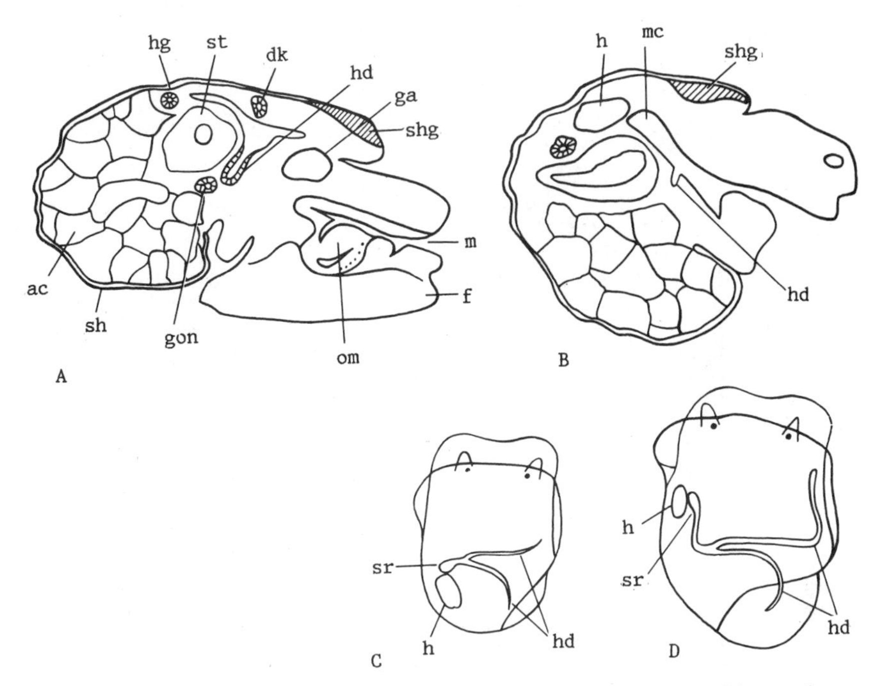

Fig. 5.22. Various views of the reproductive system during development and late embryogenesis in *L. stagnalis appressa* Say [60]. A) Schematic midsagittal section through a 6-day-old embryo. B) Oblique section through an 8-day-old embryo; part of the mantle cavity behind hermaphroditic duct rudiment will later become the seminal receptacle. C) Arrangement of reproductive ducts and heart in a 10-day-old embryo (viewed from above). D) Same as above, but in a 14-day-old embryo prior to hatching; development at 25°C. sh, shell; shg, shell gland; ac, albumen cells of the larval liver; hg, hind gut; st, stomach; dk, definitive kidney; hd, hermaphroditic duct (A shows its rudiment on the mantle wall); gon, gonad; ga, ganglion; om, oral mass or buccal complex; m, mouth; f, foot; h, heart; mc, mantle cavity; sr, seminal receptacle.

large ciliated cells. The heart is located in the sagittal plane approximately under the middle part of the shell.

Stage 24: Late Veliger or "Hippo."

This stage is again divided into two stages, as follows:

24/1. The rotation of the embryo is irregular. The foot in the posterior part has not yet reached the shell; it may extend in the direction parallel to the cranial–caudal body axis. The prototroch is still present, but has been displaced more caudally. The shell covers the whole posterior half of the body. The first evidence of its helical coiling is present. The eyes have a reddish color. A lens begins to form from the secretion supplied by the eye vesicle cells. The contracting heart, which is located in the center dorsal region, can be seen through the wide aperture of the pulmonary cavity. The heart ventricle has a multilayered wall. The pericardium has not yet completely formed. The kidney and pericardium are connected by the renal-pericardial channel which is formed at this stage. The protonephridia still function; their flame cells are no longer in contact with the cerebral ganglia. The buccal ganglia are connected by a commissure. The pleural ganglia are connected to the parietal and cerebral ganglia through connectives, which are longer on the right side of

TABLE 5.3. Sequence of Organ Formation and Development in the Pond Snail at 25°C [49]

Organs	Stage									
	18	20	22	23/1	24/1	25	26	27	28	29
	Age in days									
	2	3	4	5	6	7	8	9	10	11
Prototroch (velum)	–	+	+15	+15	+15	+15	–	–	–	–
Cilia	+	+	+	+	+	+	–	–	–	–
Protonephridium	(+)	+8	+	+	+	9	9	–	–	–
Kidney	16	16	16	16	+8	+	+	+	+	+
Renopericardial channel	–	–	–	–	+8	+	+	+	+	+
Nuchal cells	–	+	+	+	+	+1	+9	+9	+9	+9
Pericardium	–	–	–	(+)	(+)	+	+	+	+	+
Heart	–	–	–	+	+	+	+	+	+	+
Cerebral ganglia	–	14	16	16	16	+	+	+	+	+
Pedal ganglia	–	–	14	16	16	+	+	+	+	+
Buccal ganglia	–	–	–	(+)	+	+	+	+	+	+
Pleural ganglia	–	–	14	14	16	+	+	+	+	+
Parietal ganglia	–	–	14	14	16	+	+	+	+	+

Visceral ganglion	–	–	14	16	16	+	+	+	+	+
Osphradial ganglion	–	–	–	14	16	16	+	+	+	+
Osphradium	–	–	–	14	10	10	+	+	+	+
Eyes	–	10	(+)	(+)	(+)	+	+	+	+	+
Pulmonary cavity	–	–	16	10-12	10-12	10-12	+13	+	+	+
Respiratory orifice	–	–	–	–	–	–	+	+	+	+
Stomodeum	+	+	+	+2	+2	+	+	+	+	+
Jaw	–	–	–	(+)	(+)	+	+	+	+	+
Radula	16	5	3	3	3	4	+	+	+	+
Esophagus	+7	+7	+7	+7	+7	+7	+7	+7	+7	+
Salivary glands	–	–	–	–	–	(+)	+	+	+	+
Stomach	–	(+)	+	+	+	+	+	+	+	+
Hindgut	–	5/6	+6	+6	+6	+8	+	+	+	+
Definitive liver	–	–	–	–	5	5	5	5	5	5
Albumen sac	x	x	x11	x11	xx11	xx11	xx11	xx11	xx11	xx11
Tentacles	–	–	–	(+)	+	+	+	+	+	+
Oral lobes	–	–	–	(+)	+	+	+	+	+	+
Shell	–	+	+	+	+	+	+	+	+	+
Mantle cavity	–	–	–	–	–	+10	+10	+10	+	+
Genital system	–	–	–	–	–	(+)	(+)	(+)	(+)	(+)

Key to symbols: 1, decrease in number; 2, extension; 3, closing; 4, opening; 5, growth; 6, anal cells; 7, cilia; 8, rupture; 9, degeneration; 10, invagination; 11, lumen present; 12, open cavity; 13, closed cavity; 14, proliferation; 15, resorption; 16, undifferentiated cellular complex; x, one albumen sac; xx, two albumen sacs; (+), in the process of formation; –, absence; +, formed complete [e.g., +15 = the completely formed organ (velum) begins to resorb; 10-12 = pulmonary cavity is open and undergoes invagination].

the body than on the left. The osphradial ganglion has formed. Visceral–parietal and osphradial–parietal (right) connectives are present.

According to Kruglyanskaya and Sakharov [101], biogenic amines can now be detected in the cerebral and pedal ganglia for the first time.

Radular teeth begin to form. The stomach has turned even further to the right along the longitudinal axis; the large aperture is now located laterally on the left side. The epithelium has started to grow at the borders of these apertures and large and small lobes of the definitive liver have appeared (the small lobe grows very slowly). These lobes divided the albumen sac into two parts, the left being greater than the right.

24/2. The foot reaches the shell at the posterior end. The eyes are dark.

Stage 25: Veliconcha.

The embryo sometimes turns slowly in the proteinaceous fluid which is present in the capsule, and sometimes uses its foot to move along the inner capsule wall. It is still unable to withdraw into the shell. The shell now covers the whole visceral complex. A small mantle cavity has been formed by ectodermal cell proliferation and ectoderm wrapping under the shell edges. The tentacles as well as the mouth lobes are well developed. The prototroch is in the final stage of reduction. Resorption of the protonephridia and head vesicles has begun. At the beginning of the next stage the head vesicle is completely replaced by a small-cell epithelium. The eyes are dark, and the lens is formed and yet pigmented. The heart has been displaced to the left body side and beats 56-60 times a minute. The ventricle is larger than the atrium, and the pericardium is completely formed. The distances between the ganglia have decreased on the right side of the body owing to torsion, and the whole ganglionar ring possesses a plane of symmetry. The radular cavity is connected to the oral cavity. The jaw has formed completely. Salivary glands have been laid down as diverticula of the oral cavity walls.

From this stage onward torsion affects all organs of the visceral complex. The larger stomach aperture, together with the large liver lobe, has been displaced to the ventral side and the small one to the dorsal side. The larger albumen sac is located under the heart and stomach, while the smaller one is in the upper right region of the body. The hindgut perforates the pulmonary cavity and opens outside as the anal orifice. The oviduct has been laid down.

Stage 26: Transition to Foot Movement.

Owing to the presence of a larger mantle cavity, the embryo can now withdraw into its shell and use only its foot for movement. The shell is coiled. There is intensive pigmentation of the occiput, tentacles, and mouth lobes. The prototroch has been completely resorbed. The degrading protonephridia are still visible. The kidney is located on the roof of the pulmonary cavity and reaches the pericardium on the left side. Its position does not change any further before hatching. The nuchal cells are located as a single group and are active in holocrine secretion (see [149]). The heart has been further displaced to the left, its main axis located horizontally. When the embryo moves, the contraction rate of the heart is 60 beats/min, while at rest it is 39 beats/min. The osphradium cavity is still open outside.

The pulmonary cavity extends from the right to the left body wall, its orifice still open. The lens is devoid of pigmentation. The caudal part of the stomach is bent toward the left. The caudal tip of the large liver lobe borders the shell epithelium. The left albumen sac protrudes into the depression formed by the first turn of the helix. The hermaphroditic gland appears to the right of the stomach in the ventral body region (see Fig. 5.22A). The gonad begins forming after only two primary germ cells have arisen [44].

Stage 27.

The first helical turn of the shell is twice as high as at the previous stage. The kidney is bent over the pericardium to the ventral side of the embryo. The main axis of the heart is no longer horizontal, the ventricle now being located above the atrium. The heartbeat rate is 82 beats/min. The osphradium sac is closed and begins to divide into two parts. Diverticula form in the large lobe of the liver. The left albumen sac fills the entire caudal–ventral region of the visceral complex and is located perpendicularly to the main body axis. Some albumen cells divide.

Stage 28.

The first helical turn of the shell is twice as high as at stage 26. The heartbeat rate is 88 beats/min. The pulmonary cavity undergoes a marked extension in the caudal direction, its anterior wall bordering the shell. The small orifice of the stomach is on the left side owing to torsion, which generally reaches completion at this stage. Three diverticula are present in the large liver lobe. A dilatation appears in the caudal part of the esophagus. The right albumen sac is displaced to the left of the body. A penis appears at the caudal border of the right tentacle.

The hermaphroditic gland opens into a short hermaphrodite duct.

Stage 29. Hatching.

The embryo occupies the entire capsule. The capsular membrane breaks owing to movements of the young mollusc's radula. The nuchal cells are present in roughly a third of their initial number (which was maximal at stage 24). The degeneration of nuchal cells does not reach completion until the third week of postembryonic development. The main axis of the heart is vertical and the heart is positioned well away from the stomach. The dorsal thickening with cilia, previously present, disappears from the esophagus. The large liver lobe consists of four diverticula. After hatching, the cells of the albumen sacs transform into epithelial cells, becoming part of the definitive liver. The left sac participates in forming the large liver lobe, and the right contributes toward the smaller lobe. The shell has one and a half turns and its retractor is well developed. The hermaphrodite duct is connected to the rudiment of the oviduct. Generally the genital system at the time of hatching is at an early stage of development. According to Cumin [49], the pulmonary cavity is closed at this stage; the respiratory orifice first opens into the atmosphere when the mollusc creeps up to the surface of the water. On the other hand, we have observed a wide open respiratory orifice in young molluscs well under water.

When embryos are cultured in capsules on agar, all molluscs hatch within 12 h. In our experiments capsules were incubated in water and the interval between the hatching of the first and last mollusc was also 12 h.

After hatching, young snails feed on slime and the envelope of the cocoon, but even during the first day they can eat green plants [49]. According to van der Steen [163], the mortality rate of young snails during the first 11 weeks after hatching is 25%, whereas during the subsequent 3 weeks it does not exceed 10% at 20°C.

Table 5.3 shows the formation sequence of the main organs in the pond snail at 25°C (after [49]).

REFERENCES

1. W. A. Anderson and P. Personne, "The localization of glycogen in the spermatozoa of various invertebrate and vertebrate species," *J. Cell Biol.* **44**, 29–51 (1970).

2. P. Arni, ""Zur Feinstruktur der Mitteldarmdrüse von *Lymnaea stagnalis* L. (Gastropoda, Pulmonata)," *Z. Morphol. Tiere* **77**, 1–18 (1974).
3. P. Arni, "Licht- und elektronenmikroskopische Untersuchungen an Embryonen von *Lymnaea stagnalis* L. (Gastropoda, Pulmonata) mit besonderer Berücksichtigung der frühembryonalen Ernährung," *Z. Morphol. Tiere* **78**, 299–323 (1974).
4. L. Arvy, "Particularités de l'évolution nucléolaire au cours de l'ovogenèse chez *Lymnaea stagnalis* L," *C. R. Acad. Sci. Paris* **228**, 1983–1985 (1949).
5. R. Aubry, "La structure de l'acinus et la lignée femelle dans la glande hermaphrodite de *Lymnaea stagnalis* adulte," *C. R. Soc. Biol.* **148**, 1498–1500 (1954).
6. R. Aubry, "Les éléments nourriciers dans la glande hermaphrodite de *Lymnaea stagnalis* adulte," *C. R. Soc. Biol.* **148**, 1626–1629 (1954).
7. R. Aubry, "La lignée male dans la glande hermaphrodite de *Lymnaea stagnalis* adulte," *C. R. Soc. Biol.* **148**, 1856–1858 (1954).
8. R. Aubry, "Sur le cycle d'élaboration sexuelle dans la glande hermaphrodite de *Lymnaea stagnalis* adulte," *C. R. Soc. Biol.* **148**, 2075–2077 (1954).
9. R. Aubry, "De la possibilité d'une autofécondation dans l'acinus hermaphrodite de *Lymnaea stagnalis* adulte," *C. R. Soc. Biol.* **149**, 390–392 (1955).
10. R. Aubry, "La structure du canal hermaphrodite chez *Lymnaea stagnalis* adulte," *C. R. Soc. Biol.* **148**, *C. R. Soc. Biol.* **150**, 1786–1789 (1956).
11. L. C. Beadle, "Salt and water regulation in the embryos of freshwater pulmonate molluscs. I. The embryonic environment of *Biomphalaria sudanica* and *Lymnaea stagnalis*," *J. Exp. Biol.* **50**, 473–479 (1969).
12. R. Bekins, "The circulatory system of *Lymnaea stagnalis* (L.)," *Neth. J. Zool.* **22**, 1–58 (1972).
13. G. V. Berezkina, "Certain questions of development and function of the limnaeid reproductive system," in: *Ekologiya Zhivotnykh Smolenskoi i Sopredelnykh Oblastei* [in Russian], Smolensk (1980).
14. G. V. Berezkina and Ya. I. Starobogatov, "Morphology of egg batches of certain molluscs from the genus *Lymnaea* (Gastropoda, Pulmonata)," *Zool. Zh.* **60**, 1756–1768 (1981).
15. A. Berrie, "On the life cycle of *Lymnaea stagnalis* (L.) in the west of Scotland," *Proc. Malacol. Soc. London* **36**, 283–295 (1965).
16. J. J. Bezem and C. P. Raven, "Computer simulation of early embryonic development," *J. Theor. Biol.* **54**, 47–61 (1975).
17. J. J. Bezem, H. A. Wagemaker, and J. A. M. van den Biggelaar, "Relative cell volumes of the blastomeres in embryos of *Lymnaea stagnalis* in relation to bilateral symmetry and dorsoventral polarity," *Proc. K. Ned. Akad. Wet.* **C84**, 9–20 (1981).
18. J. A. M. van den Biggelaar, "Timing of the phases of the cell cycle with tritiated thymidine and Feulgen cytophotometry during the period of synchronous division in *Lymnaea*," *J. Embryol. Exp. Morphol.* **26**, 351–366 (1971).
19. J. A. M. van den Biggelaar, "Timing of the phases of the cell cycle during the period of asynchronous division up to the 49-cell stage in *Lymnaea*," *J. Embryol. Exp. Morphol.* **26**, 367–391 (1971).
20. J. A. M. van den Biggelaar, "RNA synthesis during cleavage of the *Lymnaea* egg," *Exp. Cell Res.* **67**, 207–210 (1971).

21. J. A. M. Van den Biggelaar, "Development of dorsoventral polarity preceding the formation of the mesentoblast in *Lymnaea stagnalis*," *Proc. K. Ned. Akad. Wet.* **C79**, 112–126 (1976).
22. J. A. M. van den Biggelaar, "The fate of maternal RNA containing ectosomes in relation to the appearance of dorsoventrality in the pond snail, *Lymnaea stagnalis*," *Proc. K. Ned. Akad. Wet.* **C79**, 421–426 (1976).
23. J. A. M. van den Biggelaar, "Significance of cellular interactions for the differentiation of the macromeres prior to the formation of the mesentoblast in *Lymnaea stagnalis*," *Proc. K. Ned. Akad. Wet.* **C80**, 1–12 (1977).
24. J. A. M. van den Biggelar and E. K. Boon-Niermeijer, "Origin and prospective significance of division asynchrony during early molluscan development," in: *The Cell Cycle in Development and Differentiation*, Cambridge University Press, Cambridge (1973).
25. J. A. M. van den Biggelaar, A. W. C. Dorresteijn, S. W. de Laat, and J. G. Bluemink, "The role of topographical factors in cell interactions and determination of cell lines in molluscan development," *Int. Cell Biol.*, 526–538 (1980–1981).
26. J. G. Bluemink, "The subcellular structure of the blastula of *Lymnaea stagnalis* L. (Mollusca) and the mobilization of the nutrient reserve," Thesis, Utrecht (1967).
27. D. A. Boag and P. S. M. Pearlstone, "On the life cycle of *Lymnaea stagnalis* (Gastropoda, Pulmonata) in southwestern Alberta," *Can. J. Zool.* **57**, 353–362 (1979).
28. H. H. Boer and T. Sminia, "Sieve structure of slit diaphragms of podocytes and pore cells of gastropod molluscs," *Cell Tissue Res.* **170**, 221–229 (1976).
29. H. H. Boer, C. Groot, M. de Jong-Brink, and C. J. Cornelisse, "Polyploidy in the freshwater snail *Lymnaea stagnalis* (Gastropoda, Pulmonata). A cytophotometrical analysis of the DNA in neurons and some other cell types," *Neth. J. Zool.* **27**, 242–252 (1977).
30. S. Bohlken and J. Joosse, "The effect of photoperiod on female reproductive activity and growth of the freshwater pulmonate snail *Lymnaea stagnalis* kept under laboratory breeding conditions," *Int. J. Invertebr. Reprod.* **4**, 213–222 (1982).
31. E. K. Boon-Niermeijer, "The effect of puromycin on the early cleavage cycles and morphogenesis of the pond snail *Lymnaea stagnalis*," *Wilhelm Roux's Arch. Entwicklungsmech. Org.* **177**, 29–40 (1975).
32. E. K. Boon-Niermeijer, "Morphogenesis after heat shock during the cell cycle of *Lymnaea*: a new interpretation," *Wilhelm Roux's Arch. Entwicklungsmech. Org.* **180**, 241–252 (1976).
33. T. Bose, "Onset of alkaline phosphatase synthesis during the embryonic development of *Lymnaea* sp.," *Indian J. Exp. Biol.* **15**, 554–555 (1977).
34. W. Bottke, I. Sinha, and I. Keil, "Coated vesicle-mediated transport and deposition of vitellogenic ferritin in the rapid growth phase of snail oocytes," *J. Cell Sci.* **53**, 173–191 (1982).
35. A. E. Boycott, C. Diver, S. L. Garstang, and F. M. Turner, "The inheritance of sinistrality in *Lymnaea peregra* (Mollusca, Pulmonata)," *Philos. Trans. R. Soc. London, Ser. B.* **219**, 51–131 (1930).
36. R. L. Brahmachary, "Molecular embryology of invertebrates," in: *Advances in Morphogenesis,* Vol. 10, Academic Press, New York (1973).

37. R. L. Brahmachary, T. K. Basu, and K. P. Banerji, "Effects of ATP and EDTA on cleavage of *Lymnaea* embryos," *Curr. Sci.* **37**, 505–506 (1968).
38. R. L. Brahmachary, K. P. Banerji, and T. K. Basu, "Investigations of transcription on *Lymnaea* embryos," *Exp. Cell Res.* **51**, 177–184 (1968).
39. R. L. Brahmachary and S. R. Palchoudhury, "Further investigations on transcription and translation in *Lymnaea* embryos," *Can. J. Biochem.* **49**, 926–932 (1971).
40. R. L. Brahmachary and P. K. Tapaswi, "A technique for separating the early blastomeres of *Lymnaea*," *Curr. Sci. (India)* **43**, 494 (1974).
41. L. H. Bretschneider, "Insemination in *Lymnaea stagnalis*," *Proc. K. Ned. Akad. Wet.* **51**, 358–362 (1948).
42. L. H. Bretschneider and C. P. Raven, "Structural and topochemical changes in the egg of *Lymnaea stagnalis*," *Arch. Néerl., Zool.* **10**, 1–31 (1951).
43. P. Brisson and I. Régondaud, "Observations relatives à l'origine dualiste de l'appareil génital chez quelques gastéropodes pulmonés basommatophores," *C. R. Acad. Sci. Paris* **D273**, 2339–2341 (1971).
44. P. Brisson and C. Besse, "Étude ultrastructurale de l'ébauche gonadique chez l'embryon de *Lymnaea stagnalis* L. (gastéropode pulmoné basommatophore)," *Bull. Soc. Zool. Fr.* **100**, 345–349 (1975).
45. K. M. Brown, "Effects of experimental manipulations on the life history patterns of *Lymnaea stagnalis appressa* Say (Pulmonata: Limnaeidae)," *Hydrobiologiya* **65**, 165–176 (1979).
46. T. Camey and W. L. M. Geilenkirchen, "Cleavage delay and abnormal morphogenesis in *Lymnaea* eggs after pulse treatment with azide of successive states in a cleavage cycle," *J. Embryol. Exp. Morphol.* **23**, 385–394 (1970).
47. M. R. Carriker, "Morphology of the alimentary system of the small *Lymnaea stagnalis appressa* Say," *Trans. Wis. Acad. Sci. Arts Lett.* **38**, 1–88 (1947).
48. J. N. Cather, N. H. Verdonk, and G. Zwaan, "Cellular interactions in the early development of the gastropod eye, as determined by deletion experiments," *Malacol. Rev.* **9**, 77–84 (1976).
49. R. Cumin, "Normentafel zur Organogenese von *Lymnaea stagnalis* (Gastropoda, Pulmonata) mit besonderer Berücksichtigung der Mitteldamndrüse," *Rev. Suisse Zool.* **79**, 709–774 (1972).
50. M. Dawkins, "Behavioral analysis of coordinated feeding movements in the gastropod *Lymnaea stagnalis* (L.)," *J. Comp. Physiol.* **92**, 255–271 (1974).
51. C. Diver and I. Andersson-Kottö, "Sinistrality in *Lymnaea peregra* (Mollusca, Pulmonata): the problem of mixed broods," *J. Genet.* **35**, 447–525 (1938).
52. A. A. Dogterom, "The effect of the growth hormone of the freshwater snail *Lymnaea stagnalis* on biochemical composition and nitrogenous wastes," *Comp. Biochem. Physiol.* **B65**, 163–167 (1980).
53. G. E. Dogterom, "Spontaneous and neurohormone-induced ovulation in *Lymnaea stagnalis* kept under various experimental conditions," *Haliotis* **10**, 49 (1980).
54. A. A. Dogterom and A. Doderer, "A hormone-dependent calcium-binding protein in the mantle edge of the freshwater snail *Lymnaea stagnalis*," *Calcif. Tissue Int.* **33**, 505–508 (1981).
55. M. R. Dohmen and J. M. A. van de Mast, "Electron microscopical study of RNA-containing cytoplasmic localizations and intercellular contacts in early

cleavage stages of eggs of *Lymnaea stagnalis* (Gastropoda, Pulmonata)," *Proc. K. Ned. Akad. Wet.* **C81**, 403–414 (1978).
56. A. W. Dorresteijn, J. A. M. van den Biggelaar, J. G. Bluemink, and W. J. Hage, "Electron microscopical investigations of the intercellular contacts during the early cleavage stages of *Lymnaea stagnalis* (Mollusca, Gastropoda)," *Wilhelm Roux's Arch. Dev. Biol.* **190**, 215–220 (1981).
57. P. F. Elbers and J. G. Bluemink, "Pinocytosis in the developing egg of *Lymnaea stagnalis*," *Exp. Cell Res.* **21**, 619–621 (1960).
58. P. F. Elbers, "The primary action of lithium chloride on morphogenesis in *Lymnaea stagnalis*," *J. Embryol. Exp. Morphol.* **22**, 449–463 (1969).
59. J. Faber, "Induced rotation of cleavage spindles in *Lymnaea stagnalis* L.," *Proc. K. Ned. Akad. Wet.* **C53**, 1490–1497 (1950).
60. L. A. Fraser, "The embryology of the reproductive tract of *Lymnaea stagnalis appressa* Say," *Trans. Am. Microsc. Soc.* **65**, 279–298 (1946).
61. G. Freeman and J. W. Lundelius, "The developmental genetics of dextrality and sinistrality in the gastropod *Lymnaea peregra*," *Wilhelm Roux's Arch. Dev. Biol.* **191**, 69–83 (1982).
62. W. L. M. Gellenkirchen, "Differences in lithium effects in *Lymnaea* after treatment of whole egg masses and isolated egg capsules," *Proc. K. Ned. Akad. Wet.* **C55**, 192–196 (1952).
63. W. L. M. Gellenkirchen, "The action and interaction of calcium and alkali chlorides on eggs of *Lymnaea stagnalis* and their chemical interpretation," *Exp. Cell Res.* **34**, 463–487 (1964).
64. W. L. M. Geilenkirchen, "Programming of gastrulation during the second cleavage cycle in *Lymnaea stagnalis*:: a study with lithium chloride and actinomycin D," *J. Embryol. Exp. Morphol.* **17**, 367–374 (1967).
65. J. C. George, "Experimental fusion of embryos of *Lymnaea stagnalis*," *Proc. K. Ned. Akad. Wet.* **C61**, 595–597 (1958).
66. W. P. M. Geraerts and L. H. Algera, "The stimulating effect of the dorsal-body hormone on cell differentiation in the female accessory sex organs of the hermaphrodite freshwater snail, *Lymnaea stagnalis*," *Gen. Comp. Endocrinol.* **29**, 109–118 (1976).
67. F. Giusti, "L'ultrastructura dello spermatozoo nella filogenesi e nella sistematica dei molluschi gasteropodi," *Atti Soc. Ital. Sci. Nat.* **112**, 381–402 (1971).
68. E. M. Goudsmit, "Galactogen catabolism by embryos of the freshwater snails, *Bulimnaea megasoma* and *Lymnaea stagnalis*," *Comp. Biochem. Physiol.* **B53**, 439–442 (1976).
69. B. Grygon and K. Weychert, "Z badań nad gonad obojnacaz *Helix pomatia* L. i *Lymnaea stagnalis* L," *Prz. Zool.* **18**, 72–77 (1974).
70. A. Gubicza, "Cytotopographical studies on the central nervous system of *Lymnaea stagnalis* L.," *Magy. Tud. Akad. Tihanyi Biol. Kutatoin Téz Évk.* **37**, 3–15 (1970).
71. P. Guerrier, "Origine et stabilité de la polarité animale végétative chez quelques Spiralia," *Ann. Embryol. Morphogen.* **1**, 119–139 (1968).
72. T. Guha and R. L. Brahmachary, "Tunicamycin-induced shape transformation in rabbit erythrocytes and loss of adhesion in *Lymnaea* embryonic cells," *Exp. Cell Biol.* **49**, 278–282 (1981).
73. R. E. Harris and W. A. G. Charleston, "Some temperature responses of

Lymnaea tomentosa and *L. columella* (Mollusca: Gastropoda) and their eggs," *N. Z. J. Zool.* **4**, 45–49 (1977).
74. O. Hess, "Die Entwicklung von Halbkeimen bei dem Süsswasser-Pulmonaten *Lymnaea stagnalis* L.," *Wilhelm Roux's Arch. Dev. Biol.* **150**, 124–145 (1957).
75. O. Hess, "Entwicklungsphysiologie der Mollusken," *Fortschr. Zool.* **14**, 130–163 (1962).
76. O. Hess, "Freshwater gastropoda," *in: Experimental Embryology of Marine and Freshwater Invertebrates*, G. Reverberi (ed.), North-Holland, Amsterdam–London (1971).
77. L. W. Holm, "Histological and functional studies of the genital tract of *Lymnaea stagnalis appressa* Say," *Trans. Am. Microsc. Soc.* **65**, 45–68 (1946).
78. H.-J. Horstmann, "Untersuchungen zur Physiologie der Bagattung und Befruchtung der Schlammschnecke *Lymnaea stagnalis*," *Z. Morphol. Oekol. Tiere* **4**, 222–268 (1955).
79. H.-J. Horstmann, "Ein einfacher Brutapparat für die Eier von Wasserlungenschnecken," *Zool. Anz.* **158**, 129–130 (1957).
80. H.-J. Horstmann, "Sauerstoffverbrauch und Trockengewicht der Embryonen von *Lymnaea stagnalis*," *Z. Vgl. Physiol.* **41,** 390–404 (1958).
81. J. C. Jager, N. Middelburg-Frielink, J. W. Mooij-Vogelaar, and W. J. van der Steen, "Effects of oxygen and food location on behavior in the freshwater snail *Lymnaea stagnalis* (L.)," *Proc. K. Ned. Akad. Wet.* **C82**, 177–180 (1979).
82. Br. Jockusch, "Protein synthesis during the first three cleavages in pond snail eggs (*Lymnaea stagnalis*)," *Z. Naturforsch.* **23b**, 1512–1516 (1968).
83. M. de Jong-Brink, L. P. C. Schot, H. J. N. Shoenmakers, and M. J. M. Bergamin-Sassen, "A biochemical and quantitative electron microscope study on steroidogenesis in ovotestis and digestive gland of the pulmonate snail *Lymnaea stagnalis*," *Gen. Comp. Endorcrinol.* **45**, 30–38 (1981).
84. J. Joosse, "Dorsal bodies and dorsal neurosecretory cells of the cerebral ganglia of *Lymnaea stagnalis*," Thesis, Utrecht (1964).
85. J. Joosse, "Structural and endocrinological aspects of hermaphroditism in pulmonate snails, with particular reference to *Lymnaea stagnalis* (L.)," in: *Intersexuality in the Animal Kingdom*, Berlin (1975).
86. J. Joosse and J. Lever, "Techniques of narcotization and operation for experiments with *Lymnaea stagnalis* (Gastropoda, Pulmonata)," *Proc. K. Ned. Akad. Wet.* **C62**, 145–149 (1959).
87. J. Joosse, H. H. Boer, and C. J. Cornelisse, "Gametogenesis and oviposition in *Lymnaea stagnalis* as influenced by γ-irradiation and hunger," *Symp. Zool. Soc. London* **22**, 213–235 (1968).
88. J. Joosse and D. Reitz, "Functional anatomical aspects of the ovotestis of *Lymnaea stagnalis*," *Malacologia* **9**, 101–109 (1969).
89. J. Joosse, M. A. Hemminga, and H. Loenhout, "The effect of temperature and daylength on glycogen storage in the pond snail *Lymnaea stagnalis*," *Haliotis* **10**, 73 (1980).
90. C. Jura and J. C. George, "Observations on the jelly mass of the eggs of three molluscs, *Succinea putris, Lymnaea stagnalis,* and *Planorbis corneus* with special reference to metachromasia," *Proc. K. Ned. Akad. Wet.* **C61**, 590–594 (1958).

91. N. N. Kamardin and T. P. Tsirulis, "Electron microscopic study of the pond snail osphradium," *Tsitologiya* **22**, 266–270 (1980).
92. R. G. Kessel and H. W. Beams, "10–12 nm filaments in the sperm of *Lymnaea stagnalis*," *J. Submicrosc. Cytol.* **13**, 551–560 (1981).
93. V. V. Khlebovitch and V. V. Lukanin, "Survival of spermatozoa of certain molluscs in sea water of different salinity," *Dokl. Akad. Nauk SSSR* **192**, 203–204 (1970).
94. L. Kielbowna and B. Koscielski, "A cytochemical and autoradiographic study of oocyte nucleoli in *Lymnaea stagnalis* L.," *Cell Tissue Res.* **152**, 103–111 (1974).
95. K. S. Kits, "States of excitability in ovulation hormone producing neuroendocrine cells of *Lymnaea stagnalis* (Gastropoda) and their relation to the egg-laying cycle," *J. Neurobiol.* **11**, 397–410 (1980).
96. K. S. Kits, "Electrophysiology of the caudo-dorsal neuroendocrine cells in *Lymnaea stagnalis* (Gastropoda)," *Comp. Biochem. Physiol.* **A72**, 91–97 (1982).
97. W. P. W. van der Knaap, T. Sminia, F. G. M. Kroese, and R. Dikkeboom, "Elimination of bacteria from the circulation of the pond snail *Lymnaea stagnalis*," *Dev. Comp. Immunol.* **5**, 21–32 (1981).
98. E. Kniprath, "Das Wachstum des Mantels von *Lymnaea stagnalis* (Gastropoda)," *Cytobiologie* **10**, 260–267 (1975).
99. E. Kniprath, "Zur Ontogenese des Schalenfeldes von *Lymnaea stagnalis*," *Wilhelm Roux's Arch. Dev. Biol.* **181**, 11–30 (1977).
100. A. Kordylewska, "Effect of phenol on the cell structure of the embryos of *Lymnaea stagnalis* L. (Gastropoda, Pulmonata). Light and electron microscopic study," *Acta Biol. Cracov. Ser. Zool.* **22**, 89–98 (1980).
101. Z. Ya. Kruglyanskaya and D. A. Sakharov, "Appearance of biogenic monoamines in the developing nervous system of *Lymnaea stagnalis* embryos," *Ontogenez* **6**, 194–197 (1975).
102. V. Labordus, "The effect of ultraviolet light on developing eggs of *Lymnaea stagnalis* (Mollusca, Pulmonata). I. The pattern of the effect on mitotic cycles," *Proc. K. Ned. Akad. Wet.* **C73**, 366–381 (1970).
103. S. Letelier and O. Ciolpan, "Contributions to the study of prolificity of the molluscs *Lymnaea stagnalis* (L.) and *Planorbarius corneus* (L.) (Gastropoda, Pulmonata) in natural conditions," *Trav. Mus. Hist. Nat. Grigore Antipa* **22**, 229–233 (1980).
104. O. V. Levina, "Fertility of the freshwater molluscs *Lymnaea stagnalis* and *Radix ovata*," *Zool. Zh.* **52**, 676–684 (1973).
105. D. L. Luchtel, "An ultrastructural study of the egg and early cleavage stages of *Lymnaea stagnalis*, a pulmonate mollusc," *Am. Zool.* **16**, 405–419 (1976).
106. A. ter Maat, J. C. Lodder, J. Veenstra, and J. T. Goldschmeding, "Suppression of egg-laying during starvation in the snail *Lymnaea stagnalis* by inhibition of the ovulation hormone producing caudo-dorsal cells," *Brain Res.* **239**, 535–542 (1982).
107. L. Merenyi, "A kannibalizmus egy megfigyelt esete akváriumban tartott *Lymnaea stagnalis* (L.) egyedek között," *Soosiana* **8**, 3–4 (1980).
108. V. N. Meshcheryakov, "Invariability of dissymmetrical cytokinesis in the centrifuged zygotes of pulmonate molluscs," *Ontogenez* **8**, 435–441 (1977).

109. V. N. Meshcheryakov, "Isolation of the mitotic apparatus from the eggs and embryos of pulmonate mollusc *Lymnaea stagnalis* L.," *Tsitologiya* **20**, 1211–1215 (1978).
110. V. N. Meshcheryakov, "Orientation of cleavage spindles in pulmonate molluscs. I. Role of blastomere form in orientation of the second cleavage spindles," *Ontogenez* **10**, 558–566 (1978).
111. V. N. Meshcheryakov, "Orientation of cleavage spindles in pulmonate molluscs. II. Role of architecture of intercellular contacts in orientation of the third and fourth cleavage spindles," *Ontogenez* **9**, 567–575 (1978).
112. V. N. Meshcheryakov, "Dynamics of the cell surface during early cleavage and its relation to the polarity of the zygote in Gastropoda," *Zh. Obshch. Biol.* **39**, 916–926 (1978).
113. V. N. Meshcheryakov, "Isolation of the egg cortical layer in pulmonate molluscs," *Ontogenez* **12**,177–186 (1981).
114. V. N. Mescheryakov, "Simulation of spiral cytokinesis by membrane ghosts of the pond snail zygotes," *Ontogenez* **14**, 82–85 (1983).
115. V. N. Meshcheryakov and L. N. Beloussov, "Spatial organization in cleavage," *in: Morphology of Man and Animals. Anthropology*, Vol. 8 [in Russian], VINITI, Academy of Sciences of the USSR, Moscow.
116. V. N. Meshcheryakov and G. V. Veryasova, "Orientation of cleavage spindles in pulmonate molluscs. III. Form and location of mitotic apparatus in binucleate zygotes and blastomeres," *Ontogenez* **10**, 24–35 (1979).
117. V. N. Meshcheryakov and L. V. Beloussov, "Changes in the spatial organization of early cleavage of molluscs *Lymnaea stagnalis* L. and *Physa fontinalis* L. under the effect of trypsin," *Ontogenez* **4**, 359–372 (1973).
118. V. N. Meshcheryakov and L. V. Beloussov, "Asymmetrical rotations of blastomeres in early cleavage of Gastropoda," *Wilhelm Roux's Arch. Entwicklungsmech. Org.* **177**, 193–203 (1975).
119. V. N. Meshcheryakov and L. G. Filatova, "Contractile phenomena in a model nonmuscle system," *Biofizika* **26**, 1057–1062 (1981).
120. V. N. Meshcheryakov and L. G. Filatova, "Comparative electron-microscopic study of the snail egg microfilaments and muscle F-actin," *Stud. Biophys.* **107**, No. 1, 5–12 (1985).
121. J. W. Mooij-Vogelaar, J. C. Jager, and W. J. Steen, "The effect of density changes on the reproduction of the pond snail *Lymnaea stagnalis* (L.)," *Neth. J. Zool.* **20**, 279–288 (1970).
122. J. W. Mooij-Vogelaar and W. J. van der Steen, "Effect of density on feeding and growth in the pond snail *Lymnaea stagnalis* (L.)," *Proc. K. Ned. Akad. Wet.* **C76**, 61–68 (1973).
123. J. B. Morrill, C. A. Blair, and W. J. Larsen, "Regulative development in the pulmonate gastropod, *Lymnaea stagnalis*, as determined by blastomere deletion experiments," *J. Exp. Zool.* **183**, 47–55 (1973).
124. J. B. Morrill and F. O. Perkins, "Microtubules in the cortical region of the egg of *Lymnaea* during cortical segregation," *Dev. Biol.* **33**, 206–212 (1973).
125. J. B. Morrill, R. W. Rubin, and M. Grandi, "Protein synthesis and differentiation during pulmonate development," *Am. Zool.* **16**, 547–561 (1976).
126. J. B. Morrill and L. E. Macey, "Cell surface changes during the early cleavages of the pulmonate snail, *Lymnaea palustris*," *Am. Zool.* **19**, 246 (1979).

127. J. B. Morrill, "Development of the pulmonate gastropod, *Lymnaea*," in: *Developmental Biology of Freshwater Invertebrates*, F. W. Harrison and R. R. Cowden (eds.), A. R. Liss, Inc., New York (1982).
128. L. Mulherkar, S. Goel, and M. V. Joshi, "Effects of cytochalasin H on the cleaving eggs of *Lymnaea acuminata*," *Indian J. Exp. Biol.* **15**, 1089–1092 (1977).
129. V. S. Musienko and M. A. Kostenko, "Effect of cAMP upon morphological differentiation of adult molluscs neurons in culture," *Tsitologiya* **24**, 264–269 (1982).
130. A. A. Neyfakh, "Radiation investigation of nucleocytoplasmic interrelations in morphogenesis and biochemical differentiation," *Nature (London)* **201**, 880–884 (1964).
131. A. A. Neyfakh, "Morphogenetic nuclear function in the early development of the common pond snail (*Lymnaea stagnalis*)," *Ontogenez* **1**, 630–632 (1976).
132. O. I. Patrusheva and F. M. Sokolina, "Certain aspects of keeping the lymneids in aquariums and their reproduction," in: *Problems of Malacology of Siberia* [in Russian], Tomsk (1969).
133. B. Plesch, "An ultrastructural study of the musculature of the pond snail *Lymnaea stagnalis* (L.)," *Cell Tissue Res.* **180**, 317–340 (1977).
134. B. Plesch, M. de Jong-Brink, and H. H. Boer, "Histological and histochemical observations on the reproductive tract of the hermaphrodite pond snail *Lymnaea stagnalis* (L.)," *Neth. J. Zool.* **20** (1971).
135. N. B. Pullan, F. M. Clino, and C. B. Mansfield, "Studies on the distribution and ecology of the family Lymnaeidae (Mollusca: Gastropoda) in New Zealand," *J. R. Soc. N. Z.* **2**, 393–405 (1972).
136. C. P. Raven, in: *Experimental Embryology in the Netherlands, 1940–1945*, M. W. Woerdeman and C. P. Raven (eds.), Elsevier, New York–Amsterdam (1946).
137. C. P. Raven, "Morphogenesis in *Lymnaea stagnalis* and its disturbance by lithium," *J. Exp. Zool.* **121**, 1–78 (1952).
138. C. P. Raven, "Abnormal development of the foregut in *Lymnaea stagnalis*," *J. Exp. Zool.* **139**, 189–245 (1958).
139. C. P. Raven, *Oogenesis: The Storage of Developmental Information*, Macmillan (Pergamon Press), New York (1961).
140. C. P. Raven, *Morphogenesis: The Analysis of Molluscan Development*, 2nd edn., Pergamon Press, New York (1966).
141. C. P. Raven, "The cortical and subcortical cytoplasm of the *Lymnaea* egg," Int. Rev. Cytol. **28**, 1–44 (1970).
142. C. P. Raven, "Chemical embryology of molluscs," in: *Chemical Zoology*, Vol. 7, Mollusca, M. Florkin and B. T. Scheer (eds.), Academic Press, New York (1972).
143. C. P. Raven, "Further observations on the distribution of cytoplasmic substances among the cleavage cells in *Lymnaea stagnalis*," *J. Embryol. Exp. Morphol.* **31**, 37–59 (1974).
144. C. P. Raven, "Morphogenetic analysis of spiralian development," *Am. Zool.* **16**, 395–403 (1976).
145. C. P. Raven, J. J. Bezem, and J. Isings, "Changes in size relation between macromeres and micromeres of *Lymnaea* under the influence of lithium," *Proc. K. Ned. Akad. Wet.* **C55**, 248–258 (1952).

146. C. P. Raven and U. P. van der Wal, "Analysis of the formation of the animal pole plasm in the eggs of *Lymnaea stagnalis*," *J. Embryol. Exp. Morphol.* **12**, 123–139 (1964).
147. J. Régondaud, "Development de la cavité pulmonaire et de la cavité palléale chez *Lymnaea stagnalis*," *C. R. Acad. Sci. Paris* **252**, 179–181 (1961).
148. J. Régondaud, "Origine embryonnaire de la cavité pulmonaire de *Lymnaea stagnalis*. Considerations particulières sur la morphogenèse de la commisure viscérale," *Bull. Biol. Fr. Belg.* **98**, 433–471 (1964).
149. J. Régondaud and P. Brisson, "Données radioautographiques aprés incorporation de leucine tritiée dans les cellules nucales de l'embryon de *Lymnaea stagnalis* L. (gastéropode pulmoné basommatophore)," *Bull. Soc. Zool. Fr.* **101**, 477–480 (1976).
150. A. Richards, "The development rate and oxygen consumption of snail eggs at various temperatures," *Z. Naturforsch.* **20b**, 347–349 (1965).
151. J. E. Rigby, "The fine structure of the oocyte and follicle cells of *Lymnaea stagnalis*, with special reference to the nutrition of the oocyte," *Malacologie* **18**, 377–380 (1979).
152. E. W. Roubos, W. P. M. Geraerts, G. H. Boerrigter, and G. P. J. Kampen, "Control of the activities of the neurosecretory light green and caudo-dorsal cells and of the endocrine dorsal bodies by the lateral lobes in the freshwater snail *Lymnaea stagnalis* (L.)," *Gen. Comp. Endocrinol.* **40**, 446–454 (1980).
153. E. W. Roubos, A. N. de Keijzer, and P. Buma, "Adenylate cyclase activity in axon terminals of ovulation-hormone-producing neuroendocrine cells of *Lymnaea stagnalis* (L.)," *Cell Tissue Res.* **220**, 665–668 (1981).
154. T. Salih, O. Al-Habbib, W. Al-Habbib, S. Al-Zako, and T. Ali, "The effects of constant and changing temperatures on the development of eggs of the freshwater snail *Lymnaea auricularia* (L.)," *J. Therm. Biol.* **6**, 379–388 (1981).
155. E. A. Sapaev, "Concerning intraspecific differentiation of *Chaetogaster lymnaei* (Baer, 1827)," *in: Evolutionary Morphology of Invertebrate Animals* [in Russian], Nauka, Leningrad (1976).
156. J. E. M. Scheerboom and A. Doderer, "The effects of artificially raised haemolymph-glucose concentrations on feeding, locomotory activity, growth, and egg production of the pond snail *Lymnaea stagnalis* (L.)," *Proc. K. Ned. Akad. Wet.* **C81**, 377–386 (1978).
157. J. E. W. Scheerboom and R. van Elk, "Field observations on the seasonal variations in the natural diet and the haemolymph-glucose concentration of the pond snail *Lymnaea stagnalis* (L.)," *Proc. K. Ned. Akad. Wet.* **C81**, 365–376 (1978).
158. L. C. Schlichter, "Ion relations of haemolymph, pallial fluid, and mucus of *Lymnaea stagnalis*," *Can. J. Zool.* **59**, 605–613 (1981).
159. C. T. Slade, J. Mills, and W. Winlow, "The neuronal organization of the paired pedal ganglia of *Lymnaea stagnalis* (L.)," *Comp. Biochem. Physiol.* **A69**, 789–803 (1981).
160. T. Sminia, N. D. de With, J. L. Bos, M. E. van Nieuwmegen, M. P. Witter, and J. Wondergem, "Structure and function of the calcium cells of the freshwater pulmonate snail *Lymnaea stagnalis*," *Neth. J. Zool.* **27**, 195–208 (1977).

161. Y. I. Starobogatov, "Taxonomy and phylogeny of Lymnaeidae (Gastropoda, Pulmonata, Basommatophora), in: *Problems of Ecology* [in Russian], Nauka, Leningrad (1976).
162. V. I. Starostin, N. V. Potapina, and G. P. Satdykova, "Certain aspects on histogenesis of blood and connective tissue in gastropod molluscs," in: *Cytological Mechanisms of Histogenesis* [in Russian], Nauka, Moscow (1979).
163. W. J. van den Steen, N. van Hoven, and J. Jager, "A method for breeding and studying freshwater snails under continuous water change, with some remarks on growth and reproduction in *Lymnaea stagnalis* L.," *Neth. J. Zool.* **19**, 131–139 (1969).
164. W. J. van der Steen, "Periodic oviposition in the freshwater snail *Lymnaea stagnalis*: a new type of endogenous rhythm," *Acta Biotherm.* **19**, 87–93 (1970).
165. W. J. van der Steen, J. C. Jager, and D. Tiemersma, "The influence of food quantity on feeding, reproduction, and growth in the pond snail *Lymnaea stagnalis* (L.) with some methodological comments," *Proc. K. Ned. Akad. Wet.* **C76**, 47–60 (1973).
166. C. J. Stoll, P. Sloep, Y. Veerman-van Duivenbode, and H. A. van der Woude, "Light-sensitivity in the pulmonate gastropod *Lymnaea stagnalis*: peripherally located shadow-receptors," *Proc. K. Ned. Akad. Wet.* **C79**, 510–516 (1976).
167. R. Storey, "The importance of mineral particles in the diet of *Lymnaea peregra* (Müller)," *J. Conchol.* **27**, 191–195 (1970).
168. A. H. Sturtevant, "Inheritance of direction of coiling in *Lymnaea*," *Science (Wash.)* **58**, 1501, 269–270 (1923).
169. P. K. Tapaswi, "RNA synthesis during oogenesis to the onset of fertilization in the mollusc *Lymnaea*," *Z. Naturforsch.* **27b**, 581–582 (1972).
170. H. H. Taylor, "The ionic properties of the capsular fluid bathing embryos of *Lymnaea stagnalis* and *Biomphalaria sudanica* (Mollusca, Pulmonata)," *J. Exp. Biol.* **59**, 543–564 (1973).
171. H. H. Taylor, "The ionic and water relations of embryos of *Lymnaea stagnalis*, a freshwater pulmonate mollusc," *J. Exp. Biol.* **69**, 143–172 (1977).
172. L. P. M. Timmermans, "Studies on shell formation in molluscs," *Neth. J. Zool.* **19**, 417–523 (1969).
173. N. N. Tretyak and M. A. Kostenko, "Dependence of RNA synthesis and formation of processes in isolated pulmonate neurons in culture on ionic composition and osmolarity of the nutrient medium," *Tsitologiya* **20**, 643–650 (1978).
174. G. A. Übbels, "A cytochemical study of oogenesis of the pond snail *Lymnaea stagnalis*," Thesis, Utrecht (1968).
175. N. H. Verdonk, "Morphogenesis of the head region in *Lymnaea stagnalis* L.," Thesis, Utrecht (1965).
176. N. H. Verdonk, "The determination of bilateral symmetry in the head region of *Lymnaea stagnalis*," *Acta Embryol. Morphol. Exp.* **10**, 211–227 (1968).
177. N. H. Verdonk, "Gene expression in early development of *Lymnaea stagnalis*," *Dev. Biol.* **35**, 29–35 (1973).
178. C. H. Waddington, M. M. Perry, and E. Okada, "'Membrane knotting' between blastomeres of *Lymnaea*," *Exp. Cell. Res.* **23**, 631–633 (1961).
179. U. P. van der Wal, "The mobilization of the yolk of *Lymnaea stagnalis*

(Mollusca). I. A structural analysis of the differentiation of the yolk granules," *Proc. K. Ned. Akad. Wet.* **C79**, 393–404 (1976).
180. U. P. van der Wal, "The mobilization of the yolk of *Lymnaea stagnalis*. II. The localization and function of the newly synthesized proteins in the yolk granules during early embryogenesis," *Proc. K. Ned. Akad. Wet.* **C79**, 405–420 (1976).
181. U. P. van der Wal and M. R. Dohmen, "A method for the orientation of small and delicate objects in embedding media for light and electron microscopy," *Stain Technol.* **53**, 56–57 (1978).
182. W. A. Warneck, "Uber die Bildung und Entwicklung des Embryos bei Gasteropoden," *Bull. Soc. Nat. Moscou* **23**, 90–194 (1850).
183. N. D. de With, "Water turnover, ultrafiltration, renal water reabsorption and renal circulation in fed and starved specimens of *Lymnaea stagnalis*, adapted to different external osmolarities," *Proc. K. Ned. Akad. Wet.* **C83**, 109–120 (1980).
184. N. D. de With and A. A. Dogterom, "Calcium regulation in the freshwater snail *Lymnaea stagnalis*," *Haliotis* **10**, 45 (1980).
185. O. V. Zaitseva, V. A. Kovalev, and L. S. Bocharova, "Study of morphofunctional relationships between sensor epithelium and statoconia in the statocyst of *Lymnaea stagnalis*," *Zh. Evol. Biokhim. Fiziol.* **14**, 307–309 (1978).
186. O. V. Zaitseva, "Skin innervation of the pulmonate molluscs," *Arkh. Anat., Gistol. Embriol.* **78**, 32–39 (1980).
187. V. I. Zhadin, "Molluscs occurring in freshwater and brackish habitats in the USSR," in: *Handbook of USSR Fauna*, Vol. 46 [in Russian], Nauka, Moscow–Leningrad (1952).

Chapter 6

THE MIDGE *Chironomus thummi*

I. I. Kiknadze, O. E. Lopatin,
N. N. Kolesnikov, and L. I. Gunderina

6.1. INTRODUCTION

The larvae of chironomids, along with those of drosophilids and sciarids, are widely used in cytological and biochemical studies of cell differentiation. The description of the puffing pattern in the polytene chromosomes of these larvae at various stages of metamorphosis has enabled a concept to be formulated regarding changes in the spectrum of the most active sites on chromosomes in the course of morphogenesis [4. 8, 11-14, 31, 32]. Along with the results of biochemical and cytogenetic studies, these data (for reviews see [9, 19]) provided the basis for current views about gene activity in the course of cell differentiation.

Studies of the cell protein spectrum and changes in this spectrum during dipteran development have confirmed the idea of differential gene activity and enabled the functional significance of several puffs to be elucidated, in particular those coding for tissue-specific proteins [3, 5, 21, 22, 26-28, 40, 41, 43, 47-50, 71, 77, 78, 85, 93, 96].

In diptera, as in many other organisms, RNA templates with a long half-life have been found; such RNA species are synthesized a long time before the genetic stage at which they are translated [15, 18, 56, 110].

It should be remembered that both transcription and replication of DNA in various organs of diptera are discontinuous in the course of larval development: this activity drops or may even cease completely during molts [2, 14, 17, 37, 38, 69, 76]. Therefore, just as with the general problem of regulating metamorphosis, great attention is being paid to hormones since they play a major role in controlling chromosomal activity [4, 14, 16, 51, 73-75].

Compared with *Drosophila*, chironomids have only comparatively recently been introduced as experimental species which can be reared under laboratory conditions [6, 12, 24, 30, 44, 46, 52]; this is only possible for a few species. These include *Chironomus thummi* Kieff. (subspecies *thummi* and *piger*), *C. tentans, C. pallidivitatus*, and some others. The large size of chironomid larvae, as well as the high degree of chromosomal polyteny in a number of larval tissues, make these species particularly promising for developmental biology.

Chironomid larvae are aquatic organisms with integuments that are easily permeable to many compounds. They are very suitable for studying the effects of various substances on the whole animal or its individual organs. It is important to emphasize that under vital conditions chironomids can be studied easily. This is because their skin is translucent and the state of the cells in almost all the organs can be examined directly under the microscope. The presence of polytene chromosomes with a markedly higher degree of polyteny in the larval organs of chironomids than in drosophilids makes them an indispensable species for studying many problems related to chromosomal structure and function. Another important problem lies in elucidating the relationship between chromosomal activity and cell physiology during such critical periods as larval molts, after molt, and intermolt growth.

Unfortunately, the life cycle of chironomids has not been fully studied and more detailed information regarding the embryonic stages is particularly necessary. Data on the developmental physiology of chironomids are very scarce; there are few genetic and cytogenetic investigations. This is because it is difficult to obtain progeny from individual pairs, which cannot be traced during swarming. However, methods designed to mate males and females outside swarming have been described: this should facilitate future genetic and cytogenetic studies. The development of methods for rearing chironomids on a laboratory scale may in the future provide the basis for rearing vast numbers of them. This is economically promising because they make excellent fish food.

This section contains basic information on the main features of the life cycle of *Chironomus thummi* Kieff. in laboratory culture under conditions of artificial rearing. Up-to-date studies on chironomid development place great emphasis on analyzing the larval stage and metamorphosis, so there is plenty of information on these. We deal primarily with problems of the morphology and physiology of these developmental stages. Little work has been done on the embryogenesis of chironomids, and this is discussed briefly.

6.2. TAXONOMY AND DISTRIBUTION

The gnatlike midge *Chironomus thummi* Kieff. belongs to the order Diptera, family Chironomidae, genus *Chironomus* Meig. Chironomids belong to the group of holometabolous insects (Holometabola); their life cycle includes embryonic, larval, pupal, and adult stages. Classifying chironomids is very difficult and confusing, since the presence of larval, pupal, and adult forms requires the development of a taxonomic system for the various developmental stages. Up-to-date keys for chironomids have been compiled taking all these requirements into account [20, 45, 63, 65, 66, 79, 80, 86-89, 92, 104-107].

The main diagnostic features of chironomid larvae include the size and pigmentation of the body and the head capsule, characteristics of the mentum ventrolabial plate and mandibulae, the number and size of antennal segments, and the presence and shape of abdominal appendages on the VIIth and VIIIth abdominal segments (see Figs. 6.2 through 6.5). In adults the size and pigmentation of the body and wings, and the structure of hypopygium, antenna, forelegs, and eyes serve as diagnostic criteria (Fig. 6.1A, B, C; 6.5).

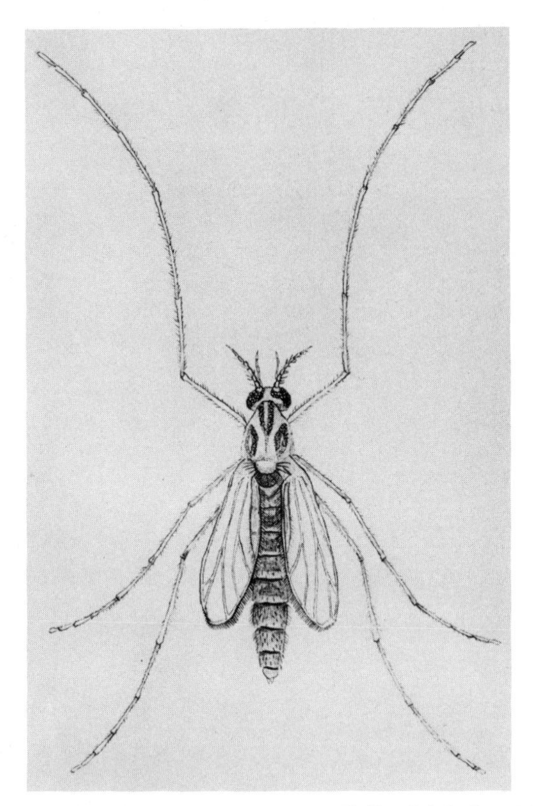

Fig. 6.1A. *Chironomus thummi* Kieff. Adult female.

C. thummi can be easily reared under laboratory conditions. Laboratory stocks of this species are maintained and successfully used for studying developmental biology and cytology in a number of laboratories in various countries [30, 52, 59, 102, 103]. In the Soviet Union the laboratory stock of *C. thummi* (see Figs. 6.1 to 6.5) was obtained from the natural population in the Saratov region by Professor A. S. Konstantinov [44, 46]. However, he initially determined the species as *C. dorsalis*. Cytological analysis of polytene chromosomes of this stock has shown that, according to the Keyl system [30], the specimens actually belong to *Chironomus thummi thummi* (Figs. 6.6 and 6.7). In 1960, Professor Konstantinov made his stock of *Chironomus* available to the Institute of Cytology and Genetics, Siberian Branch, Academy of Sciences of the USSR, where it is still being maintained. Rearing chironomids under laboratory conditions is also being performed at the Department of Invertebrate Zoology, Saratov University, under the supervision of Professor Konstantinov.

C. thummi is widespread in nature [89]. Larvae of this species may be found in different freshwater habitats. Together with other species of chironomids they generally comprise the dominant invertebrates in each habitat.

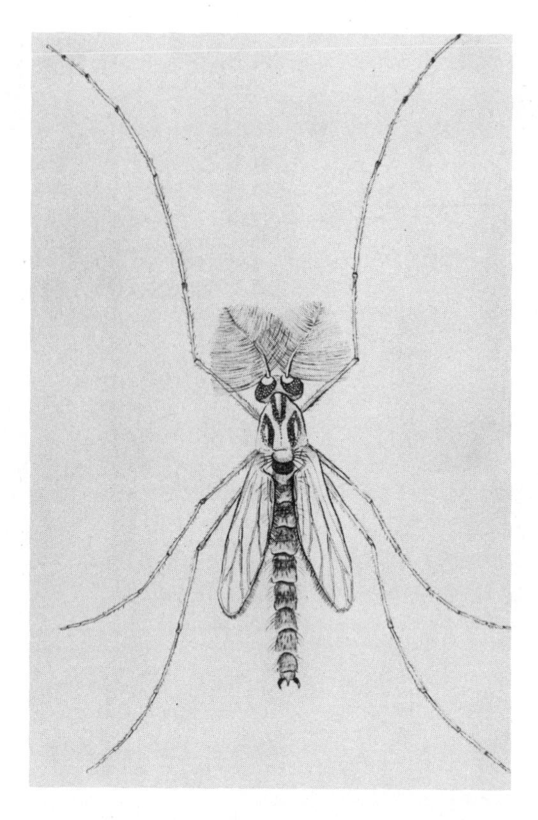

Fig. 6.1B. *Chironomus thummi* Kieff. Adult male (original sketches by Valleva).

6.3. REPRODUCTIVE BIOLOGY AND THE MAIN FEATURES OF THE LIFE CYCLE

The biology of reproduction and the developmental cycle of chironomids have been described in numerous studies [1, 64, 68, 72, 89, 91, 97], in particular by Konstantinov [46].

Adult males and females of *C. thummi* differ distinctly in the length and shape of the antennae (males have large, wide, plumed antennae), width of the wings (they are narrower in males), abdominal shape (in females the posterior part of the abdomen is more constricted), and genitalia (see Fig. 6.1).

Adults live only 4 or 5 days. Males form the mating swarm 1.0-1.5 days after emerging. On windless days these swarms can be easily observed over bushes and trees near water. During swarming the males produce a characteristic thin sound. Females sit either on bushes or on grass and fly individually into the swarm; mating takes place during flight, and the mating pairs leave the swarm. According to Konstantinov [46], males and females mate only once during their lifetime. Between

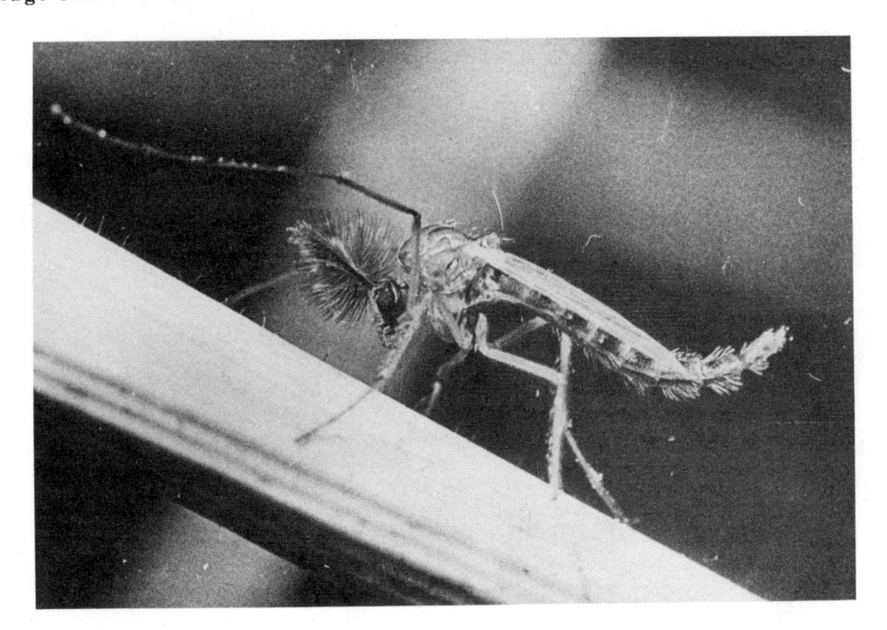

Fig. 6.1C. *Chironomus thummi* Kieff. Adult male.

1.5-2.0 days after mating the females lay egg masses on water (see Fig. 6.2). Oviposition takes place either at sunrise or sunset. The egg mass is invested in a gelatinous matrix. The number of eggs in the egg mass of a *C. thummi* laboratory specimen varies from 300 to 900. The length of the egg mass is from 7 to 20 mm. The eggs in the mass are arranged regularly with 10-17 eggs in each row; the gelatinous matrix is divided into corresponding segments. The central fibers consist of thicker gelatinous material, and their shape is significant as a taxonomic trait. The egg mass is attached to any solid object on the water surface by a gelatinous cord.

The eggs of *C. thummi* are elongated (length 0.3-0.35 mm, width 0.12-0.14 mm). They have a dark-brown pigmentation depending on the color of the yolk. The chorion is colorless, and the micropyle is located on the rounder anterior pole.

The embryonic development of chironomids at 20-22°C lasts 2.5-3 days. A characteristic feature is the early segregation of primary germ cells.

When embryonic development has been completed, the larvae hatch from the egg but do not leave the gelatinous matrix for 24 h after hatching. The larvae which have left the gelatinous matrix (Fig. 6.3) are capable of building specialized structures in which they live. These are tubes consisting of small particles of mud and detritus glued together by a salivary gland secretion. Each tube is open on both ends, and water is drawn through the tube by anteroposterior undulation of the

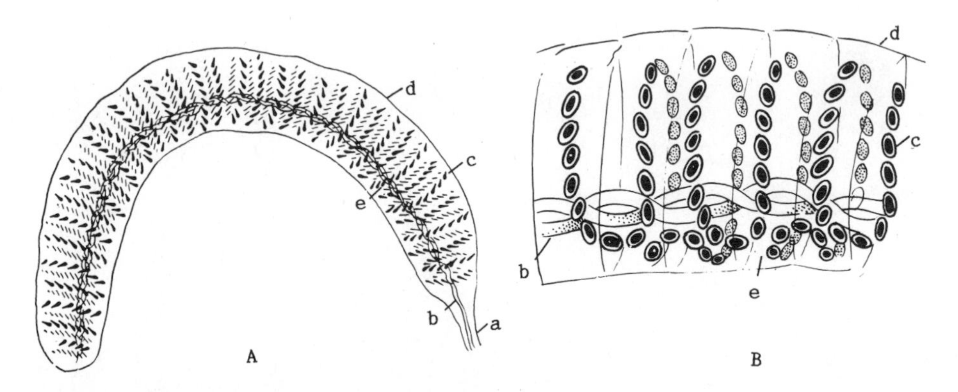

Fig. 6.2. Egg mass of *Chironomus thummi*. A) General view; B) part of the egg mass at higher magnification. a) Gelatinous stalk for attachment; b) central fibers of gelatinous matrix; c) eggs; d) gelatinous envelope; e) gelatinous matrix.

larva's body. Particles of food suspended in the water are deposited on the sticky walls of the tube and may be eaten by the larva. In addition, the larvae can emerge from their tubes, spread salivary gland secretion with their anterior forelegs, and then withdraw into the tube again, dragging along the secretion with any adhering particles.

The larvae are microphagous, feeding on algae, protozoa, small invertebrates, bacteria, and detritus. The larval stage is the longest of the chironomid life cycle and lasts 35 days on average. However, the duration of larval development depends on environmental conditions, for example in winter the larval stage can continue without metamorphosis for several months.

The pupae of *C. thummi* (see Fig. 6.4) do not feed; when mature they leave their tubes and move to the surface of the water. The ascent of the pupa in water is aided by air accumulated under the pupal skin around the adult's thorax. The adult (Fig. 6.5) emerges from a dorsal slit in the thorax of the pupal skin.

6.4. CULTURE

A procedure for rearing *C. thummi* under laboratory conditions has been developed by Konstantinov. We have used this method with some modifications throughout 20 years of continuously rearing this species.

The larvae of *Chironomus* are reared in trays filled with moist mud or tap water which has been standing for several hours in the laboratory. The mud is collected in summer in the shallow parts of a habitat such as a river or pond. Before filling the trays, the mud should be washed well and boiled.

The egg masses taken from jars of water which have been placed in rooms where the *Chironomus* are being reared are then transferred to Petri dishes with a diameter of 10 cm for 2-3 days; then the egg masses are placed in trays. Six egg masses are usually placed into a 35 × 23 × 4 cm tray. The trays are placed into specially constructed multilevel racks (Fig. 6.8). It is important to emphasize that the

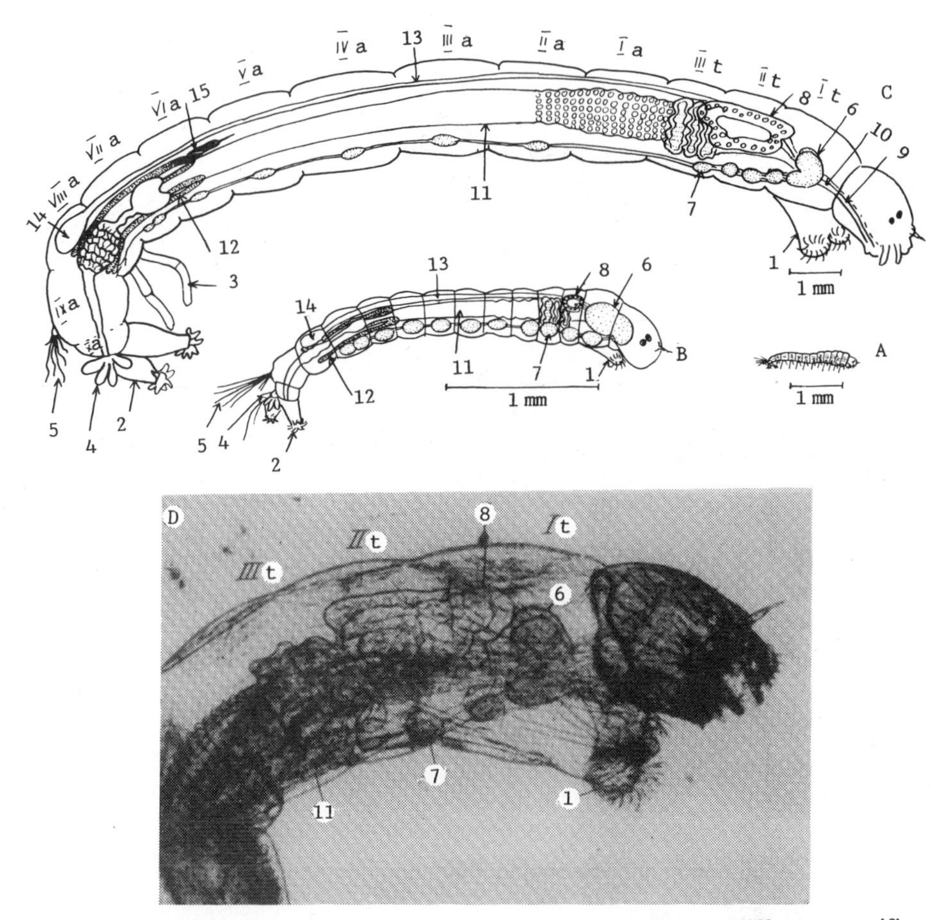

Fig. 6.3. Larval stages of *Chironomus thummi*. A, B) 1st Instar larvae (different magnifications); C) 4th instar larva; D) photograph of late 4th instar showing head and thoracic segments (living state). It–IIIt) Thoracic segments; Ia–Xa) abdominal segments; 1, 2) appendages with a circle of hooklets (for locomotion) on the ventral side of the first and last body segments; 3) two pairs of appendages (tubuli) of VIIIth abdominal segment; 4) four anal papillae; 5) bristle tuft; 6) brain; 7) ventral nerve cord; 8) salivary gland; 9) pharynx; 10) esophagus; 11) gut; 12) Malpighian tubules; 13) aorta; 14) heart; 15) ovaries.

trays should be placed horizontally into these racks. This is because even slight changes in position result in a nonuniform distribution of mud and water over different parts of the tray, thereby adversely affecting the larva development. The trays should be watered daily to keep the mud wet, but the water level should not exceed 1-2 mm above the mud. A constant water level of 2-3 mm adversely affects development. The larvae are fed with dry yeast powder which is spread evenly onto the surface of the tray. This can be easily performed by sieving the dry yeasts through 2-3 layers of cheesecloth, so that only the finest particles reach the mud surface. The larvae should be fed every other day. It is better to spread the food

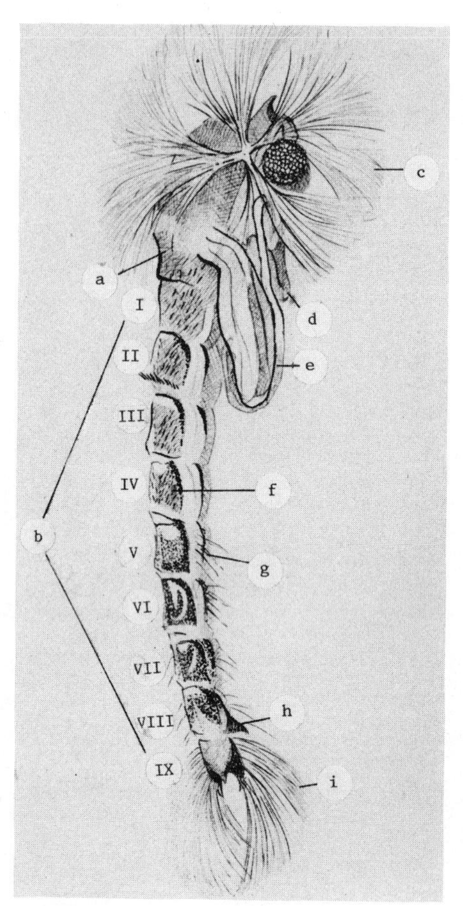

Fig. 6.4. Pupa of *Chironomus thummi*. a) Cephalothorax; b) I–IX, abdominal segments; c) thoracic horns; d) leg sheaths; e) wing sheaths; f) shagreen; g) bristles; h) spur of segment VIII; i) swim fin.

before watering the trays. The dry yeasts should be heated at 50°C in a thermostatically controlled oven before use.

The mud should be washed and boiled after each generation. Larvae are very sensitive to oxygen deficit during development. If this happens, the tubes extend high above the surface of the water and the larvae leave the tubes for a long time. The racks with the trays should be placed in a separate room or in a cage (180 × 80 × 80 cm) and tightly covered with fine mesh gauze. Large jars of water are placed on the floor near the racks. The females lay eggs into these jars. Egg masses are collected daily.

The larvae can also be reared in water. This method of rearing is used to maintain laboratory stocks of *C. thummi* and other chironomid species – *C. tentans* and *C. pallidivitatus* [6, 12, 24]. The water should be well aerated when using this method.

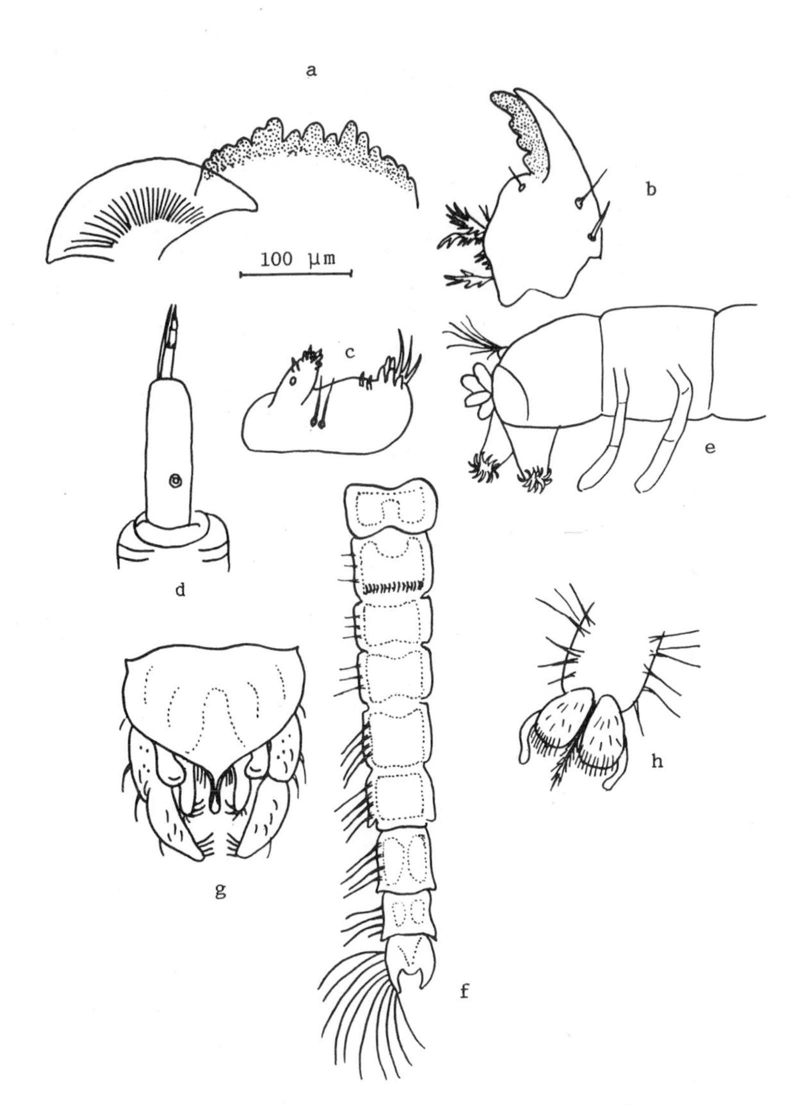

Fig. 6-5. Diagnostic features of *Chironomus thummi*. a) Submentum with submental plate; b) mandible; c) maxilla; d) antenna; e) appendages of VIIIth abdominal segment; f) shagreen pattern and characteristic features of the pupal abdominal segments; g) hypopygia of adult male; h) tip of adult leg.

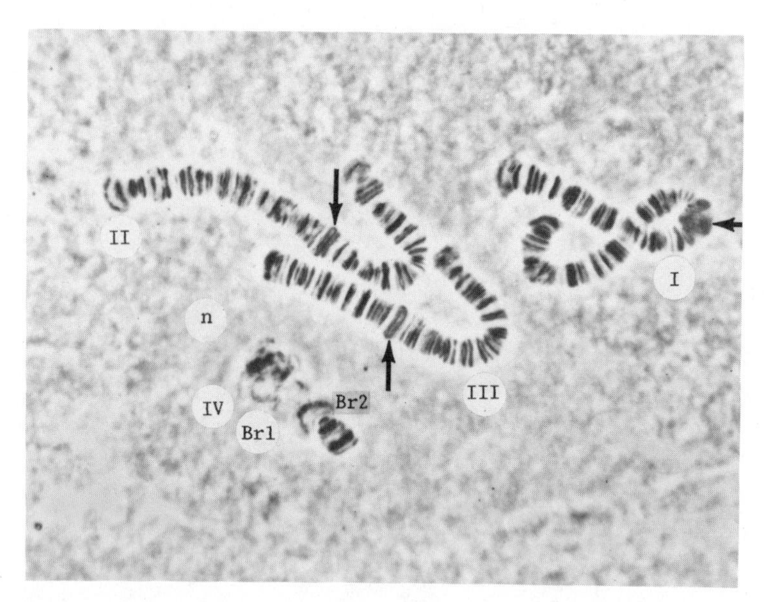

Fig. 6.6. Karyotype of *Chironomus thummi*. Polytene chromosomes from larval salivary glands. I–IV, chromosome number; n, nucleolus; Br, Balbiani rings. Centromeric regions are shown by arrows.

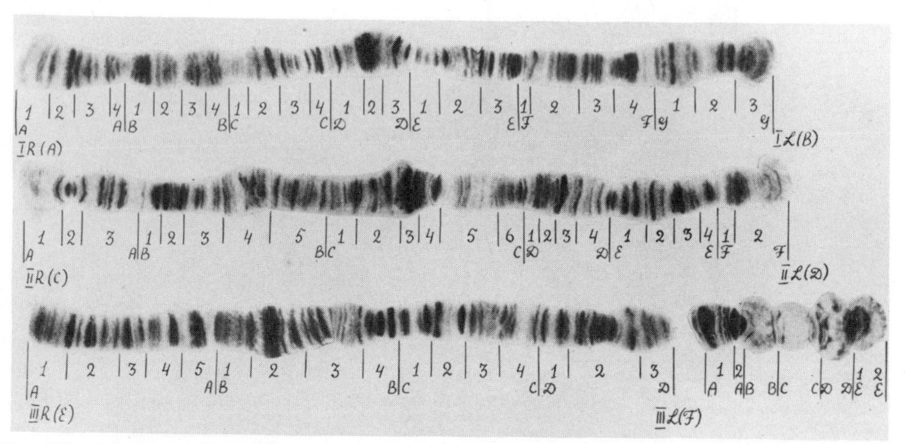

Fig. 6.7. Photographic map of polytene chromosomes from *Chironomus thummi* salivary glands. Letters and numbers correspond to the chromosomal regions from Hägele [24].

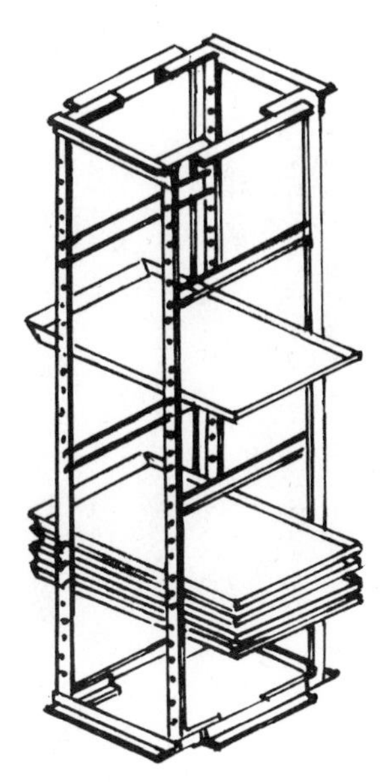

Fig. 6.8. Rack of trays for rearing chironomids [46].

The larvae of chironomids usually develop asynchronously even if they originate from the same egg mass. To synchronize larval development, the aeration and feeding conditions should be strictly standardized. The population density is of fundamental importance. By changing it one may obtain larval populations developing at different rates and with a different degree of asynchrony. We have obtained the optimal uniformity of larval development [23] at a population density of 0.6 larvae/cm^2. A drop in the population density by about one-third did not bring

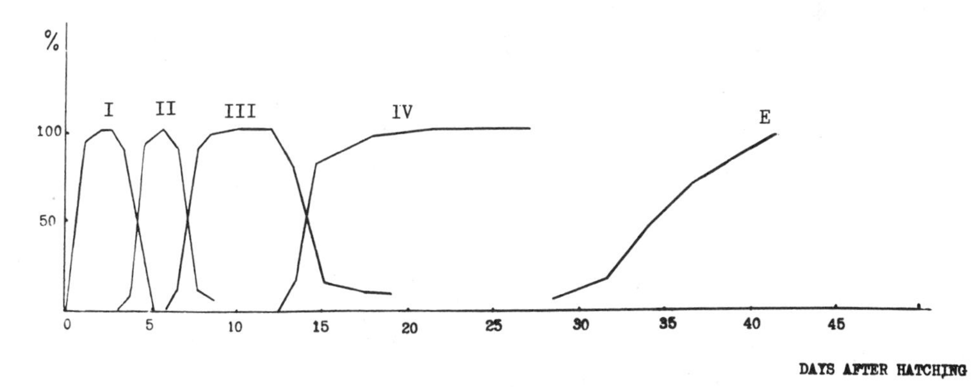

Fig. 6.9. Duration of larval development of *Chironomus thummi* under laboratory conditions (15–25°C). Abscissa shows time after larval hatching from the gelatinous envelope; ordinate gives the number of a given instar as a percentage of the total. I–IV, larval instars; E, emergence of adults.

TABLE 6.1. Characteristics of *Chironomus thummi* Developmental Stages

Stage	Time after oviposition (hours, days)	Diagnostic features	
		External features	Internal features
Embryonic development at 20°C (arbitrary stages):	Total duration 2.5-3 days		
1	Immediately after oviposition	Eggs elongated, dark brown, length 0.3-0.35 mm	Completion of the maturation divisions
2	2.5 h	Segregation of two primary germ cells	Four nuclei embryos
3	10-15 h		Completion of blastoderm formation and beginning of the second and third embryonic leaflet formation
4	End of first day	Beginning of embryo rotation	
5	End of second day		Formation of mouth organ rudiments
6	End of third day	Completion of embryo formation and its emergence from egg membranes	
Postembryonic development at 23-25°C:	Total duration 43-46 days		
Ist instar	Days 3-4	Actively moving larva feeding on mucus	Head ganglion in the head capsule
	Days 4-8	Larvae immediately after leaving the egg mass 0.88 ± 0.07 mm in length with width of the head capsule 0.104 ± 0.002 mm. Mouth parts are incompletely formed as compared with later larval instars: only one pair of bristles on labrum, frontal and oral fields are smooth, premandibulae devoid of bristles. Triple median notch in the submentum, antennae are short (44 μm), relation between the first and second antenna segments is 0.77. No	Hemolymph is colorless due to low hemoglobin concentration. Tracheae are not filled with air, corpus allatum consists of two cells

		appendages on the VIII ventral segment. The base of bristle tuft is low	
IInd instar	Days 8-10	Larval length at the beginning is 1.73 ± 0.08 mm. Head capsule width 0.197 ± 0.003 mm. Formation of mouth parts completed. Submentum acquires the shape characteristic of the species. Appendages on the VIIIth ventral segments appear	Head ganglion displaced to the thoracic segment. Hemolymph acquires pink color. Tracheae are filled with air. The number of cells in corpus allatum is 4-12
IIIrd instar	Days 11-17	Larval body length at the beginning of the instar 3.32 ± 0.08 mm. Head capsule width 0.326 ± 0.008 mm. Appendages on ventral segment undergo elongation	Pink color of hemolymph intensifies. The number of cells in corpus allatum is 12-40
IVth instar (phase 1-5)	Days 17-36	Larval body length from 6.74 ± 0.24 mm at the instar beginning to 12.32 ± 0.48 mm at the instar end. Head capsule width 0.523 ± 0.01 mm	Intensive gonad development. Hemolymph is dark red in color. Progressive development of imaginal disks proceeds in the course of the instar (see Fig. 7-13). The number of cells in corpus allatum is 40-180
Phase 6-9 (prepupa)	Days 36-38	Swelling of thoracic segments. Body contracts and tapers, false legs decrease in size	Imaginal disks continue to form (see Fig. 7-14). Characteristic large cells associated with the epidermis appear on the ventral body side. Metamorphosis puffs in salivary gland chromosomes appear. Formation of oocytes begins in female gonads, in males spermatids and spermatozoa are abundant
Pupa	Days 38-41	Change of body pigmentation from red to dark grey. Fusion of the head and three thoracic segments in cephalothorax. Formation of pupal integuments. Formation of prothoracic horns	Gradual lysis of larval organs (salivary glands, intestine, fat body, and others, with the exception of Malpighian tubules and neural chain). Formation of adult organs from the imaginal disks completed
Adult	Days 43-46	–	–

any marked improvement in the synchrony. Conversely, a four- to sixfold increase in the population density resulted in a marked desynchronization and a decreased rate of larval development in the population.

C. thummi should be reared under a regime of 15 h of light to 9 h of dark at a standard temperature of 20-22°C.

6.5. NORMAL DEVELOPMENT

The general characteristics of the normal developmental stages of *C. thummi* are given in Table 6.1. This table shows that the average lifespan of *C. thummi* under laboratory conditions at 20-25°C is 45-48 days. The larval period of development last 35 days on average, and three molts divide it into four larval instars. Since larval development is asynchronous, the data on instar duration can be conveniently expressed as a diagram (Figs. 6.9 and 6.10). The mean duration of a larval instar in this diagram is taken to be equal to the length of the line corresponding to the 50% diagram cross section at any given instar (the prepupa is included in the IVth larval instar). It follows from Fig. 6.10 that the mean duration of the first larval instar is equal to 3.5 days; of the IInd instar, three days; of the IIIrd, seven days; of the IVth (including the prepupa), 21 days. Prepupae live 4-5 days.

As previously mentioned, the population density greatly affects the rate of larval development. The emergence of adults in populations with a density of 0.2-0.6 larvae/cm^2 begins 28 days and ends 35 days after oviposition. In populations of a higher density, adults begin to emerge 35 days after oviposition although the majority of the population does not complete metamorphosis for at least another month.

These data show that the duration of the last two larval instars, particularly that of the IVth instar, is quite long. The larval development becomes particularly asynchronous during this period. Therefore, the use of data on the mean duration of an instar and of diagrams appears to be more suitable for cytological and developmental biological studies compared with the use of estimates of a minimal instar duration [46].

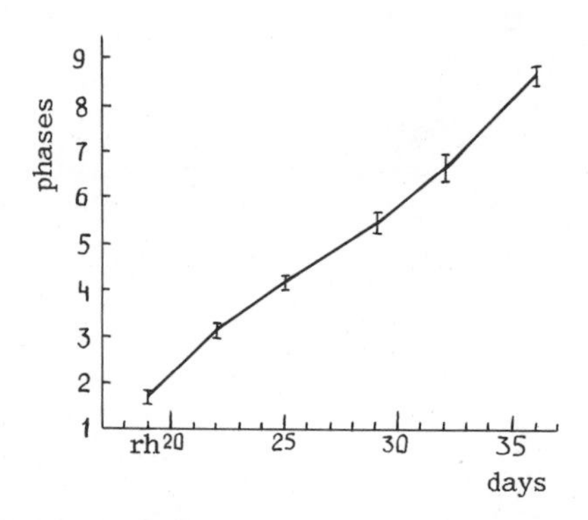

Fig. 6.10. Duration of the different phases of the 4th larval instar of *Chironomus thummi*. Abscissa, time (in days) after hatching; ordinate, 4th instar phases; rh, red-head phase.

6.6. CHARACTERISTIC FEATURES OF THE LARVAL INSTARS

The animals grow rapidly during the larval period, and the larvae of different instars differ markedly in size. The mean length of the body and salivary glands (in millimeters) at the beginning of each larval instar (after the molt) and by the end of larval development is shown below:

Larval instar	I	II	III	IV	End of IVth instar	References
Body-length *l*	0.88 ± 0.07	1.73 ± 0.08	3.32 ± 0.08	6.74 ± 0.24	12.32 ± 0.46	Our data
	–	2.9	3.7	6.9	15.5	[46]
Length of salivary gland	0.08 ± 0.01	0.15 ± 0.01	0.29 ± 0.01	0.60 ± 0.05	1.07 ± 0.08	

The length of the larvae, however, cannot serve as a reliable criterion of instar determination since the length varies, depending on developmental conditions such as temperature, feeding conditions, and oxygen supply [46, 91]. The width of the head capsule serves as the main morphological criterion and enables different instars to be distinguished. The mean width of the head capsule is characteristically different for each larval instar. The mean size of the head capsules at different larval instars in millimeters is presented below:

Larval instar	I	II	III	IV	References
Head capsule					
Length	0.15 ± 0.008	0.25 ± 0.003	0.43 ± 0.12	0.69 ± 0.12	Our data
Width	0.10 ± 0.002	0.20 ± 0.003	0.33 ± 0.008	0.52 ± 0.01	Our data
	0.125	0.20-0.25	0.33-0.38	0.48-0.68	[44]

Since the width of the head capsule increases only after the molts and shows practically no changes during the intermolt periods, the measurement of this characteristic enables the larval instar to be reliably determined. One should bear in mind, however, that the absolute size of the head capsule may vary somewhat depending on the conditions for larval development [46]. Despite this, the size ratio for different larval instars remains unchanged. This ratio should be used for determining larval instars.

Although the increase in the larval size during larval development is a characteristic trait, other morphological and physiological characteristics can also be used to identify individual larval instars. These include certain changes in the body structure. During the first larval instar there are no appendages (tubuli) on the VIIIth abdominal segment. They appear after the first molt and then become enlarged; bristle tufts undergo elongation. The relative length of the antenna feelers decreases with age, and there is a change in the length ratio of antenna segments

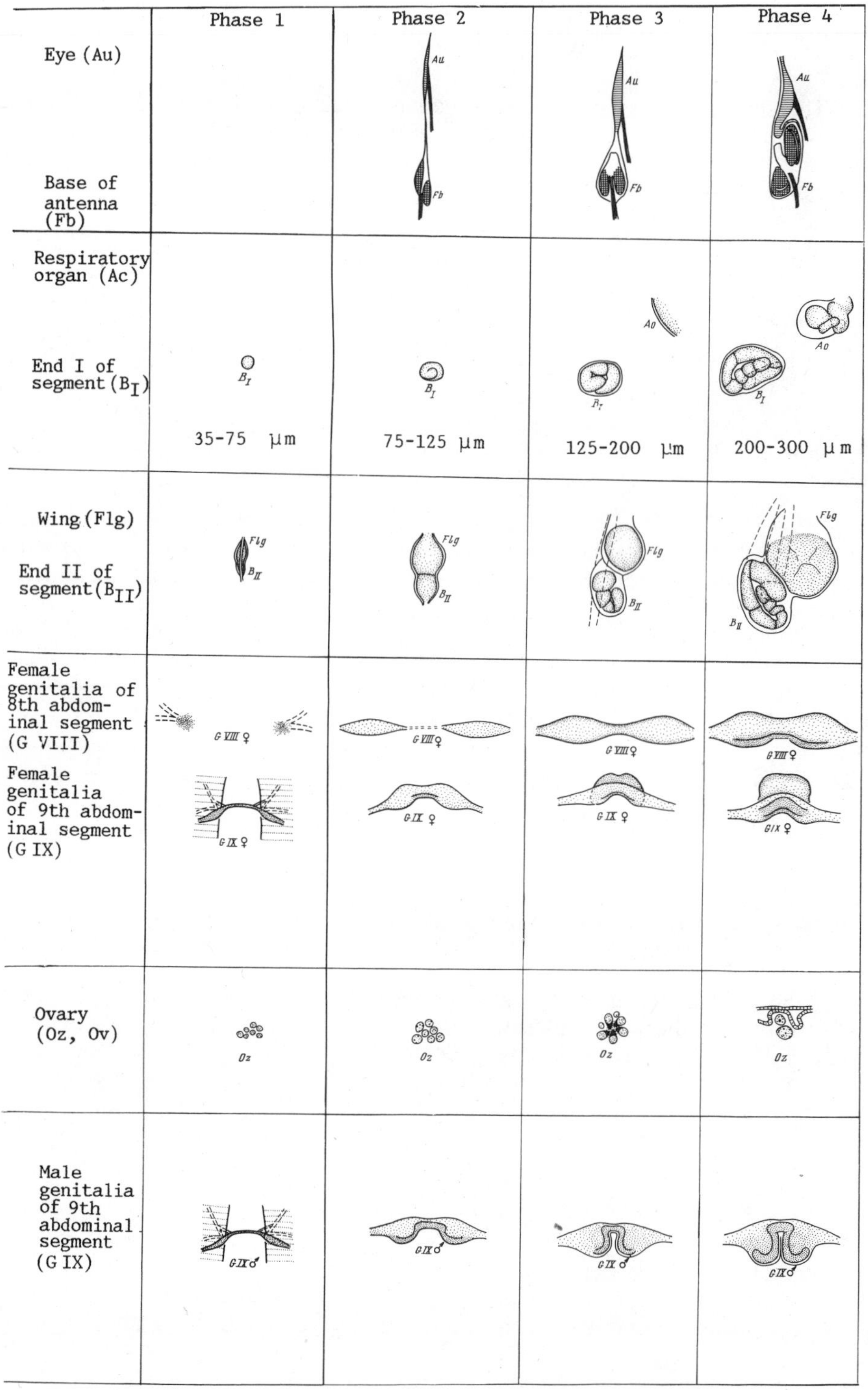

	Phase 1	Phase 2	Phase 3	Phase 4
Eye (Au) Base of antenna (Fb)		Au Fb	Au Fb	Au Fb
Respiratory organ (Ac) End I of segment (B_I)	B_I 35-75 µm	B_I 75-125 µm	Ao B_I 125-200 µm	Ao B_I 200-300 µm
Wing (Flg) End II of segment (B_{II})	Flg B_{II}	Flg B_{II}	Flg B_{II}	Flg B_{II}
Female genitalia of 8th abdominal segment (G VIII) Female genitalia of 9th abdominal segment (G IX)	G VIII ♀ G IX ♀	G VIII ♀ G IX ♀	G VIII ♀ G IX ♀	G VIII ♀ G IX ♀
Ovary (Oz, Ov)	Oz	Oz	Oz	Oz
Male genitalia of 9th abdominal segment (G IX)	G IX ♂	G IX ♂	G IX ♂	G IX ♂

Fig. 6.11. Development of the imaginal disks in *Chironomus thummi*.

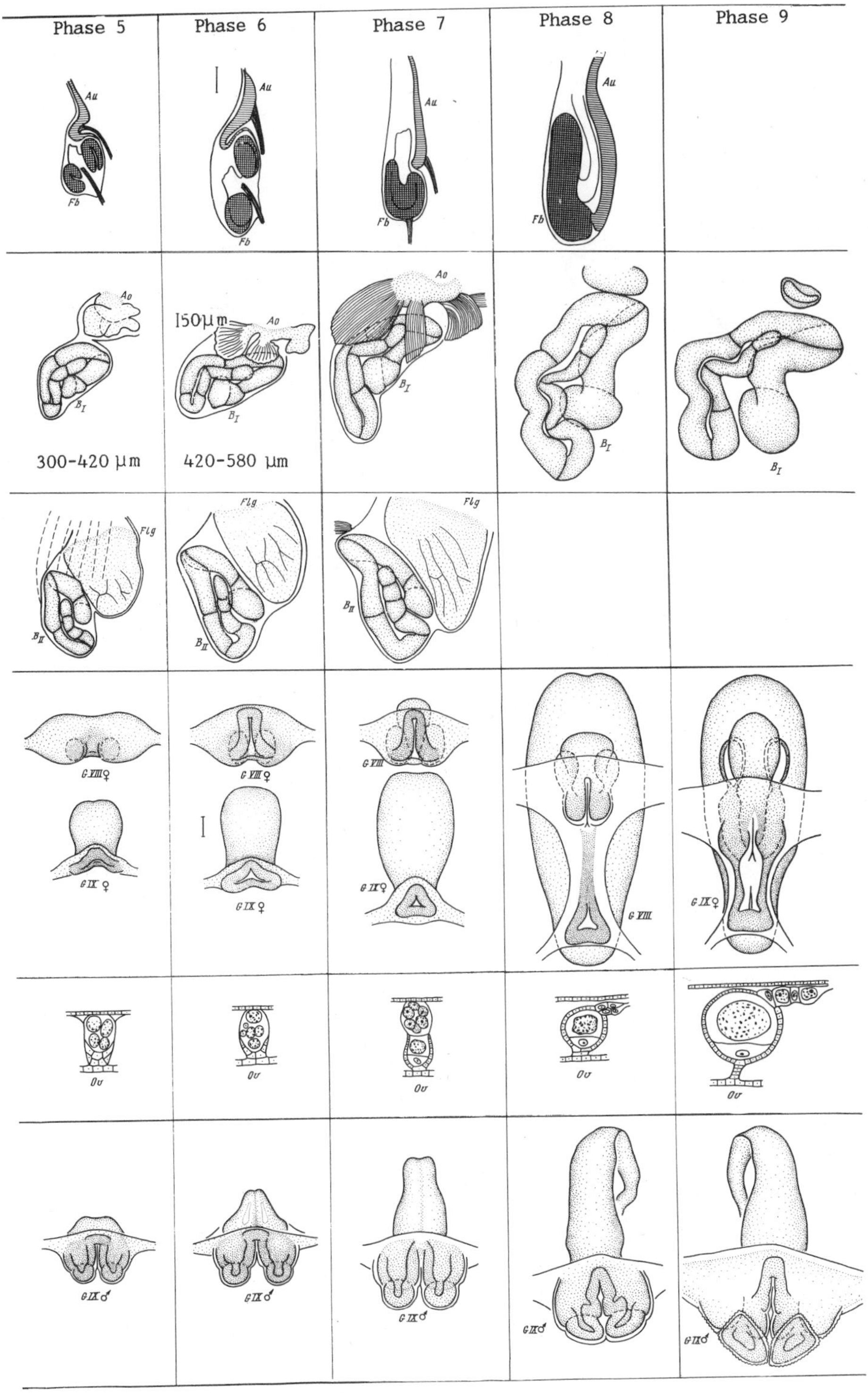

Phase 5	Phase 6	Phase 7	Phase 8	Phase 9
Au Fb	Au Fb	Au Fb	Au Fb	
Ao B_I 300-420 µm	150µm Ao B_I 420-580 µm	Ao B_I	B_I	B_I
Flg B_{II}	Flg B_{II}	Flg B_{II}		
G VIII ♀ G IX ♀	G VIII ♀ G IX ♀	G VIII G IX ♀	G VIII	G IX ♀
Ov	Ov	Ov	Ov	Ov
G IX ♂	G IX ♂	G IX ♂	G IX ♂	G IX ♂

Fig. 6.12. Developmental changes in certain larval features. Numbers correspond to instars I–IV. a) Mentum and ventramental plate; b) antenna; c) mandible; d) appendages of 8th abdominal segment in instars II–IV; d_4, d_4') beginning and end of the 4th instar.

and in the mentum shape (Fig. 6.12). The intensive development of the imaginal disks begins during the IVth larval instar (Fig. 6.11).

Body pigmentation of the larva changes during this period, too. This is related to an increased hemoglobin concentration in the hemolymph. The first instar larvae appear colorless, but during the second and third larval instars they gradually acquire a red color, and during the fourth larval instar they assume the dark-red color characteristic of chironomids. It is believed that hemoglobin is absent during the first larval instar [46, 72], but our own observations (Kolesnikov, unpublished) suggest that hemoglobin is present during the first larval instar, but that its concentration is too low to color the larvae.

The tracheae begin to fill with air after the first molt.

After presenting this short list of differences between the larval instars, it should be stressed that both in *C. thummi* and in other chironomid species the early larval instars have been far from sufficiently investigated. The most detailed information is available for the last larval instar which is more accessible to investigations; this instar has been widely used in various cytogenetic and biochemical studies. As the IVth larval instar lasts for a relatively long period (see Fig. 6.10), it is necessary to identify more accurately the consecutive developmental phases during this period. Such identification is also essential for experimental purposes since analyzing many developmental aspects requires a large number of synchronously developing animals. As the natural development of the larvae is rather asynchronous, it is necessary to select larvae at the specifically required developmental

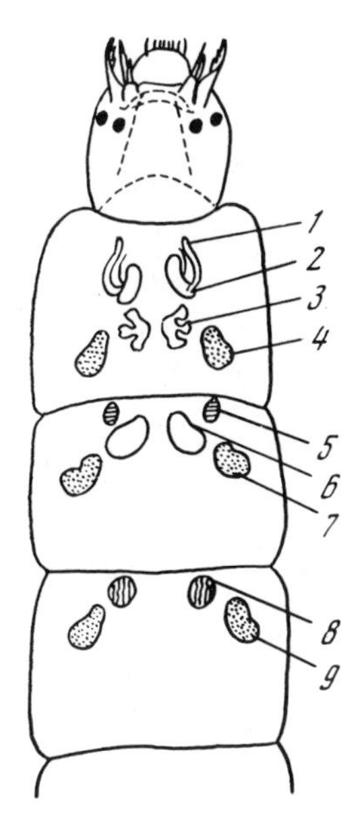

Fig. 6.13. Location of the imaginal disks in the thoracic segments of a *Chironomus* larva. 1) Antenna; 2) eye; 3) respiratory organ; 4) 1st leg; 5) salivary gland; 6) wing; 7) 2nd leg; 8) haltere; 9) 3rd leg.

phases using reliable criteria. Essential criteria for distinguishing consecutive developmental phases of the IVth instar larvae are described in the paper by Wülker and Götz [103]. They compiled a detailed description of the formation of imaginal disks corresponding to different organs and, using these criteria, divided the IVth larval instar (including the prepupal stage) into nine phases. The map showing the location of imaginal disks in a *C. thummi* larva is presented in Fig. 6.13, while the scheme of consecutive stages in their formation during the development of the IVth instar larva is shown in Fig. 6.11. The system for determining the individual phases of the IVth instar larva according to Wülker and Götz is suitable for practical use and can be easily applied to selected larvae corresponding to each of the IVth instar phases. Operationally the long-term practice was to divide the IVth instar into early, middle, and late periods using mainly such characteristics as the duration of development, larval body size, shape of imaginal disks of the forelegs, and the set of puffs [12, 32, 51-53]; there is no need to emphasize that this division was rather arbitrary and not very reliable.

It has been demonstrated that the size of the imaginal disks may vary significantly under different rearing conditions and depending on the physiological peculiarities of the larvae. For example, the size of the imaginal disks is significantly affected by the population density. As can be seen from Fig. 6.14, at a population density of 0.2-0.6 larva/cm^2, the diameter of the foreleg imaginal disks at the redhead stage (first 12 h after the molt from the IIIrd to the IVth instar) was about 80 μm. This corresponds to the second developmental phase according to Wülker and Götz rather than the first phase, as could have been expected for the beginning of the IVth instar. Upon a significant increase in the population density (2.5-3.8 larvae/cm^2) the diameter of these imaginal disks was found to be 50 μm, which completely corresponds to the value listed by Wülker and Götz for the first phase of development. The rate of larval development also affects the size of the imaginal disk. If the size of the imaginal disks at the redhead stage is compared to the larvae developing from one egg mass which take a different time to reach the stage of molt from the IIIrd instar to the IVth instar, one notes that the most rapidly developing larvae have imaginal disks of about 80 μm in diameter, while the larvae completing the molt have imaginal disks with a diameter of 50 μm. The diameter of the imaginal disks always corresponds to their morphological structure.

Thus, when using the characteristics of the imaginal disks as criteria for selecting larvae at a given stage, one should always remember that this requires

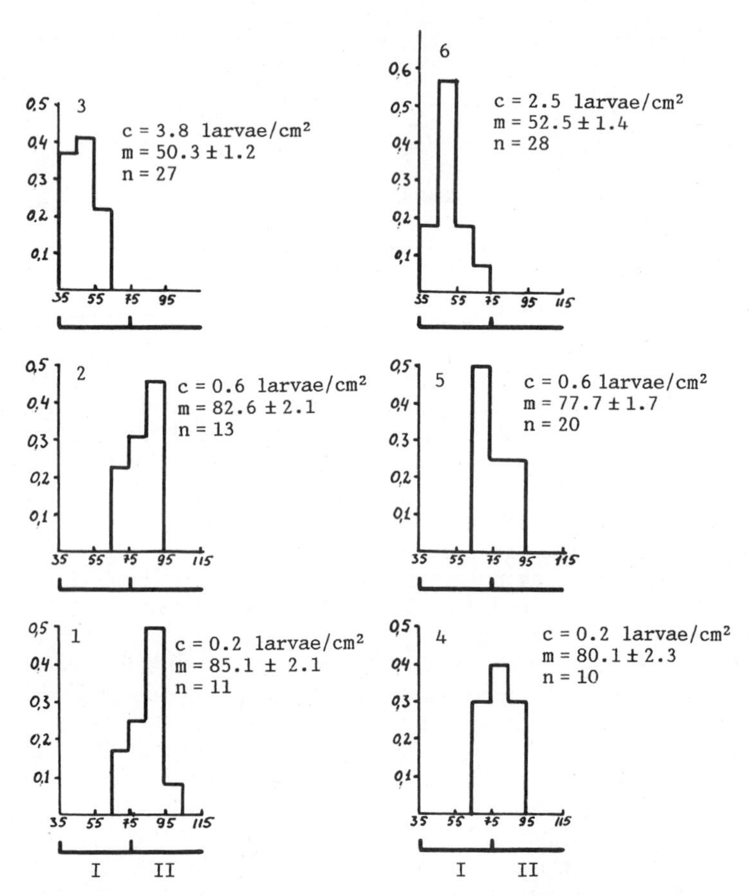

Fig. 6.14. Size changes in the 1st leg imaginal disk in the 4th instar. Different population sizes of *Chironomus* larvae at the redhead stage were used. Abscissa, diameter of imaginal disk (μm); ordinate, frequency of occurrence. I, II) Developmental phases. c, population density; m, mean disk size; n, number of larvae examined. 1–3 and 4–6 are the results of two independent experiments.

keeping a certain population density of larvae and standardizing other rearing factors.

While describing larval development, we would like to mention the differentiation of two organs, the gonads and salivary glands. Gonads comprise one of the most important characteristics of the developing animal, while salivary glands are widely used in developmental cytogenetics owing to the presence of chromosomes with extremely high degrees of polyteny.

6.6.1. Gonads

As previously mentioned, the primary germ cells in *C. thummi* segregate at the posterior egg pole at the stage of the tetranuclear embryo. The two primary germ cells undergo two divisions and the resulting group, consisting of 8-16 cells, migrates inside the cleaving embryo. These cells do not change until the larva hatches from the egg; this is followed by the beginning of gonad growth. It proceeds very slowly between the Ist and IIIrd larval instars. Intensive gonad growth and oocyte differentiation begin only in IVth instar larvae [103].

During the first and second phases of the IVth instar larvae, female gonads consist of apparently uniform multiplying oogonia. During the third phase the groups of cells become organized to acquire a rosettelike shape; during phases 4-6 the ovarian follicles are gradually formed (see Fig. 6.11). By the beginning of pupation (phase 7) the oocyte, nurse cell, follicular cells, and interstitial cells adjacent to the ovarian epithelium can be distinguished in the follicle (see Fig. 6.11). Interstitial cells are located between the follicles. The follicular cells continue to divide and are suitable for studying the metaphase chromosomes. The latter feature is very important, since nearly all the larval tissues in chironomids have polytene chromosomes and do not divide mitotically. In addition to gonads, mitotic division figures can be conveniently observed in the imaginal disks during the IVth larval instar as well as in the brain and the epidermal cells during molts.

Prior to pupation the oocytes are at the slow growth stage; they begin to grow rapidly at prepupa and pupa stages and cease at the adult stage. Prior to pupation the ovary can be clearly distinguished from the testis, owing to its more elongated shape and larger size; it is located in the VIth abdominal segment. Examination *in vivo* enables oocytes of different sizes to be distinguished.

By the end of larval development the testis is markedly smaller than the ovary. It has an oval shape and is also located in the VIth abdominal segment.

During phases 1-4 of the IVth instar larva the testes contain mainly spermatogonia which are clustered in rosettelike groups. Spermatocytes appear during phases 5-6 (Fig. 6.15). Prior to pupation the testes may contain spermatocytes and the pachytene, diplotene, and metaphase I stages, as well as spermatids and spermatozoa. In the prepupae (phases 7-9) metaphase II and telophase II stages, as well as spermatids and spermatozoa, are the most frequent stages.

6.6.2. Salivary Glands

The salivary glands complete their formation by the time of larval hatching. The salivary gland of *C. thummi* comprises a flat sac whose wall consists of 28-36 glandular cells (Fig. 6.16). The cells are arranged as one layer around a central gland lumen. The number of cells in the gland does not change throughout larval life. There are two salivary glands and their two ducts unite in the narrow chi-

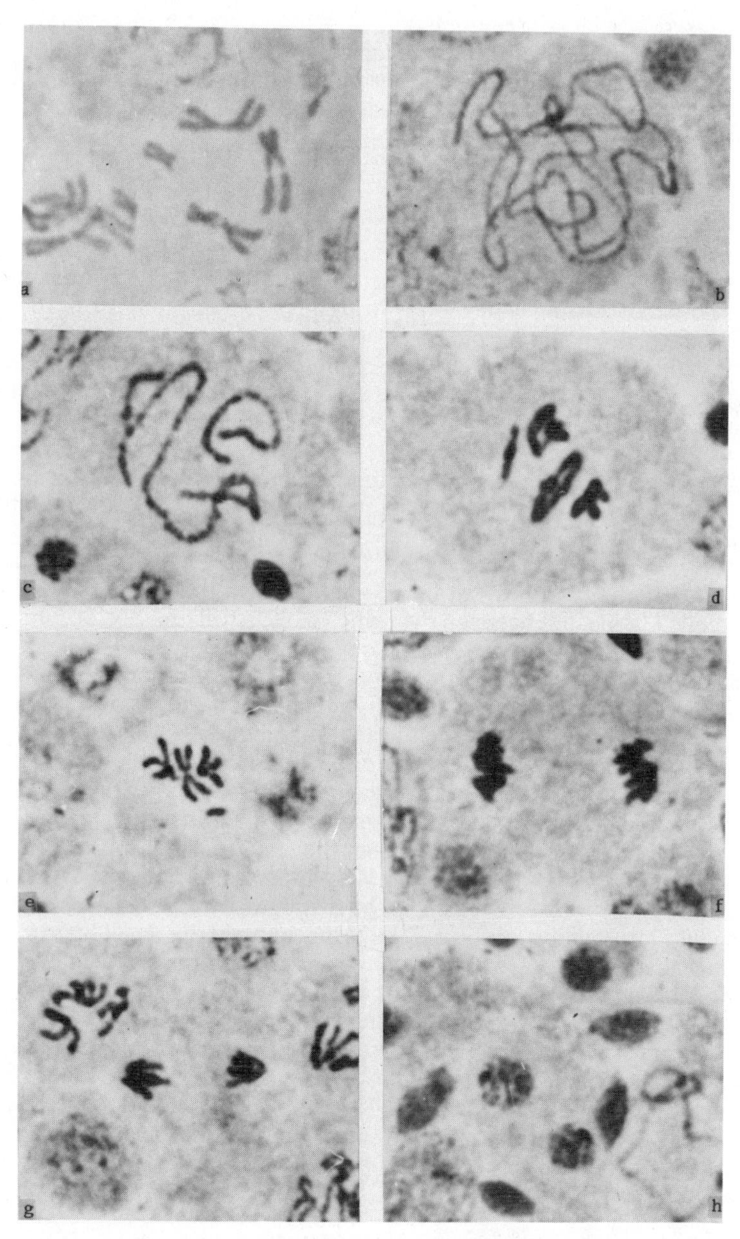

Fig. 6.15. Meiosis in *Chironomus thummi* testis. a) Spermatogonial metaphase; b) pachytene; c) diplotene; d) diakinesis; e) metaphase II; f) anaphase I; g) anaphase II; h) spermatids. Squashed aceto-orcein preparations (original photographs by T. M. Panova).

tinized duct, which carries secretion to the mouth of the larva. The imaginal disk of a mature salivary gland is located at the source of the duct within the gland. The glandular cells produce glycoprotein secretion, which is formed in Golgi bodies and then released to the central secretory lumen of the gland.

The gland is differentiated into lobes (Fig. 6.16a). The four cells forming the special lobe produce granular secretion of a particular type. Electrophoretic study of the secretion proteins produced by this lobe has shown (Fig. 6.16b, c) the presence of an additional fraction with a molecular weight of 160 kD which was lacking in the secretion synthesized by the cells of other lobes. The secretion of the special lobe also differs from that of other lobes in carbohydrate composition and the intensity of carbohydrate synthesis [43, 84].

It is now established that the genes coding for the secretory proteins of chironomids are located in the Balbiani rings. In *C. thummi* the Balbiani rings (BR) are located in the short chromosome IV (Fig. 6.16d). In the cells of all lobes, two Balbiani rings are present: BR1(c) and BR2(b). In the special lobe an additional Balbiani ring BR3(a) is present. It is most probably responsible for the synthesis of sp 160, the additional fraction of secretory protein referred to above.

The salivary gland cells contain extremely large polytene chromosomes which are very suitable for cytogenetic analysis (Figs. 6.6 and 6.18). The formation of polytene chromosomes in salivary gland cells begins quite early during their differentiation. At the beginning of the first instar the salivary gland cell nuclei look like typical interphase nuclei, but their content of DNA attains 32-64 C. Two cycles of DNA replication take place during the Ist larval instar, 2-3 during the IInd larval instar, and 1-2 in larvae of the IIIrd and IVth instars; 11-12 DNA replication cycles proceed during larval development, and the completely formed polytene chromosomes are between 4000 and 8000 C (Fig. 6.17). An increase in the quantity of DNA templates by endoreplication in salivary gland chromosomes provides the genetic basis for gland cell growth. The process of endoreplication proceeds in relation to body growth.

A total of 336 puffs has been identified in the giant polytene chomosomes of *C. thummi* salivary glands [33, 34]. As shown in Figs. 6.18 and 6.25, the puffing pattern in salivary gland chromosomes is quite stable during the intermolt period of the last two larval instars (studies of puffing are impossible during earlier larval instars owing to the low polyteny of the chromosomes).

The larval molts are important periods in insect larval development. It is known that molts are accompanied by significant changes in the concentration of hormones such as ecdysone and juvenile hormone in the hemolymph, leading to characteristic changes in the physiological state of the larval tissue (for review see [70, 90, 100]). These changes are primarily observed in the cells of the epidermis, forming the cuticle (Fig. 6.23). However, as already mentioned, synthesis of DNA, RNA, and protein in a number of larval organs decreases or even stops completely during the molt [15, 17, 18, 69, 76, 101].

The molting physiology, however, is far from being completely understood even for such classic experimental animals of invertebrate physiology as butterflies, ticks, and crustaceans [98, 99, 108]. This process has been investigated in *Chironomus* to only a small extent, although this species is one of the few experimental models to provide an opportunity for comparing the morphological and biochemical characteristics of molting larvae with visually observable changes in chromosomes. The fragmentary data available suggest that further work in this direction could be quite promising. Using the salivary gland as a model, one can expect to observe

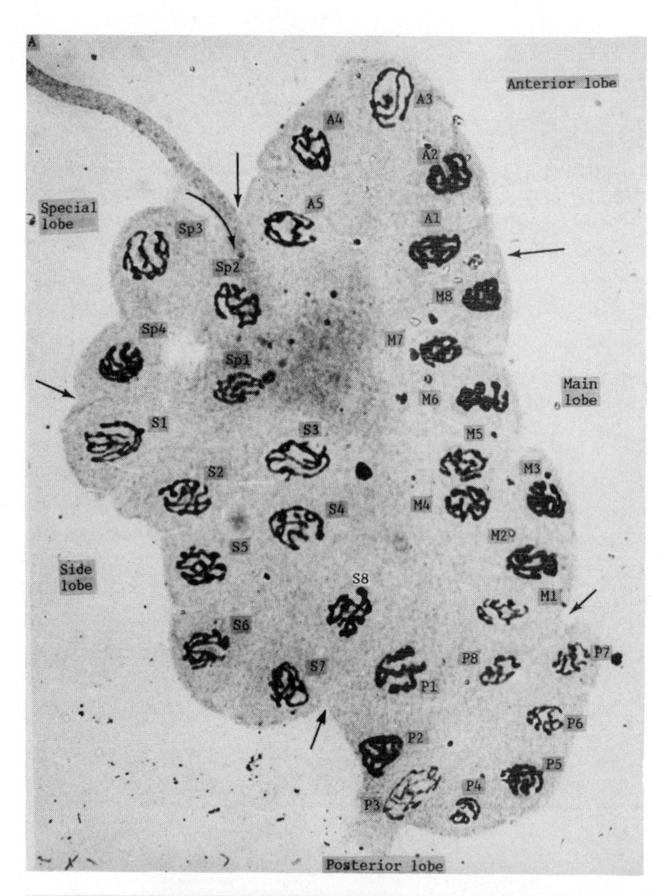
A
Anterior lobe
A3
A4
A2
A5
A1
Special lobe
Sp3
Sp2
M8
Sp4
M7
Sp1
Main lobe
M6
S1
S3
M5
S2
M3
M4
S4
M2
S5
Side lobe
S8
M1
S6
P8
P7
S7
P1
P6
P2
P5
P4
P3
Posterior lobe

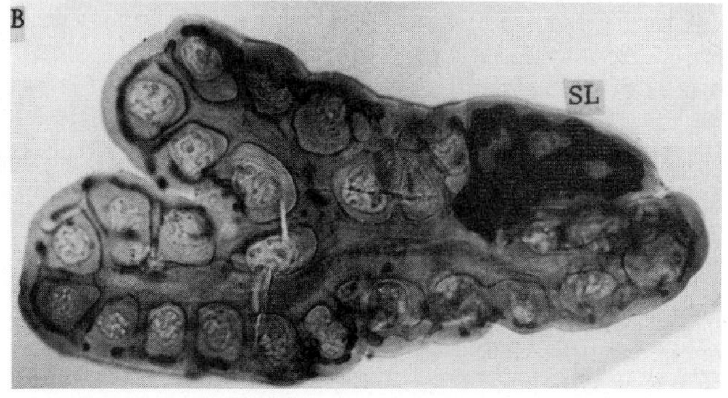
B
SL

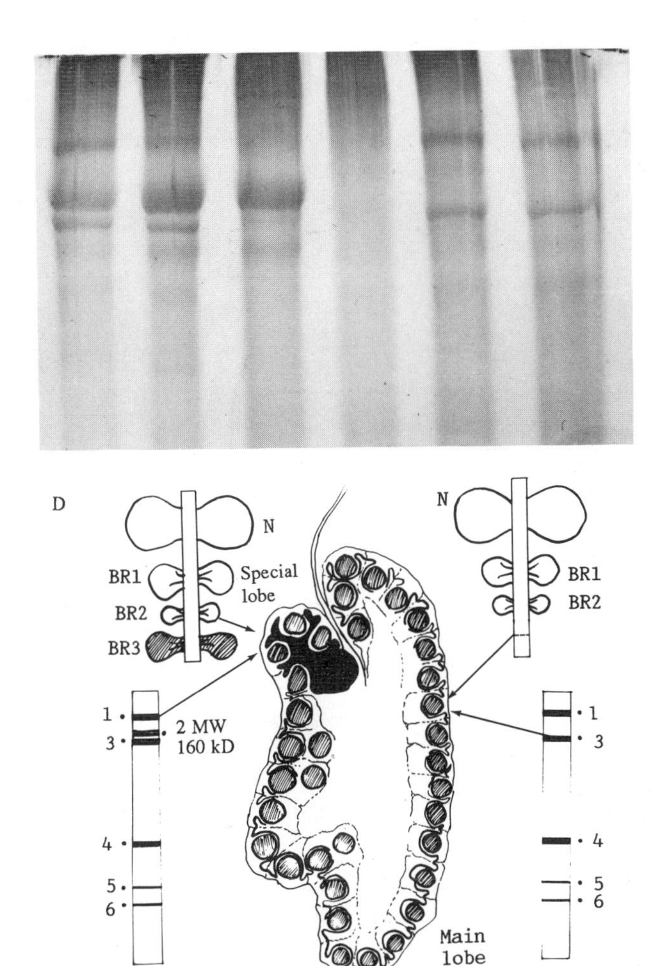

Fig. 6.16. Salivary gland of *Chironomus thummi* 4th larval instar. A) Whole mount and Feulgen preparations. Arrows indicate lobe boundaries. Letters and numbers indicate cells in different lobes. B) Cytochemical differences of the secretion from the special lobe and other lobes of the salivary gland. Paraldehyde–fuchsin staining. SL, special lobe. C) Differences in the electrophoretic patterns of secretory proteins from the special (SSL) and the main (SML) lobes. Polyacrylamide gel-electrophoresis. D) Sketch illustrating differences between the special lobe and other lobes of the salivary gland in the pattern of Balbiani rings and electrophoretic patterns of secretory proteins [85].

interesting correlations between changes in the puffing pattern and a number of physiological characteristics reflecting the activity of the gland.

Studies of puffing during the larval development of *C. thummi* have shown that, as in other Diptera, the most significant changes in the puffing pattern take place during the metamorphosis molt (Figs. 6.18 and 6.19), when metamorphic puffs appear and some of the larval puffs are reduced [31, 33-35, 52-55, 60-62].

It has been demonstrated that larval molts are also associated with considerable changes in the puffing pattern [33, 34]. The main puffs, previously regarded as

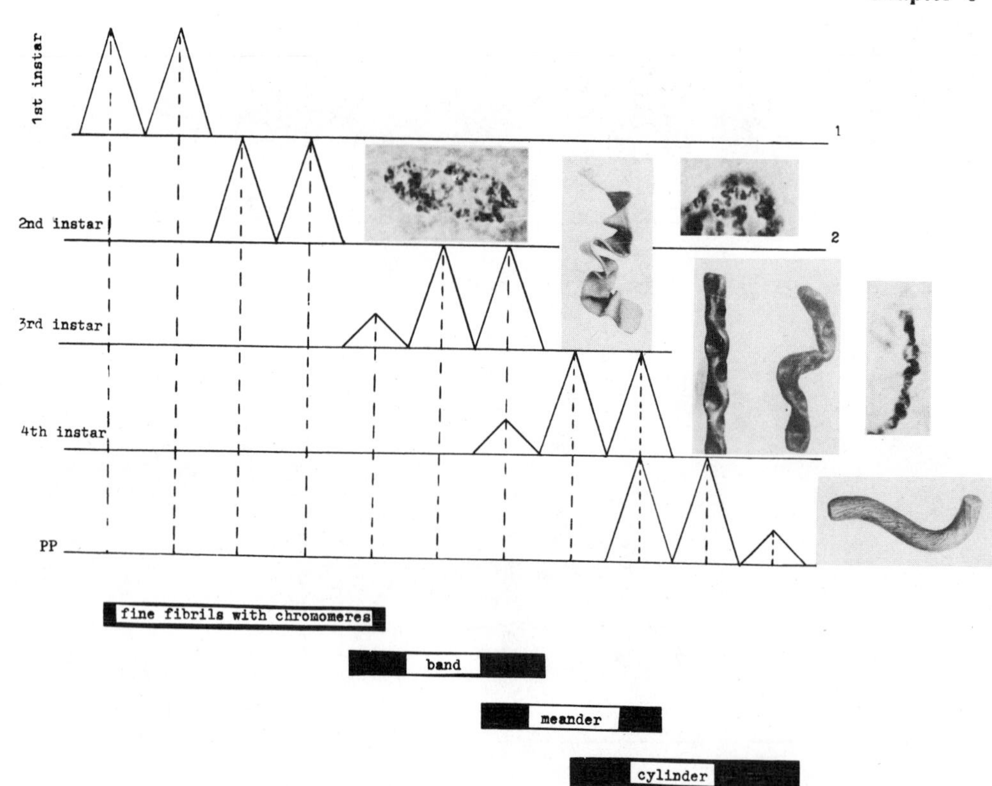

Fig. 6.17. Formation of polytene chromosomes during larval development. Abscissa, degree of polyteny indicated by cytophotometry of the Feulgen-stained chromosomes; ordinate, larval instars. (For each instar, photometry was conducted at the redhead stage, when DNA synthesis is very low.) PP, prepupa. Photographs of chromosomes with different degrees of polyteny and corresponding plastic models are shown on the right-hand side of the figure.

typical only for the metamorphosis molt (Fig. 6.18), also appear during these periods. Only the late puffs of metamorphosis do not develop during larval molts. A characteristic feature of the puffing pattern during larval molts is the absence of regression of the main larval puffs, with the exception of tissue-specific puffs which become inactive. The Balbiani rings, in particular, undergo regression during larval molts. This regression is reversed after the molt has been completed (Fig. 6.19). The BR reduction during larval molts correlates with decreased ribosomal RNA synthesis. The incorporation of ^{3}H-uridine into nucleoli is much reduced and in many cells is completely absent during the larval molts; however, it is fully restored when the molts are completed (Fig. 6.19).

In order to explain the functional aspect of puffing properly we must emphasize that the reduction of RNA synthesis in the BR and the nucleoli is accompanied by a cessation of secretion production. Electron microscopy apparently demonstrates that, as a consequence of ecdysone action, the Golgi bodies become inactive during molts, secretory granules disappear from the cytoplasm, and the secretion synthesized earlier is actively released into the lumen of the gland. Ultrastructural changes

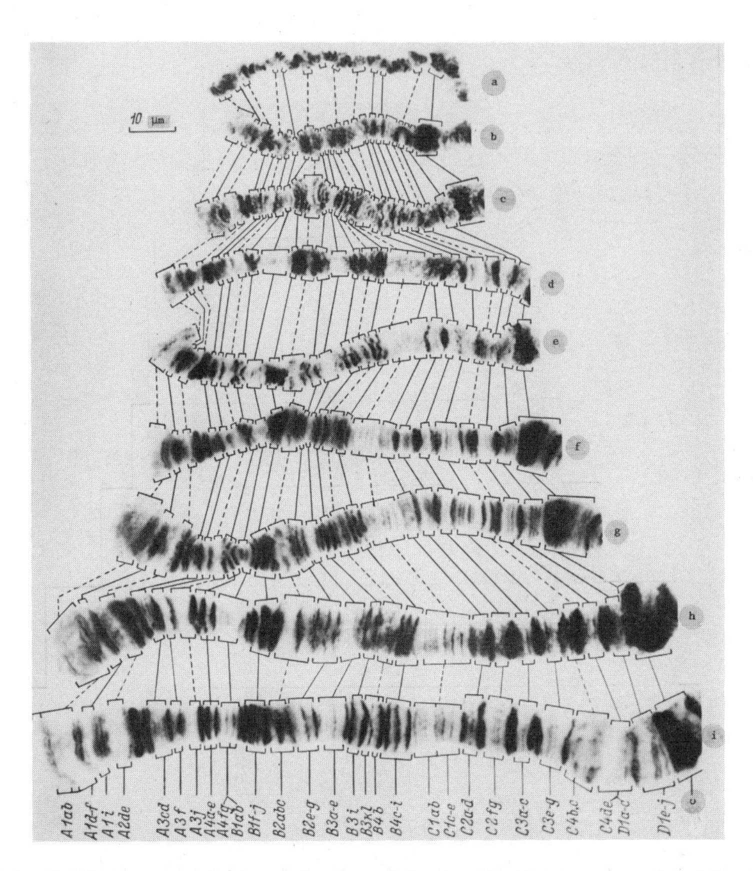

Fig. 6.18A. Puffing patterns in the right arm of the 1st chromosome. a–c) 3rd Instar – 3, 5, and 7 days after molting, respectively; d) 4th instar molt; e–g) 4th instar, phases 1 (redhead stage), 3, and 6; h) prepupa, phase 7; i) red pupa. c, centromeric region. In this and the following figures the locations of the main puffs are connected with lines.

of the cytoplasm of the salivary gland cells during larval molts can also be detected (Fig. 6.20). The cessation of secretion formation is accompanied by the regression of a large area of endoplasmic reticulum (ER) with attached polysomes. This regression predominantly involves the areas of ER located in the apical parts of the cells which are directed toward the lumen. Mitochondria migrate *en masse* from the distal cell regions where they are usually concentrated in the apical parts.

Thus, during the larval molts when the synthesis of tissue-specific BR RNA and ribosomal RNA is reduced, the salivary gland cell cytoplasm undergoes significant ultrastructural changes. These changes include the disappearance of secretory granules, changes in the shape of the Golgi bodies, migration of mitochondria from the distal zones to the zone of microvilli turned toward the lumen, and the formation of vast regions of destroyed rough endoplasmic reticulum [2, 35, 39].

Fig. 6.18B. Puffing patterns in the left arm of the 1st chromosome. a–c) 3rd Instar – 3, 5, and 7 days after molting, respectively; d–g) 4th instar, phase 1 (redhead stage), 1–3, and 5–6, respectively; h, i) prepupa phases 8 and 9; j) red pupa. c, centromeric region.

After the molt has been completed, the normal cytoplasm ultrastructure recovers along with the normal puffing pattern and the resumption of RNA synthesis in the Balbiani rings and the nucleolus.

Periodic reduction in the activity of the BR and the nucleolus, and simultaneous activation of the main set of metamorphic puffs during larval molts, can be explained by an increased titer of ecdysone in the larval hemolymph during these periods [95]. Changes of the puffing pattern and cytoplasmic ultrastructure can be experimentally induced by injecting ecdysone into the intermolt larvae [39].

The described transformations of the puffing pattern and cytoplasm ultrastructure can be seen as the periodic reprogramming of terminally differentiated cells during larval molts. A similar periodic switching on and off of the tissue-specific genes and corresponding changes in the cytoplasm ultrastructure during larval molts is described by Suzuki in the silk gland of the silkworm [90].

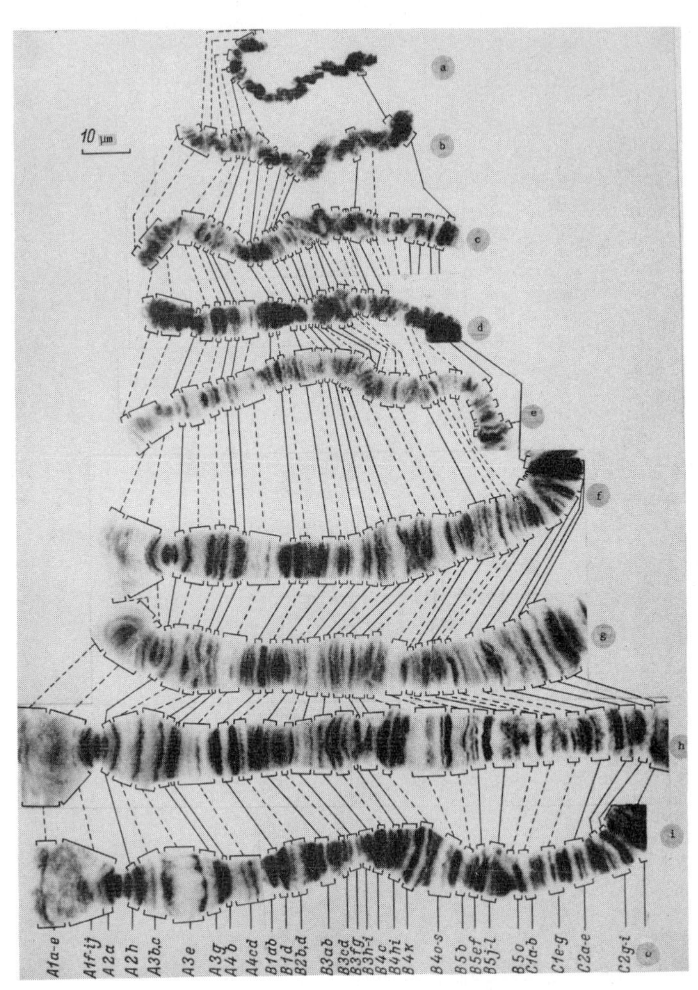

Fig. 6.18C. Puffing patterns in the right arm of the 2nd chromosome. a–c) 3rd Instar – 1, 3, and 5 days after molting; d) molt; e, f) 4th instar, phases 1 (redhead stage) and 6; g–i) prepupal phases 7, 8, and 9. c, centromeric region.

Fig. 6.18D. Puffing patterns in the left arm of the 2nd chromosome. a, b) 3rd Instar – days 5 and 7 after molt; c) 4th instar molt; d, e) 4th instar, phases 1 and 6; f–h) prepupa phases 7, 8, and 9; i) red pupa.

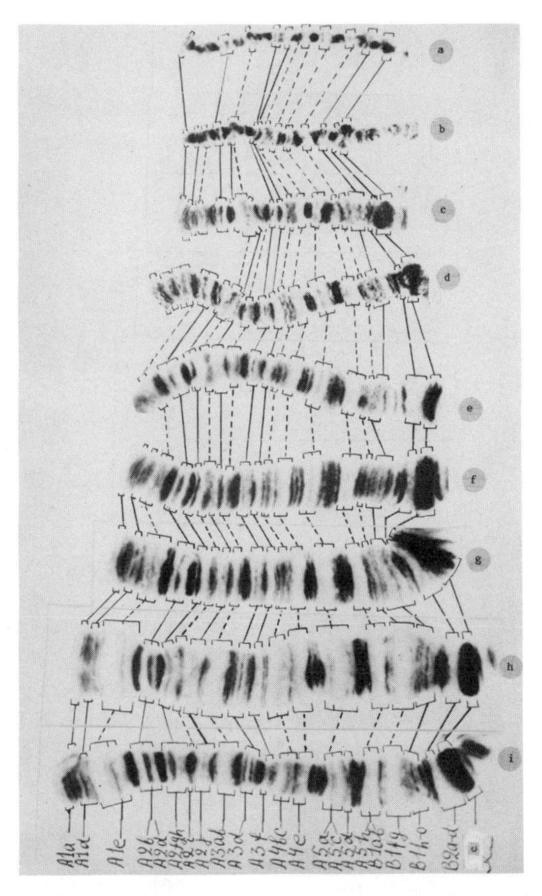

Fig. 6.18E. Puffing patterns in the right arm of the 3rd chromosome. a, b) 3rd Instar – days 1 and 5 after molt; c) molt; d–f) 4th instar, phase 1 (redhead stage), phases 1, 3, and 6; g, h) prepupa, end of phase 9; i) red pupa. c, centromeric region.

An almost complete arrest of DNA replication is another characteristic feature of the molt period (Fig. 6.21).

In order to study the sequential changes of larvae during the molt, one has to find a group of traits enabling larvae to be identified and selected at vital phases, such as the preparation to molt, the molt itself, and immediately after the molt. These traits include the state of the epidermis and cuticle, the position of eye spots on the head capsule, and pigmentation of the head capsule and mouth parts. Some of these traits have been described briefly [13, 18, 46, 64]. We have studied the changes of these characters in the course of the last two larval molts and the metamorphosis molt. During the molt from the IIIrd to the IVth instar, for example, there are marked morphological changes about 1.5 days prior to ecdysis (Figs. 6.22 and 6.23). At this time, the eye spots have already shifted to the center of the head capsule, while the cuticle begins to separate from the epidermal cells. A narrow

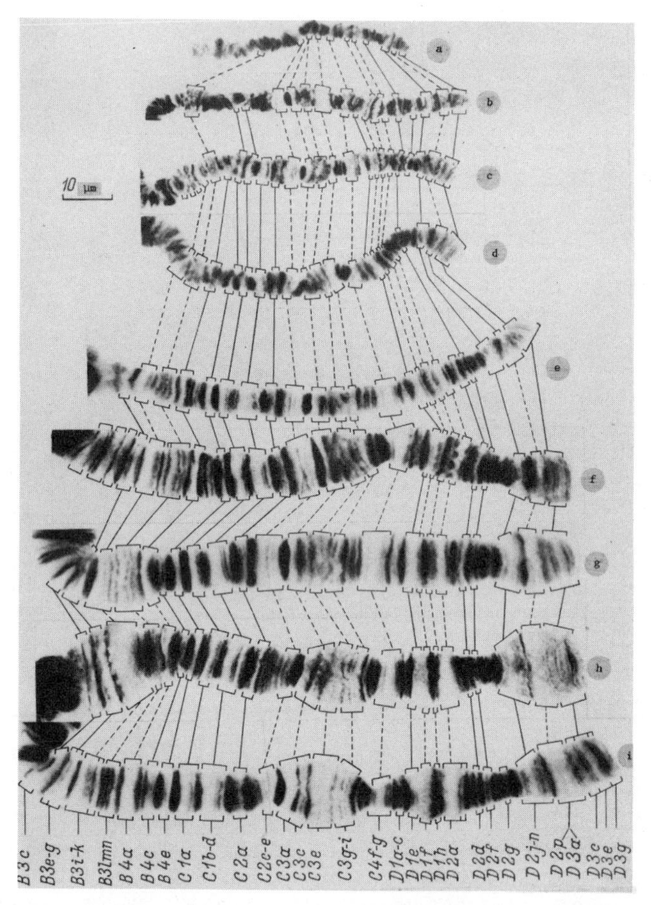

Fig. 6.18F. Puffing patterns in the left arm of the 3rd chromosome. a, b) 3rd Instar – days 3 and 7 after molt; c–f) 4th instar, phase 1 (redhead stage), phases 1, 2, and 6; g, h) prepupal phases 7 and 9 (pupation); i) red pupa.

exuvial space may be detected in the anterior body segments and in the appendages with hooklets on the last body segments, and the thoracic segments become thicker. Twelve hours prior to the molt the eye spots eventually shift to the base of the head capsule and the exuvial space can be seen clearly in all segments (see Fig. 6.22). At this time the salivary glands stop supplying secretion, and swell. The larvae, having completed ecdysis, can be easily identified by the bright red color of the head capsule; hence the description "redhead larvae." The true redhead stage in the IVth instar larva continues for 8-10 h, then the head capsule becomes gradually darker, assuming a brown pigmentation by the twelfth hour (see Fig. 6.22, b_1-b_3). Pigmentation of the mentum and mandibula changes in parallel with the pigmentation of the head capsule (see Fig. 6.22, c_1-c_3). Use of these characteristics results in a more reliable determination of the consecutive stages of larval development and of the first phase of the IVth larval instar. Correspond-

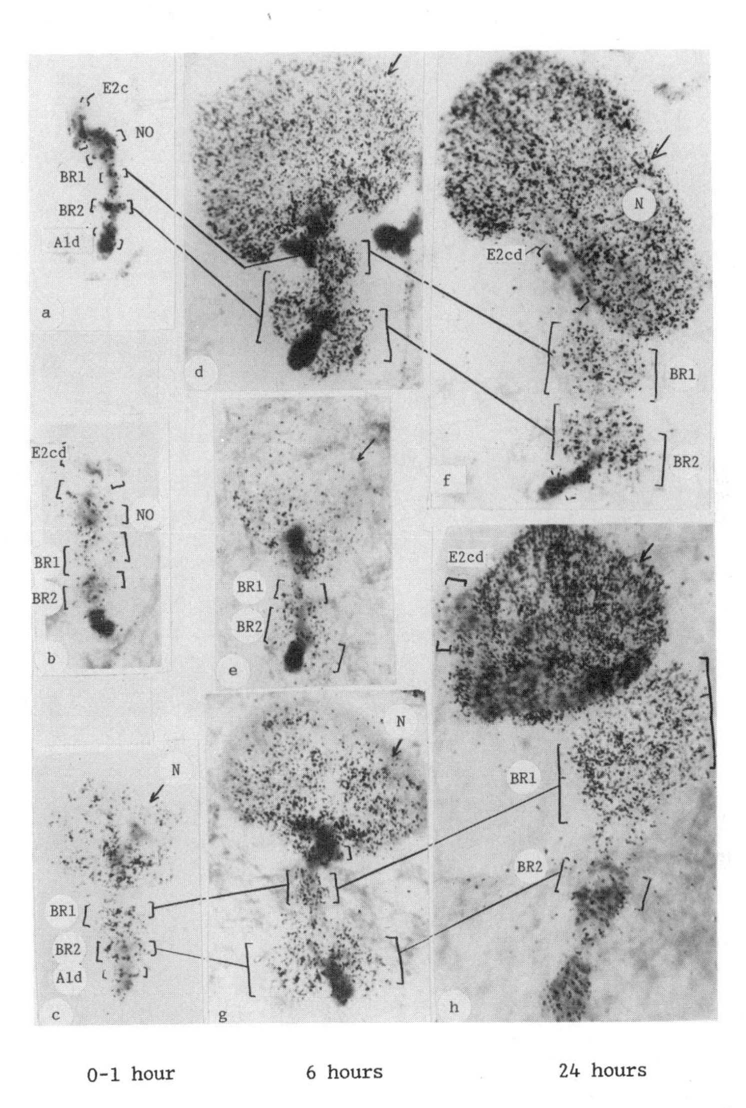

Fig. 6.19. Reversible decrease in transcription activity in Balbiani rings and nucleolus during larval molt. a–c) Decrease in incorporation of ^{3}H-uridine during and immediately after the larval molt; d–f) recovery of ^{3}H-uridine incorporation into Balbiani rings and nucleolus during the redhead stage, 6 h after the molt. The activity of BR1 during this period is still low; g–h) complete recovery of transcriptional activity of both Balbiani rings by the end of the redhead stage.

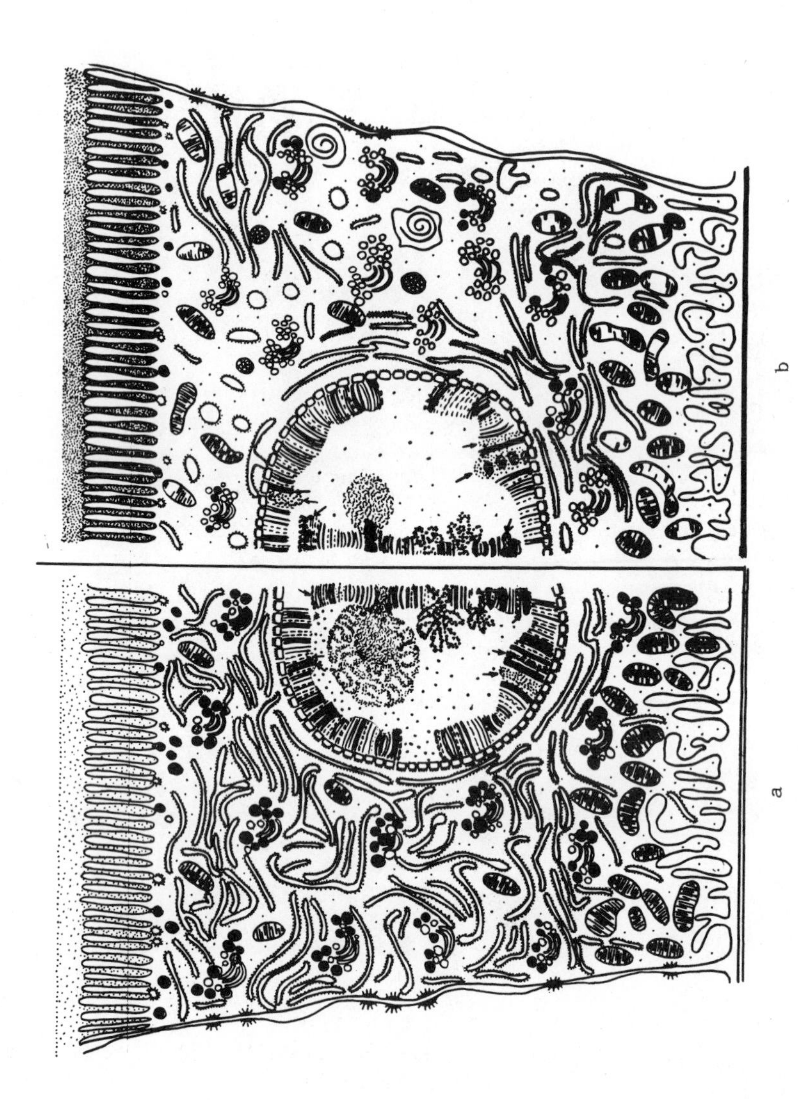

Fig. 6.20. Sketch illustrating ultrastructural changes in a salivary gland cell during larval molt. a) Ultrastructure of cytoplasm during intermolt period; b) during larval molt. For explanation see text.

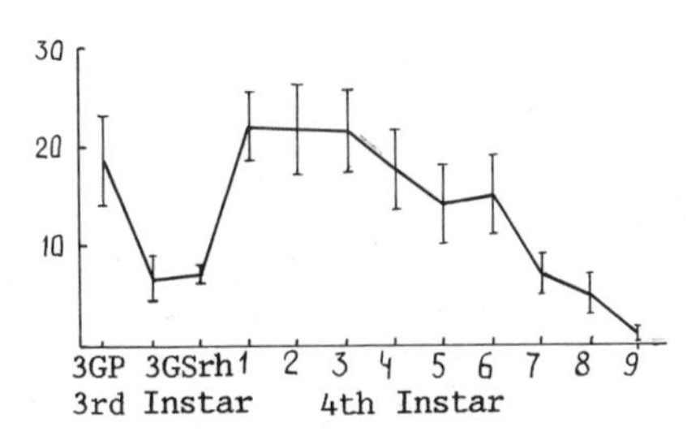

Fig. 6.21. Frequency of ^{3}H-thymidine-labeled nuclei in salivary gland cells of the last larval instars. 3rd Instar: 3GP, 3GS, stages of preparation to molt; 4th instar: 1–9, phases. rh, redhead stage.

ingly, the larvae of different stages can be selected for experiments. The latter is quite important since the redhead stage is suitable for various studies. It is particularly suitable for determining the quantity of DNA in the salivary gland nuclei, since the intensity of DNA replication decreases at this time; this stage is also promising for the detection of factors controlling the initiation of DNA, RNA, and protein synthesis during larval development.

The complex of traits, which has been used in the case of the last larval molt, can also be employed during the molt of the second instar larvae to the third instar larvae. It should be borne in mind, however, that the duration of the redhead stage in this case is only a little more than half the molt stage from the third to the fourth instar.

6.7. PREPUPA AND PUPA

6.7.1. Prepupa

The transition to metamorphosis begins from the prepupa stage (phases 7-9 according to Wülker and Götz [103]), and the state of the endocrine system changes markedly. The epidermal cells are switched over to the synthesis of the pupal cuticle and the imaginal disks grow intensively (see Figs. 6.11, 6.23, and 6.24). Externally, the transition to the prepupal state is expressed as a marked thickening of the larval thoracic segment, gradual contraction of the head, and shortening of the body. The formation of metamorphic puffs begins during this period. The prepupal phase leads to the metamorphosis molt, resulting in the pupa.

6.7.2. Pupa

The pupa differs drastically from the larva in its morphology (see Fig. 6.4). The pupa does not feed, so its wet weight is not much more than half that of the larva [46]. The morphological peculiarities of the pupa are primarily associated with the formation of the cephalothorax following the fusion of the head and three anterior segments. The remaining ten segments form the abdomen and the leg and wing sheaths. Envelopes containing wing rudiments are located laterally on the pupal body and the prothoracic horns are formed on the dorsal side of the anterior segments.

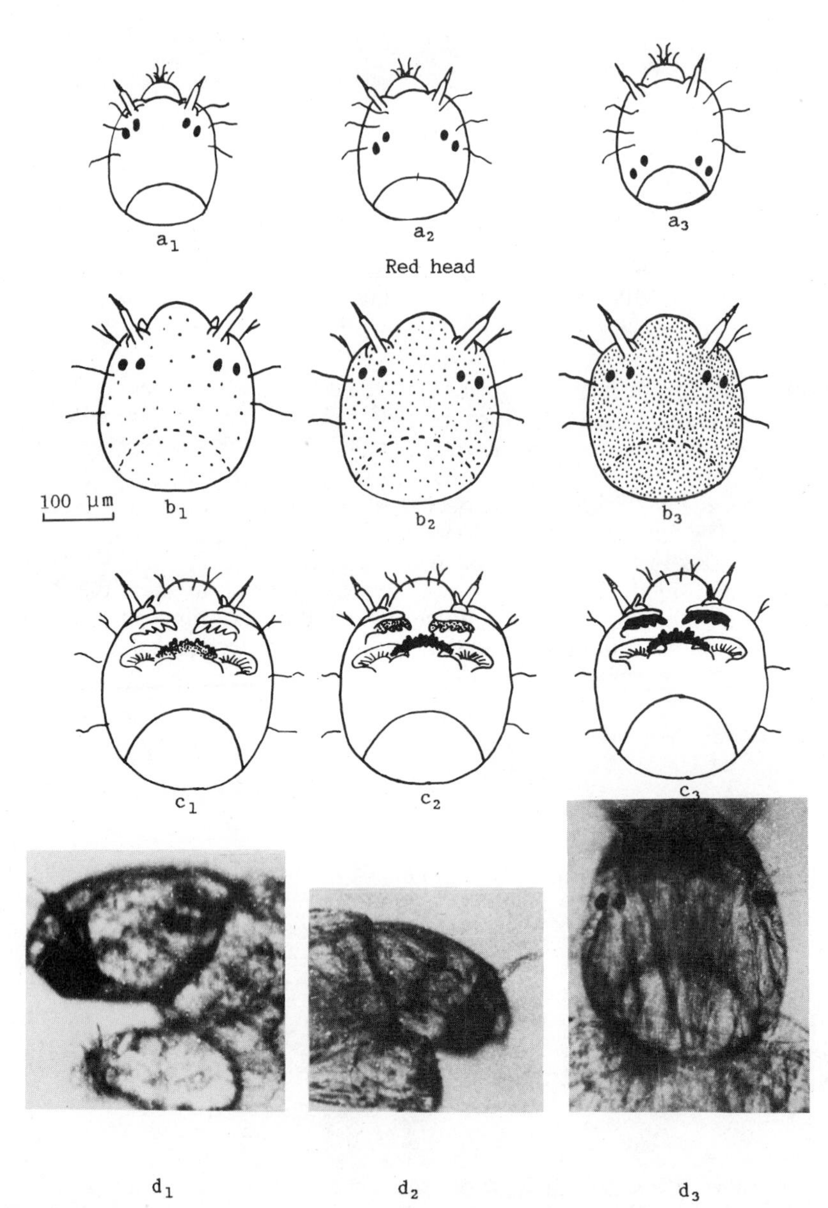

Fig. 6.22. Changes in the position of the eye spots (a_1–a_3) and pigmentation of the head capsule (b_1–b_3) and mouth parts (c_1–c_3) during molt of the 3rd larval instar. a_1) Usual position of the eye spots during the intermolt period up to 36 h prior to molt; a_2) 12–24 h prior to molt; a_3) immediately before molt; b_1) red head color immediately after molt; b_2) light-brown color 8-10 h after molt; b_3) dark brown pigmentation 12 h after molt; c_1) yellow pigmentation of the mentum and absence of mandible pigmentation after molt; c_2) dark brown mentum pigmentation and yellow pigmentation of mandible 8–10 h after molt; c_3) submentum and mandible are dark brown 12 h after molt; d_1–d_3) photographs of the eye spot of living larvae.

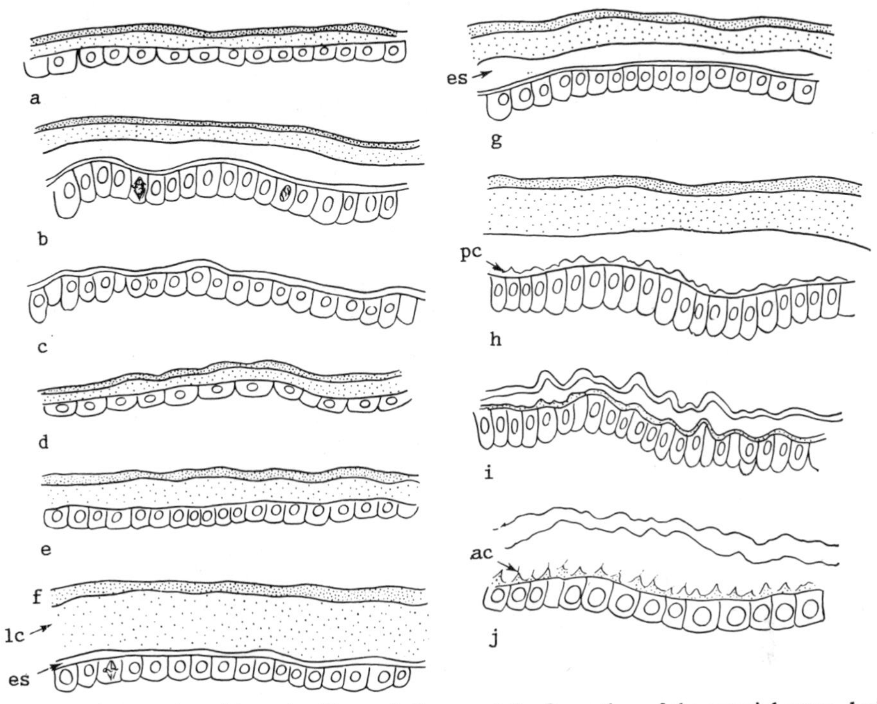

Fig. 6.23. Changes in epidermal cell morphology and the formation of the exuvial space during larval and metamorphic molts. a) Intermolt period of 3rd instar; b) prior to the molt of the 3rd to the 4th larval instar; c) 4th instar immediately after molt (redhead); d) phases 1–2; e) phase 3; f) phase 4; g) phase 6; h) phase 8; i) phase 9; j) pupa. es, exuvial space; lc, larval cuticle; pc, pupal cuticle; ac, adult cuticle.

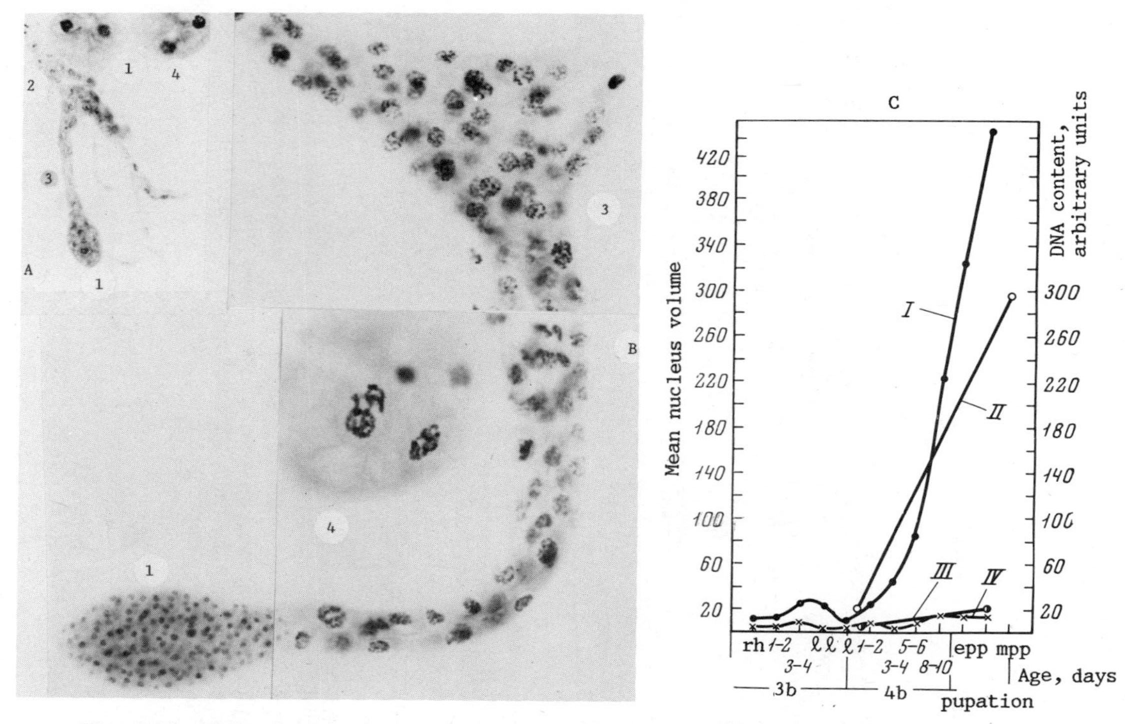

Fig. 6.24. Endocrine glands during larval development and metamorphosis of *Chironomus thummi.* A) 3rd Instar larva, redhead stage; B) late prepupa; C) change in the mean volume of the nuclei of endocrine gland cells. 1) Corpus allatum; 2) corpus cardicum; 3) peritracheal gland; 4) giant cells. I, II) Volume and DNA content of nuclei in peritracheal gland cells, respectively; III, IV) DNA content and nuclear volume of corpus allatum cells, respectively. rh, redhead stage; l, larva; ll, late larva; epp, early prepupa; mpp, middle prepupa; 3b, 3rd instar; 4b, 4th instar.

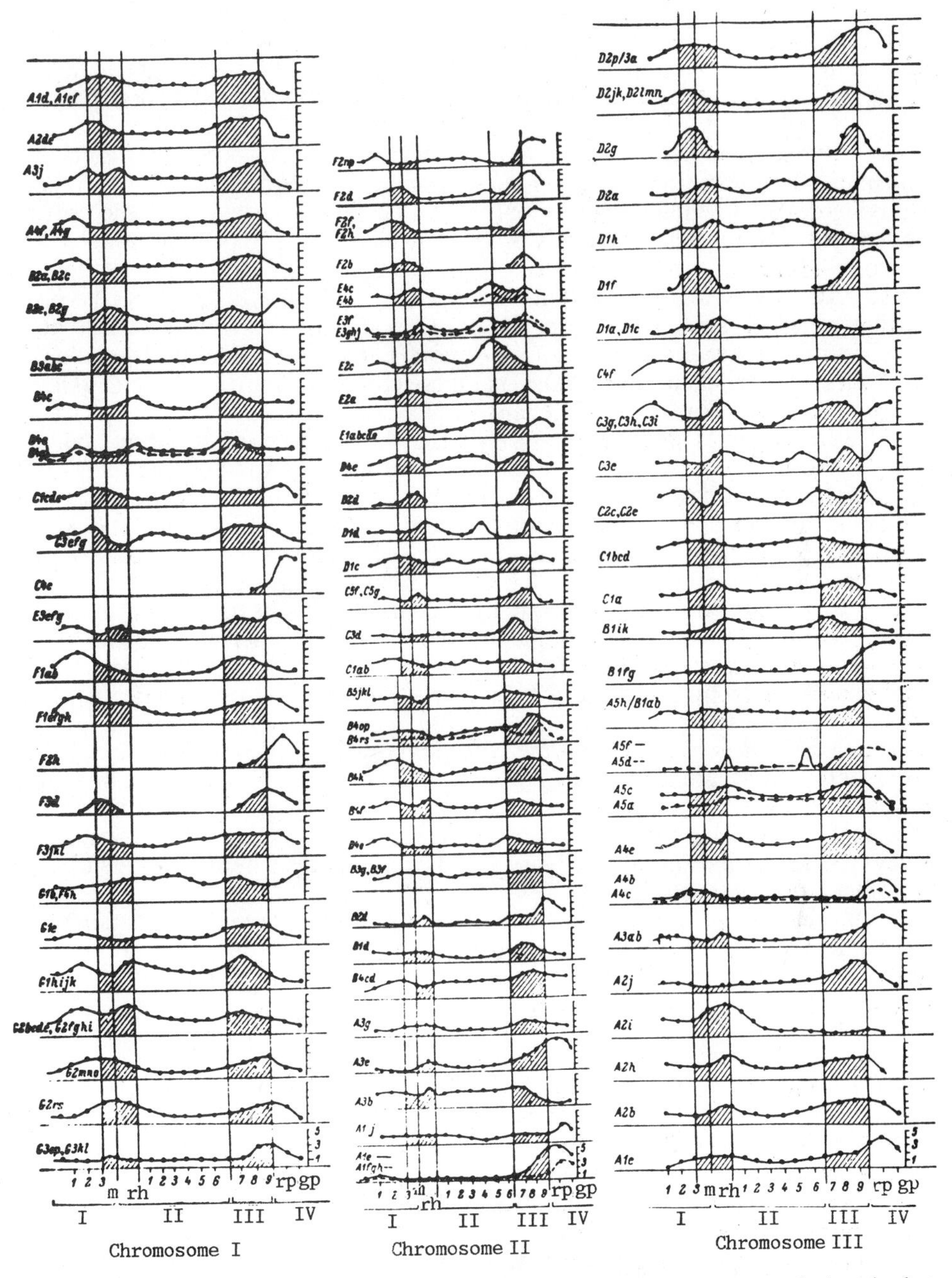

Fig. 6.25. Changes in the puff size of salivary gland polytene chromosomes during the last larval molt, intermolt period, and metamorphosis. I) 3rd Larval instar: 1) middle of 3rd instar; 2) 1–1.5 days before molting; 3) 0.5–1 day before molting; m, molt. II) Intermolt period of 4th larval instar. rh, redhead stage, phases 1–6 of 4th instar. III) Prepupa, phases 7-9. IV) Pupa. rp, red pupa; gp, grey pupa. Left, puff designation; right, scale of puff class.

The pupa stage lasts 2.5-3 days, and its pigmentation gradually changes. Immediately after pupation it is red and then becomes dark grey. In the course of pupal life the larval organs, excluding the nervous system and the Malpighian tubules, undergo histolysis. Imaginal disks of the larval organs produce the corresponding organs of the adult animal.

The presence of such complex morphogenetic transformation during metamorphosis makes this period of dipteran development extremely attractive to those studying developmental cytogenetics. Many studies have dealt with the puffing pattern in polytene chromosomes of several larval organs of chironomids and drosophilids during metamorphosis [4, 6, 8, 10, 11, 32, 109]. The authors of these papers agree that the most distinct changes of the puffing pattern take place during metamorphosis when a special complex of puffs is formed. These puffs are, as a rule, larger and synthesize RNA more actively than other puffs (excluding the Balbiani rings in chironomids). That these conclusions are valid for *C. thummi* can be seen from Figs. 6.18, 6.19, and 6.25.

Of the 336 puffs identified in the polytene chromosomes of *C. thummi* salivary glands, around 40% increase in size during the metamorphosis molt. Metamorphosis is associated with the formation of new puffs. As demonstrated above, the main set of these metamorphosic puffs appears during larval molts, too; they account for about 4% of the total number of puffs studied. They differ from the majority of larval puffs in that they are particularly large (Figs. 6.18, 6.19, and 6.25). Certain larval puffs undergo regression during the metamorphosis molt, but these constitute only about 2% of the total number. A characteristic feature of the changed puffing pattern during metamorphosis molts, just as during larval molts, involves the reduction of tissue-specific puffs and nucleoli. In contrast to larval molts, however, this reduction is irreversible in the metamorphosis molt. All the mentioned changes of puffing during the metamorphosis molt in *C. thummi*, just as in other dipterans, are induced by hormones, primarily ecdysone, although there is evidence that juvenile hormone also plays an important part [51-53, 55, 59-62]. The metamorphosis molt leads to a drastic increase in the activity of lysosomal enzymes and, as a result, the salivary glands undergo lysis in the late pupa just as do other larval organs [25, 56-58, 81-83].

The pupal stage ends in the metamorphosis molt leading to the formation of the adult (see Fig. 6.1). As previously mentioned, adult chironomids have a short lifespan, and hardly ever feed; they are only occasionally used in experimental studies.

REFERENCES

1. D. T. Aderson, "The comparative embryology of the Diptera," *Annu. Rev. Entomol.* **11**, 23–46 (1966).
2. O. A. Agapova and I. I. Kiknadze, "Puffing and tissue-specific function of *Chironomus thummi* salivary gland cells. 3. The cell ultrastructure during larval molt," *Tsitologiya* **21**, 508–513 (1979).
3. M. E. Akam, D. B. Roberts, G. P. Richards, and M. Ashburner, "*Drosophila*: the genetics of two major larval proteins," *Cell* **13**, 215–225 (1978).
4. M. Ashburner, "Function and structure of polytene chromosomes during insect development," *Adv. Insect Physiol.* **7**, 1–95 (1970).

5. S. K. Beckendorf and F. C. Kafatos, "Differentiation in the salivary glands of *D. melanogaster*. Characterization of the glue proteins and their developmental appearance," *Cell* **9**, 365–373 (1976).
6. W. Beermann, "Chromomerenkonstanz und spezifische Modificationen der Chromosomenstruktur in der Entwicklung und Organ-Differenzierung von *Chironomus tentans*," *Chromosoma* **5**, 139–198 (1952).
7. W. Beermann, "Ein Balbiani-Ring als Locus einer Speicheldrusenmutation," *Chromosoma* **12**, 1–25 (1961).
8. W. Beermann, "Riesenchromosomen," *Protoplasmatologia* **6**, 1–161 (1962).
9. W. Beermann, "Chromomeres and genes," *in: Results and Problems in Cell Differentiation*, Vol. 4, Springer-Verlag, Berlin–London–New York (1972).
10. E. S. Belyaeva, L. S. Korochkina, I. F. Zimulev, and N. K. Nazarova, "Puff characteristics in the X-chromosome of *Drosophila melanogaster* females," *Tsitologiya* **16**, 440–446 (1973).
11. H. D. Berendes, "Factors involved in expression of gene activity in polytene chromosomes," *Chromosoma* **24**, 418–437 (1968).
12. U. Clever, "Genaktivitäten in der Riesenchromosomen von *Chironomus tentans* und ihre Beziehungen zur Entwicklung," *Chromosoma* **13**, 385–436 (1962).
13. U. Clever, "Genaktivitäten in der Riesenchromosomen von Chromosomen von *Chironomus tentans*," *Chromosoma* **14**, 651–675 (1963).
14. U. Clever, "Regulation of chromosome function," *Annu. Rev. Genet.* **2**, 11–27 (1968).
15. U. Clever, H. Bultmann, and J. M. Darrow, "The immediacy of genomic control in polytenic cells," in: *RNA in Development*, University of Utah Press (1969).
16. U. Clever and P. Karlson, "Induction von Puff-Veränderung in den Speicheldrüsenchromosomen von *Chrionomus tentans* durch Ecdyson," *Exp. Cell Res.* **20**, 623–626 (1960).
17. G. A. Danielli and E. Rodino, "Larval moulting cycle and DNA synthesis in *Drosophila hydei* salivary gland," *Nature (London)* **213**, 424–425 (1967).
18. J. M. Darrow and U. Clever, "Chromosome activity and cell function in polytenic cells," *Dev. Biol.* **21**, 331–348 (1970).
19. E. H. Davidson, *Gene Activity in Early Development*, Academic Press, New York–London (1969).
20. E. I. Fittkau and F. Reiss, "*Chironomidae,*" in*: Limnofauna Europea*, J. Illies (ed.), G. Fischer, Stuttgart (1978).
21. U. Grossbach, "Chromosomen-Aktivität and biochemische Zelldifferenzierung in den Speicheldrusen von *Camptochironomus*," *Chromosoma* **28**, 136–187 (1969).
22. U. Grossbach, "The salivary gland of *Chironomus*: A model system for the study of cell differentiation," in*: Biochemical Differentiation in Insect Glands*, W. Beerman (ed.), Springer-Verlag, Berlin (1977).
23. L. I. Gunderina and I. I. Kiknadze, "A study of the possibility to synchronize the development of *Chironomus thummi* larvae," *Ontogenez* **13**, 162–168 (1982).
24. H. Hägele, "*Chironomus*," in: *Handbook of Genetics*, Plenum Press, New York–London (1975).
25. P. A. Henrickson and U. Clever, "Protease activity and cell death during metamorphosis in the salivary gland of *Chironomus tentans*," *J. Insect Physiol.* **18**, 1981–2004 (1972).

26. F. Kafatos, "Controlled synthesis of egg shell proteins in insects," *Results and Probl. Cell Differ.* **8**, 44–145 (1977).
27. E. I. Karakin, T. Y. Lerner, V. A. Kokoza, and S. M. Sviridov, "Immunochemical analysis of *Drosophila melanogaster* salivary gland secretion," *Dokl. Akad. Nauk SSSR* **233**, 698–701 (1977).
28. E. I. Karakin, S. M. Sviridov, L. I. Korochkin, and I. I. Kiknadze, "Antigens of *Drosophila melanogaster* larval organs," *Tsitologiya* **19**, 111–119 (1977).
29. E. I. Karakin and S. M. Sviridov, "The phenogenetics of *Drosophila* water-soluble antigens," in: *Biochemical Genetics of Drosophila* [in Russian], Nauka, Novosibirsk (1981).
30. H.-G. Keyl, "Untersuchungen am Karyotypus von *Chironomus thummi*. I. Karte der Speicheldrusen-Chromosomen von *C. thummi* und cytologische Differenzierung der Subspecies *C. thummi thummi* und *C. thummi* Piger," *Chromosoma* **8**, 739–756 (1957).
31. I. I. Kiknadze, "The structural and cytochemical characteristics of chromosomal puffs," in: *G. Mendel Memorial Symposium; Genetic Variations in Somatic Cells*, Academia, Prague (1965).
32. I. I. Kiknadze, *Functional Organization of Chromosomes* [in Russian], Nauka, Moscow–Leningrad (1972).
33. I. I. Kiknadze, "A comparative study of puffing larval development and metamorphosis. I. Puffing patterns in chromosome IV," *Tsitologiya* **18**, 1322–1329 (1976).
34. I. I. Kiknadze, "A comparative study of puffing patterns in *Chironomus thummi* salivary gland chromosomes during larval development and metamorphosis. II. Puffing patterns in chromosomes I, II, III," *Tsitologiya* **20**, 514–521 (1978).
35. I. I. Kiknadze, "The puffing pattern and transcriptional activity of polytene chromosomes," in: *Molecular Basis of Genetic Processes. Proceedings 14th International Congress Genetics, Book 2* (1982).
36. I. I. Kiknadze and T. M. Panova, "Puff heteromorphism in *Chironomus thummi*," *Tsitologiya* **14**, 1084–1090 (1972).
37. I. I. Kiknadze, I. E. Vlasova, and O. E. Lopatin, "DNA replication in salivary chromosomes of *Chironomus thummi* during larval development and its regulation," *Genetics* **74**(2), 139 (1973).
38. I. I. Kiknadze, T. M. Panova, and L. P. Zacharenko, "A comparative study of puffing patterns in *Chironomus thummi* salivary gland chromosomes during larval development and metamorphosis. 3. Transcriptional activity of Balbiani rings and nucleolus," *Tsitologiya* **23**, 523–531 (1981).
39. I. I. Kiknadze, T. M. Panova, and O. A. Agapova, "Repeated regression and recovery of the Balbiani rings and nucleolus in *Chironomus thummi* salivary glands during development," in: *Abstracts of International Symposium on Organization and Expression of Tissue-Specific Genes* [in Russian], Nauka, Novosibirsk (1982).
40. W. A. Kokoza, E. I. Karakin, and S. M. Sviridov, "Genetical and biochemical analysis of the secretory proteins in *Drosophila melanogaster* salivary glands," *Dokl. Akad. Nauk SSSR* **252**, 735–738 (1980).
41. W. A. Kokoza and E. I. Karakin, "Electrophoretical analysis of the secretory proteins from salivary glands in some stocks selected from natural populations of *Drosophila melanogaster*," *Genetika* **17**, 936–940 (1981).

42. N. N. Kolesnikov and I. F. Zhimulev, "Some characteristics of glycoprotein secretion synthesis in *Drosophila melanogaster*," *Ontogenez* **6**, 177–182 (1975).
43. N. N. Kolesnikov, E. I. Karakin, T. E. Sebeleva, L. Meyer, and E. Serfling, "Cell-specific synthesis and glycosylation of secretory proteins in larval salivary glands of *Chironomus thummi*," *Chromosoma* **83**, 661–679 (1981).
44. A. S. Konstantinov, "On the biology and development of *Chironomus dorsalis* Meig.," *Byull. Mosk. Ova. Ispyt. Prir.* **57**, 40–43 (1952).
45. A. S. Konstantinov, "On the taxonomy of genus *Chironomus dorsalis* Meig.," *Tr. Saratov. Otd. VNIORCH* **4**, 155–190 (1956).
46. A. S. Konstantinov, "Biology of the *Chironomus* and their cultivation," *Tr. Saratov. Otd. VNIORCH* **5**, 1–358 (1958).
47. E. P. Kopantzev, E. I. Karakin, T. M. Panova, and I. I. Kiknadze, "The antigens of the secretion of salivary glands from *Chironomus thummi* larvae," *Dokl. Akad. Nauk SSSR* **249**, 1477–1480 (1979).
48. E. P. Kopantzev, E. I. Karakin, and I. I. Kiknadze, "Immunochemical study of water-soluble proteins of *Chironomus thummi* salivary glands," *Ontogenez* **13**, 137–142 (1981).
49. G. Korge, "Chromosome puff activity and protein synthesis in larval salivary glands of *D. melanogaster*," *Proc. Natl. Acad. Sci. USA* **72**, 4550–4554 (1975).
50. G. Korge, "Genetic analysis of the larval secretion gene Sgs–4 and its regulatory chromosome sites in *D. melanogaster*," *Chromosoma* **84**, 373–390 (1981).
51. L. S. Korochkina, I. I. Kiknadze, and S. W. Muradov, "Effects of hormones on puffing pattern in *Chironomus thummi*," *Ontogenez* **3**, 177–185 (1972).
52. H. Kroeger, "Zell-physiologische Mechanismen bei der Regulation von Genaktivitäter," *Chromosoma* **15**, 36–70 (1964).
53. H. Kroeger, "Zur Normalentwicklung von *Chironomus thummi*: Allgemeine Morphologie, Speicheldrüsenentwicklung und puffing Aktivität in Riesenchromosomen," *Z. Morphol. Tierre* **74**, 65–88 (1973).
54. H. Kroeger, W. Trösch, and G. Müller, "Changes in nuclear electrolytes of *Chironomus thummi* salivary gland cells during development," *Exp. Cell Res.* **80**, 329–339 (1973).
55. H. Laufer, "Developmental interactions in the dipteran salivary gland," *Am. Zool. 8*, 257–271 (1968).
56. H. Laufer, G. Wakahase, and J. Vanderberg, "Developmental studies of the dipteran salivary gland. I. Effect of actinomycin D on larval development, enzyme activity, and chromosomal differentiation in *Chironomus thummi*," *Dev. Biol.* **9**, 367–384 (1964).
57. H. Laufer and G. Nakase, "Developmental studies of the dipteran salivary gland. 2. DNAase activity in *Chironomus thummi*," *J. Cell. Biol.* **25**, 96–102 (1965).
58. H. Laufer and G. Nakase, "Salivary gland secretion and its relation to chromosomal puffing in the dipteran *Chironomus thummi*," *Proc. Natl. Acad. Sci. USA* **53**, 511–516 (1965).
59. H. Laufer and T. K. Holt, "Juvenile hormone effects on chromosome puffing and development in *Chironomus thummi*," *J. Exp. Zool.* **173**, 341–352 (1970).
60. H. Laufer and M. Wilson, "Hormonal control of gene activity as revealed by

puffing of salivary gland chromosomes in dipteran larvae in laboratory experiment," in: *A Student's Guide to Laboratory Experiments in General and Comparative Endocrinology*, R. E. Peter and A. Gorbman (eds.), Prentice-Hall, Englewood Cliffs, New Jersey (1970).

61. H. Laufer and J. P. Calvet, "Hormonal effects on chromosomal puffs and insect development," *Gen. Comp. Endocrinol., Suppl.* **3**, 137–148 (1972).
62. M. Lezzi and M. Frigg, "Specific effects of juvenile hormone on chromosome function," *Mitt. Schweiz. Entomol. Ges.* **44**, 163–170 (1971).
63. A. A. Linevitch and E. A. Erbaeva, "On the taxonomy genus *Chironomus* Meig. from reservoirs of Baikal regions," *Izv. Biol.-Geogr. Nauchno-Issled. Inst. Irkutsk. Gos. Univ.* **25**, 127–190 (1971).
64. N. N. Lipina, *Larvae and Pupae of Chironimidae* [in Russian], *Izd. Nauchnogo Instituta Rybnogo Khozyaistva*, Moscow (1929).
65. I. Martin, "A review of the genus *Chironomus*. II. Added descriptions of *Chironomus cloacalis*," *Stud. Nat. Sci. (Portales, N. M.)* **1**, 1–21 (1971).
66. I. Martin and J. E. Sublette, "A review of the genus *Chironomus*. III. *Chironomus yoshimatsui*," *Stud. Nat. Sci. (Portales, N. M.)* **1**, 1–59 (1972).
67. I. Martin and D. T. Porter, "The salivary gland chromosomes of *Glyptotendipes barbipes*: description of inversions and comparison of Nearctic and Palearctic karyotypes," *Stud. Nat. Sci. (Portales, N. M.)* **1**, 1–25 (1973).
68. L. C. Miall and A. R. Hammond, *The Structure and Life History of the Harlequin-Fly (Chironomus)*, Oxford (1900).
69. D. Nash and J. Bell, "Larval age and the pattern of DNA synthesis in polytene chromosomes," *Can. J. Genet. Cytol.* **10**, 82–90 (1968).
70. V. J. A. Novak, *Insect Hormones*, Methuen, London (1966).
71. N. Pasteur and C. D. Kastritsis, "Developmental studies on *Drosophila*," *Experientia* **28**, 215–216 (1972).
72. J. Pause, "Beitrage zur Biologie und Physiologie der Larve von *Chironomus gregarins*," *Zool. Jahrb. Abt. 1* **36**, 339–352 (1918).
73. G. Richards, "Sequential gene activation by ecdysone in polytene chromosomes of *Drosophila melanogaster*. IV. The mid-prepupal period," *Dev. Biol.* **54**, 256–263 (1976).
74. G. Richards, "Sequential gene activation by ecdysone in polytene chromosomes of *Drosophila melanogaster*. V. The late prepupal puffs," *Dev. Biol.* **54**, 264–275 (1976).
75. G. Richards, "Sequential gene activation by ecdysone in polytene chromosomes of *Drosophila melanogaster*. VI. Inhibition by juvenile hormones," *Dev. Biol.* ***66***, 32–42 (1978).
76. T. C. Rodman, "Relationship of developmental stage to initiation in polytene nuclei," *Chromosoma* **23**, 271–287 (1968).
77. D. B. Roberts, "Antigens of developing *Drosophila melanogaster*," *Nature (London)* **233**, 394–397 (1971).
78. L. Rydlander and J.-E. Edström, "Large-sized nascent protein as dominating component during protein synthesis in *Chironomus* salivary glands," *Chromosoma* **81**, 85–99 (1980).
79. A. I. Shilova, "Systematics of genus *Tendipes* Meig. (Diptera, Tendipedidae)," *Entomol. Rev.* **37**, 434–452 (1958).
80. A. I. Shilova, *Chironomidae of the Rybinsk Reservoir* [in Russian], Nauka, Leningrad (1976).

81. K. S. Schin and U. Clever, "Lysosomal and free acid phosphatase in salivary gland of *Chironomus tentans*," *Science* **150**, 1053–1055 (1965).
82. K. S. Schin and U. Clever, "Ultrastructural and cytochemical studies of salivary gland regression in *Chironomus tentans*," *Z. Zellforsch.* **86**, 262–279 (1968).
83. K. Schin and H. Laufer, "Studies of programmed salivary gland regression during larval–pupal transformation in *Chironomus thummi*. 1. Acid hydrolase activity," *Exp. Cell Res.* **82**, 335–340 (1973).
84. T. E. Sebeleva and I. I. Kiknadze, "A cytochemical analysis of secretory glycoproteins of *Chironomus thummi* salivary glands," *Tsitologiya* **19**, 147–154 (1976).
85. T. E. Sebeleva, N. N. Kolesnikov, and I. I. Kiknadze, "A comparative analysis of secretory proteins of the salivary gland cells differing by the number of Balbiani rings," *Dokl. Akad. Nauk SSSR* **256**, 975–978 (1980).
86. J. E. Sublette and M. F. Sublette, "The morphology of *Hyptotendipes barbipes*," *Stud. Nat. Sci. (Portales, N. M.)* **1**, 1–81 (1973).
87. J. E. Sublette and M. F. Sublette, "A review of the genus *Chironomus*. V. The maturus complex," *Stud. Nat. Sci. (Portales, N. M.)* **1**, 1–41 (1974).
88. J. E. Sublette and M. F. Sublette, "A review of the genus *Chironomus*. VI. The morphology of *Chironomus stigmaterus*," *Stud. Nat. Sci. (Portales, N. M.)* **1**, 1–66 (1974).
89. K. Strenzke, "Revision der Gattung *Chromosomus* Meig.," *Arch. Hydrobiol.* **56**, 1–42 (1959).
90. Y. Suzuki and H. Maekawa, "Repeated turnoff and turnon of fibroin gene transcription during silk gland development of *Bombyx mori*," *Dev. Biol.* **78**, 394–406 (1980).
91. A. Thienemann, *Chironomus. Leben, Verbreitung, und wirtschaftliche Bedeutung der Chironomiden*, Die Binnengewässer, 20, Stuttgart (1954).
92. A. Thienemann and K. Strenzke, "Larventyp und Imaginalart bei *Chironomus*," *Entomol. Tidskr.* **72**, 1–21 (1951).
93. A. Tissieres, H. K. Mitchell, and U. M. Tracy, "Protein synthesis in salivary glands of *Drosophila melanogaster*: relation to chromosome puffs," *J. Mol. Biol.* **84**, 389–399 (1974).
94. A. A. Tchernovsky, *A Key to Larvae of Family Tendipedidae* [in Russian], Moscow–Leningrad (1949).
95. M. Valentin, W. E. Bollenbacher, L. I. Gilbert, and H. Kroeger, "Alterations in ecdysone content during the postembryonic development of *Chironomus thummi*: correlations with chromosomal puffing," *Z. Naturforsch.* **33c**, 557–560 (1978).
96. V. Velissarionu and M. Ashburner, "The secretory proteins of the larval salivary gland of *Drosophila melanogaster*. Cytogenetic correlation of a protein and a puff," *Chromosoma* **77**, 13–27 (1980).
97. R. I. D. Wensler and I. G. Rempel, "The morphology of the male and female reproductive system of the midge *Chironomus plumosus* L.," *Can. J. Zool.* **40**, 199–229 (1962).
98. V. B. Wigglesworth, "The physiology of insect metamorphosis," *Cambridge Monogr. Exp. Biol.* **1** (1954).
99. V. B. Wigglesworth, *The Principles of Insect Physiology*, Chapman and Hall, London (1972).

100. I. Whitten, "Metamorphic changes in insects," in: *Metamorphosis*, L. J. Gilbert (ed.), New York (1968).
101. I. E. Vlasova, I. I. Kiknadze, and A. I. Scherudilo, "DNA reduplication pattern during chromosome polytenization in *Chironomus thummi* ontogenesis," *Dokl. Akad. Nauk SSSR* **203**, 459–462 (1971).
102. U. Wobus, E. Serfling,, and R. Panitz, "Salivary gland protein of *Chironomus thummi* strain with an additional Balbiani ring," *Exp. Cell. Res.* **65**, 240–245 (1971).
103. W. Wülker and G. Götz, "Die Verwendung der Imaginalscheiben zur Bestimmung des Entwicklungszustandes von *Chironomus* Larven (Diptera)," *Z. Morphol. Tiere* **62**, 363–388 (1968).
104. W. Wülker, J. Sublette, M. F. Sublette, and J. Martin, "A review of the genus *Chironomus*. I. The *staegeri* group," *Stud. Nat. Sci.* (*Portales, N. M.*) **1**, 1–89 (1971).
105. W. Wülker and I. Martin, "A review of genus *Chironomus*. VI. Cytology of the maturus-complex," *Stud. Nat. Sci. (Portales, N. M.)* **1**, 1–21 (1974).
106. W. Wülker, "Basic patterns in the chromosome evolution of the genus *Chironomus* (Diptera)," *Z. Zool. Syst. Evolutionsforsch.* **18**, 112–123 (1980).
107. W. Wülker, H. M. Ryser, and A. Scholl, "Revision der Gattung *Chironomus* Meigen (Dipt.). VI. *C. holomelas* Keyl, *C. saxatilis* n. sp., *C. melanescens* Keyl," *Rev. Suisse Zool.* **88**, 903–924 (1981).
108. G. R. Wyatt, "Biochemistry of insect metamorphosis," in: *Metamorphosis*, L. Y. Gilbert (ed.), New York (1968).
109. I. F. Zhimulev, "Analysis of the puffing pattern in laboratory Batumi stock of *Drosophila melanogaster*," *Chromosoma* **46**, 59–76 (1974).
110. I. F. Zhimulev and N. N. Kolesnikov, "Synthesis and secretion of a mucoprotein glue in the salivary gland of *Drosophila melanogaster*," *Wilhelm Roux's Arch. Dev. Biol.* **178**, 15–28 (1975).

Chapter 7

THE FRUIT FLY *Drosophila*

E. V. Poluektova, V. G. Mitrofanov, G. M. Burychenko, E. N. Myasnyankina, and E. D. Bakulina

7.1. INTRODUCTION

The fact that the genetics and biology of *Drosophila* have been studied in great detail ([5, 6, 11, 21, 25, 40-42, 55], and periodic publication of *Drosophila Information Service*) and that many mutations affecting various developmental stages are known, makes *Drosophila* one of the most suitable models for analyzing the way in which genetic information is expressed in the course of development. Embryos and larvae develop rapidly and are readily available for embryological experiments [17, 33, 37, 47, 53, 73, 74, 88]. Imaginal disks undergoing differentiation into the organs and tissues of the adult provide a unique basis for studying problems of determination and differentiation [27, 28, 32, 36, 66].

7.2. TAXONOMY

The genus *Drosophila* contains over 1200 species and belongs to the family Drosophilidae, order Diptera. In accordance with Sturtevant's classification [85], this genus has been divided into six subgenera: *Hirtodrosophila, Pholodoris, Drosilopha, Phloridosa, Sophophora,* and *Drosophila.* Subsequently, Patterson and Mailand [68] and Wheeler [89] identified two other subgenera – *Siphlodora* and *Sordophila.* Each of the subgenera includes several groups of species, which in turn are divided into subgroups [13, 69] consisting of at least one species.

As a rule, developmental studies make use of *Drosophila melanogaster*, but *Drosophila virilis* and other species belonging to the *virilis* group are also being used.

7.3. DISTRIBUTION AND COLLECTION

The distribution of the genus *Drosophila* is extremely wide. With the exception of the extreme Arctic regions, species belonging to this genus can be found all over the world. Most *Drosophila* species live in the tropics, but around 200 species have been listed among the Palaearctic fauna. In nature *Drosophila* is usually found in wet, shaded places. Certain species are synanthropic and can be found in dwelling places and vegetable and fruit warehouses.

The food requirements of *Drosophila*, both at the larval and the adult stages, are closely connected with certain species of yeasts and fungi. Even members of one species group may differ in their capacity to use different yeasts. Adult *Drosophila* are best caught in the summer using special traps, which generally consist of a 0.5-1.0 liter can containing 150-200 ml of the standard medium (see p. 181). The upper part of the can is covered with thick paper with a small aperture for flies. In nature such traps are best placed near water reservoirs in shaded places. A small suction trap may be used to catch flies in food factories or warehouses. After the flies are caught, they can live for 2-3 days under conditions of total starvation and in the absence of moisture, and up to 4 days in the presence of water [43].

7.4. MANIPULATION AND CULTURE

The methods of keeping and manipulating *Drosophila* under laboratory conditions are relatively easy [22, 23, 60, 73, 75, 79, 81, 83, 84, 90].

7.4.1. Anesthetization

All manipulations with flies are performed under anesthesia, usually with diethyl ether. A small quantity of ether is applied on cotton wool in a specimen jar which consists of a glass beaker closed with a cork [60]. A funnel with a glass capsule on the end is inserted into the cork; this capsule has holes allowing ether vapor to enter. By tapping the side of the vial with fingers, the flies are shaken down to the capsule and kept there until they cease to move. Thereafter the flies are shaken out onto a white ceramic plate or paper sheet, and analyzed. Anesthesia

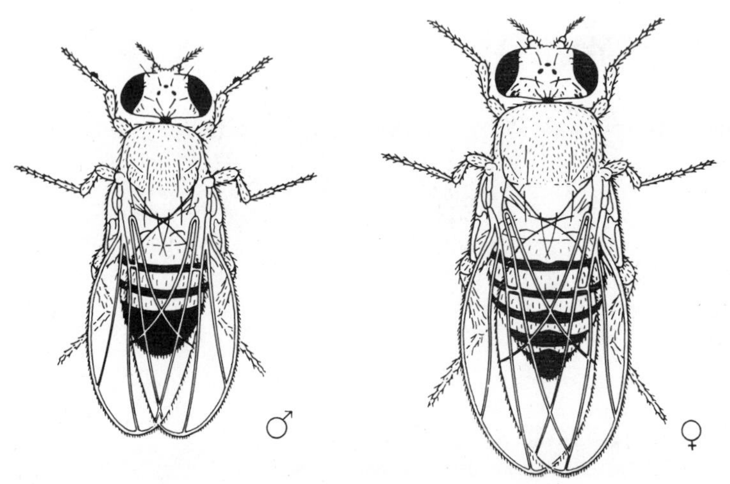

Fig. 7.1. Sexual dimorphism in *Drosophila*.

should be performed carefully since large doses of ether can kill the flies. In this case they have a characteristic strained posture, with extended legs and wings protruding above and laterally. Ether cannot always be used for anesthesia because it modifies many physiological processes; in particular, it delays oviposition. Carbon dioxide can be used as an alternative [23, 73, 74, 87]: here the gas is passed through a Buchner funnel, the lower end of which is attached through tubing and a pressure reducer to a pressure container with CO_2. The specimen jar is connected to the funnel.

7.4.2. Inspection of Flies

Anesthetized flies are inspected using a low-power dissecting microscope. Bird feathers may be used for examining and counting the flies. For a detailed examination, individual flies may be handled with fine forceps. If the flies start moving during examination, they can be anesthetized once more. Flies not required for further work are discarded into a container containing liquid such as transformer oil, alcohol, or aqueous detergent solution, in which they drown rapidly.

7.4.3. Nutrient Medium

Yeasts comprise the major component of the nutrient medium for *Drosophila*. As well as yeasts, the cultivation media contain other substrates such as sugar, raisins, and treacle. Unrefined and dark treacle is preferable. Agar is used to solidify the medium. Many recipes of nutrient media have been described [22, 23, 46, 60, 73, 74, 81, 83]. A recipe which proved quite suitable in our laboratory is as follows:

Bakers yeast	60 g (dry yeasts 18 g)
Raisins	40 g
Agar	6 g
Semolina	36 g
Treacle	50 ml
Propionic acid	0.8 ml
Tap water	1 liter

The medium is boiled for 1.5 h, then cooled to 60°C, propionic acid is added,* and the mixture distributed into vials. The solidified medium is inoculated with live yeasts using a pulverizer containing the yeast suspension. The vials are dried under a fan and then covered with cotton-wool plugs. Larvae develop quite well on this medium, but 5-6 days after mating crosses have been made, tampons made of filter paper have to be placed into the vials, because larvae liquify the medium and the newly-hatched flies may drown. This procedure is not necessary if the agar content is increased to 10-12 g/liter. Flies, however, develop far less well on a more solid medium, and their fertility decreases.

---

*Propionic acid (or nipagin) is used for preventing mould from growing on the surface of the nutrient medium. It should be borne in mind that vials for media distribution must be sterilized at a temperature of at least 180°C.

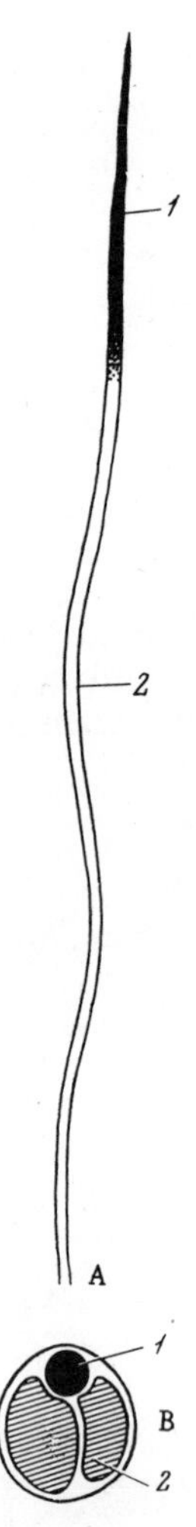

Fig. 7.2. Structure of the *Drosophila* spermatozoon. A. 1) Head (nucleus); 2) tail. B. Cross section through the tail. 1) Flagellum; 2) paracrystalline bodies.

7.4.4. Temperature

Under laboratory conditions *Drosophila* is usually cultured at 22-25°C [20, 60]. At this temperature the duration of *D. melanogaster* development is about 10 days, and the larval period lasts 4 days. At 10°C the duration of development extends to 57 days, and the development is still normal. At temperatures above 30°C the flies become partially, and even completely, sterile. Retardation of development at lower temperatures is employed for keeping the stock. It is usually maintained at 18-19°C, allowing the flies to be transferred onto a fresh nutrient medium at less frequent intervals.

7.5. SEXUAL DIMORPHISM AND CULTURING

Sexual dimorphism in *D. melanogaster* is obvious, males having smaller bodies than females. The male abdomen is cylindrical with a blunt end, while the female abdomen is more rounded with a sharper tip. Several of the most posterior abdominal tergites in males are pigmented and form the black spot which is visible to the naked eye, making them easily distinguishable from the females (Fig. 7.1). Moreover, the female abdomen has eight segments, while the abdomen of a male has only six. The distal end of the first segment of the male forelegs bears the sex comb, which consists of a number of large chitin bristles. These are absent in females.

Depending on the purpose of the experiment, *Drosophila* may be cultured either *en masse* or individually. Mass cultivation is conducted in glass vials 25 × 100 mm in size. Individual crosses are best performed in smaller vials (17 × 80 mm).

When making crosses and keeping special stocks, virgin females should be used because viable sperm can survive in the females' spermathecas for several days after mating. Virgin females can be selected as follows: all flies are removed from the cultures and then, after 6 h during which new flies are hatched, females are selected. Since the first mobile spermatozoa in *D. melanogaster* males do not appear until 7 ± 1 h after hatching [49], the virginity of the females selected during this interval is ensured.

Anesthetized flies may stick to the fresh medium and die, so after inspection they should be placed onto the sides of the vials. Later, when they begin moving, the vials may be placed vertically. During the mass cultivation of flies the population density in cultures should be controlled, since too high a density inhibits their development (lifespan decreases, body size decreases). In order to avoid overpopulation of the culture, no more than 4-5 pairs of flies should be placed into each vial.

7.6. GAMETOGENESIS

Gametes originate from a small number of pole cells, which segregate at the posterior embryo pole during the ninth cleavage division and prior to blastoderm formation [80]. The cytoplasm of these cells contains RNA-rich polar granules [18, 59]. In the course of development, the pole cells form a cap consisting of 24-48 cells which occupy the place between the vitelline membrane and the blastoderm. However, not all pole cells produce gametes; these are derived from only 2-12 of the original population of pole cells.

7.6.1. Spermatogenesis

The most complete description of the development of *D. melanogaster* male gametes is given in Cooper's review [16]. The results of Cooper's investigations are summarized in Lindsley and Tokuyasu's review [56]. All male gametes originate from a group of stem cells located in the apical part of the testes; these periodically form primary spermatogonia. Immediately after separating from a stem cell, each spermatogonium is surrounded by two somatic cells, and a cyst forms inside this aggregate. The spermatogonium undergoes four mitotic divisions and produces 16 primary spermatocytes. They enter the growth phase, which lasts about

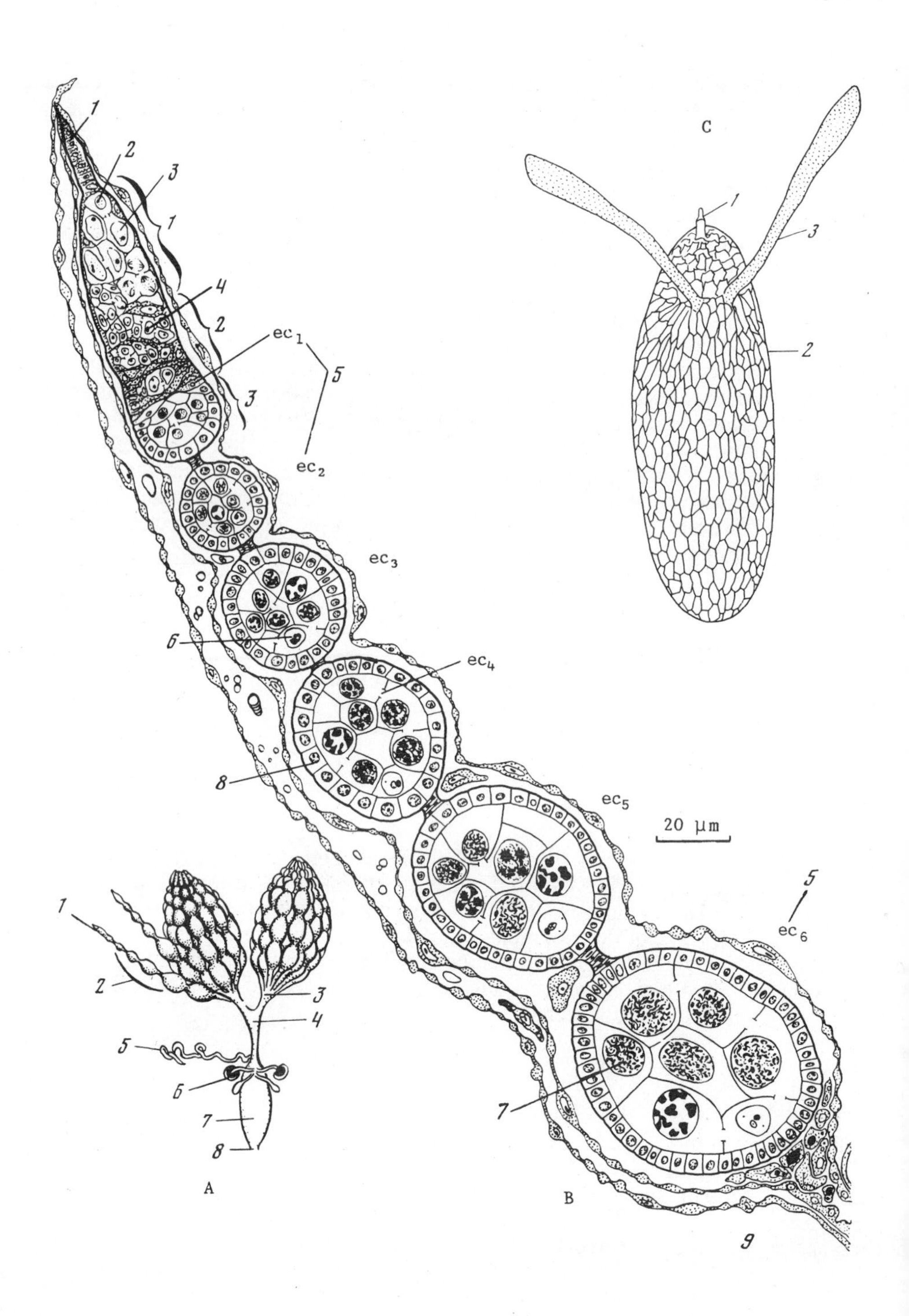
1
2
3
1
4
2
ec1
5
3
ec2
ec3
6
ec4
8
ec5
20 μm
5
ec6
7
9
C
1
3
2
1
2
3
4
5
6
7
8
A
B

90 h. During this period, the cell volume increases approximately 25-fold and the cyst migrates to the middle part of the testis. After two meiotic divisions, the cyst, which now consists of 64 spermatids, enters the stage of spermiogenesis. This process, which lasts more than 130 h and is accompanied by numerous morphological and biochemical changes, has been described in detail by Hess and Meyer [38]. Spermatids undergo elongation during spermiogenesis, and the head and tail of the future gamete is formed. Chromatin condenses, the nuclear volume decreases, and the nucleus changes shape. The bundle of spermatids undergoes extension in the testis lumen along almost its entire length, while the spermatid heads are tightly fixed by their anterior ends in the base epithelium of the testis through one of the two cells forming the cyst envelope. Prior to the completion of spermiogenesis, the cells lose most of the cytoplasm and cell organelles. The bundle of spermatozoa moves from the testis lumen to its base and, as a result of twisting, the tails form a dense hexagonal package. After maturation has ended, the anterior cell of the envelope holding the gametes in a fixed position ruptures and the released spermatozoa pass into the seminal vesicle, probably as a result of their own active movement. Gametogenesis is repeated throughout the lifespan of the male; therefore its testes always contain cysts originating from exactly the same stem cell; the cysts can be found at consecutive developmental stages. Spermatogenesis lasts a total of about 250 h.

7.6.2. Mature Spermatozoon

The sperm head consists of the acrosome and nucleus surrounded by a thin cytoplasmic layer. The flagellum extends from the level of the nucleus almost to the end of the tail, and two paracrystalline bodies are located parallel to the long axis of the flagellum (Fig. 7.2). According to Cooper [16], the length of the sperm head is 1.76 μm and its width is 0.1-0.2 μm.

It has already been mentioned that males produce sperm throughout their life. However, their fertility decreases with age, and by the end of adult life the males may become sterile [14].

7.6.3. Oogenesis

The ovaries of newly hatched larvae contain 8-12 oogonia. In the course of larval development, the germ and mesodermal cells divide and the volume of the ovaries increases approximately 50-fold. The ovaries at the adult stage consist of ovarioles (Fig. 7.3). The number of ovarioles varies considerably both for a given species and for different species. Each ovariole is a tube and consists of a series of

Fig. 7.3. *Drosophila melanogaster*, genital system of a mature female [51]. A. General structure. 1) Germarium; 2) vitellarium; 3) lateral oviduct; 4) mature oviduct; 5) seminal receptacle; 7) uterus (distended if containing mature eggs); 8) vagina. B. Sketch of an individual ovariole. 1) Anterior part; 2) stem cell; 3) cystoblast; 4) cystocyte; 5) egg chambers (ec) containing a complex of nurse cells and the oocyte; 6) oocyte nucleus; 7) nucleus of nurse cell; 8) follicular cells; 9) posterior part. C. External view of the mature *D. melanogaster* egg. 1) Micropyle; 2) chorion; 3) filaments (the number varies with species).

connected egg chambers which are located one behind another. One ovariole usually contains six egg chambers. Ovarioles begin to form at the pupal stage, approximately 24 h after puparium formation [51]. Each ovariole has an anterior part, the so-called germarium, and a posterior part, the vitellarium.

The germarium, which contains roughly 200 cells, includes two to three stem oogonia; these divide mitotically and produce two daughter cells, one cell remaining a stem cell while the other forms a cystoblast [51, 90]. The cystoblast enters mitosis and produces cells called cystocytes. Each cystocyte passes through four consecutive divisions. The cystocytes of four generations differ in the number of interconnected daughter cells. The fourth and final division yields a generation consisting of 16 cells (1 oocyte and 15 trophocytes or nurse cells). The formation of four generations of cystocytes and their differentiation into oocytes and trophocytes begins at the pupal stage and continues almost throughout the whole adult life. Oocytes differ morphologically from trophocytes in that the synaptonemal complex is present. Each oocyte and its 15 trophocytes become surrounded by a layer of follicle cells, which segregate from the germarium and form the first egg chamber, which belongs to the vitellarium, the second part of the ovariole. The vitellarium consists of several egg chambers, each containing a developing oocyte and 15 trophocytes. King [51] described fourteen consecutive stages of egg chamber development in *Drosophila* .

The cytoplasmic content of the oocyte increases during oogenesis 90,000 times. Within 1 h after eclosion the ovaries of adult *D. melanogaster* contain egg chambers of stages ranging from 1 to 7. The duration of each stage differs [51]. The stem oogonium divides every 12 h. Cystoblast progression to stage 1 egg chamber lasts approximately 5 days and it takes approximately 3 days to pass from stage 1 to stage 14 egg chamber [51, 52].

Since the oocyte nucleus is extremely minute, its cytological investigation under a light microscope is difficult. The nuclear morphology of the pre-oocyte and oocyte has, however, been studied using electron microscopy [51]. The synaptonemal complex appears in the pre-oocyte nucleus soon after the division of third-generation cystocytes (corresponding to the zygotene stage of meiotic prophase). The zygotene stage continues throughout the last generation of oocytes and includes stages 1 and 2 of the egg chamber development. At stages 3 and 4 the synaptonemal complex attains maximal development (this corresponds to the pachytene stage). The synaptonemal complex degenerates by stage 5 of egg-chamber development, and disappears by stage 7. From this time onward, up to stage 13, the oocyte nucleus is at the diplotene stage of meiotic prophase.

7.6.4. The Egg

The *Drosophila* egg is centrolecithal. Its cytoplasm contains abundant vacuoles and is full of yolk granules. The nucleus has an oval shape, somewhat flattened on the dorsal side and more convex on the ventral side. The egg is surrounded by two envelopes. The external chorion is thick and white; it is impermeable with a characteristic surface polygonal pattern rather like a honeycomb (see Fig. 7.3C). The inner envelope in direct contact with the egg is the vitelline membrane. It is thin and transparent. The filaments, twin chorion protrusions located on the dorsal side of the egg closer to its anterior pole, enable one to distinguish easily between the dorsal and ventral side of the egg as well as between the anterior and posterior pole. A small cone-shaped protuberance located at the anterior pole of the egg contains the

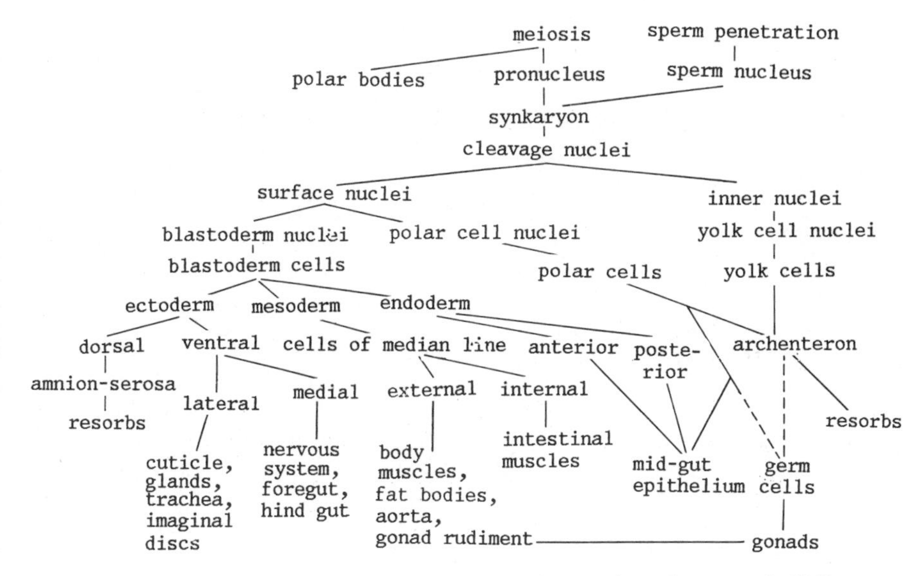

Fig. 7.4. *Drosophila melanogaster*, cell lineages in embryogenesis [71].

micropyle, through which spermatozoon penetrates the egg. Egg size is variable. The mean length and diameter of the *D. melanogaster* egg are 420 μm and 150 μm, respectively [80].

7.7. FERTILIZATION

Monospermic fertilization is the rule in *Drosophila* [39]. Spermatozoa enter the egg at the stage of meiotic division I as it passes through the uterus, where only one egg may be present at any given moment [65]. The female and male pronuclei fuse after oviposition, which occurs no later than 15 min after fertilization. During the telophase of the first meiotic division, the head of the sperm approaches the female nucleus. The second meiotic division takes place immediately after the first, yielding four haploid nuclei. Of these four, the nucleus furthest away from the surface becomes a female pronucleus, while the remaining three form the nuclei of polar bodies which are then degraded.

7.7.1. Oviposition

Females usually deposit fertilized eggs after the completion of maturation divisions, at the stage of female pronucleus formation. However, if for some reason the fertilized egg is retained for some time in the female genital tract, it still continues to develop. In this case, by the time of oviposition the embryo may be at any developmental stage up to completion of larval differentiation. As the duration of embryonic stages is measured from the moment of oviposition, the eggs must be collected from those females which perform oviposition quickly.

Females lay eggs sequentially at intervals of about 3 min [72]. *D. melanogaster* females aged between 4 and 7 days have the best fecundity. During this period a female may lay 50-70 eggs in 1 day. In other species the period of maximal oviposition intensity also coincides with the early adult life. For egg collection it is preferable that flies from cultures containing no more than 100 flies per 25 × 100 mm vial be used. Oviposition is intensive if females are kept on a moist medium containing fermentation products. To stimulate oviposition, the surface of the medium may be supplemented with a paste, which is prepared by mixing nutrient medium (100 g) with bakers yeast (1 g) and left to ferment for 48 h at room temperature. If such a paste is not available, one may use a thick suspension of yeasts in water supplemented with a few drops of ethyl alcohol and dilute acetic acid [23].

7.7.2. Collecting Eggs

Methods of collecting eggs have been described by several authors [23, 39, 72, 75, 80, 86]. When precise timing of egg laying is important, for example for embryological studies, egg collection is performed as follows. Virgin females and males are collected individually from mass cultures and kept separately for 4-5 days in vials containing fresh medium. Then the males and virgin females are transferred into a tube containing a piece of moist filter paper with a thin layer of a nutrient medium, on which they mate. Thereafter the flies are anesthetized and the males removed since they disturb sequential oviposition. Fertilized females are then placed into tubes containing a microscope slide with a strip of dark filter paper. A thin layer of the medium is applied to the filter paper; another possibility is to use microscope slides coated with a thin agar layer [72, 80, 86].

To study early developmental stages, the eggs should be collected every 15 min. For experiments at later developmental stages they may be collected every 1-2 h.

Large-scale egg collection over longer periods is conveniently performed using glass vessels (0.75-1.0 liter) into which 200-300 pairs of flies are placed. Petri dishes 5-6 cm in diameter, containing nutrient medium, are placed inside these vessels. If a large number of eggs is to be collected in a short time, special containers may prove useful [45]. For certain purposes dark medium is best, since eggs are more visible against this background. To achieve this, either treacle, grape juice, or charcoal is added to the medium before it is distributed into jars. The Petri dishes are then withdrawn and the eggs are collected from the medium surface using a needle under a low-power binocular microscope.

The eggs should not dry up during manipulations. Moisture may be retained by placing the eggs into a drop of water or Ringer solution, or onto moist filter paper. Eggs develop normally even if completely submerged in water [72]. To ensure that fixative penetrates the eggs rapidly and evenly, both the chorion and the vitelline membrane should be punctured.

7.8. NORMAL DEVELOPMENT

The life cycle of *Drosophila* consists of the following developmental stages: embryonic, larval, pupal, and adult; information about the development of this fly may be found in a number of reviews [1-3, 7, 8, 10, 16, 17, 21, 23, 25, 27, 38, 47, 51, 62, 67, 73].

TABLE 7.1. Early Developmental Stages of *D. melanogaster* at 24°C [72] (a highly inbred wild-type Florida stock)

Stage	Time after oviposition (min)	Diagnostic features
1	0-15 ± 1.21	Telophase of the second maturation division and fusion of pronuclei. Cleavage division:
2	23 ± 1.72	I
3	34 ± 1.72	II
4	47 ± 1.47	III
5	53 ± 1.05	IV
6	60 ± 0.57	V
7	70 ± 0.58	VI
8	78 ± 1.28	VII
9	93 ± 1.13	VIII
		Division of the blastema nuclei
10	99 ± 0.60	I
11	109 ± 0.51	II
12	120 ± 0.90	III

7.8.1. Embryonic Development

Drosophila embryogenesis has been investigated in some detail [17, 71, 72, 80, 91]. According to Rabinowitz [72], the early period of embryonic development may be divided into 12 stages (Table 7.1).

Stages 1-8. After the pronuclei have fused, the nuclei undergo rapid synchronous divisions in the absence of cell membrane formation. The nuclei divide every 9.5 min. The longest mitotic stage is the prophase, which lasts 4 min. The interphase continues for 3.4 min; the duration of other mitotic stages in minutes is as follows: metaphase 0.3, anaphase 1.0, and telophase 0.9 [72]. Thus the embryo of *Drosophila* at the early developmental stages is a multinuclear syncytium.

Stages 10-12. After the eighth synchronous division most of the nuclei migrate to the surface of the egg. During this developmental period three groups of nuclei may be distinguished in the embryo: blastodermal nuclei, which are located on the surface of the embryo; the nuclei of yolk cells, located in the central part of the embryo; and the nuclei of the pole cells. These three groups of nuclei produce all the cell lines which differentiate into tissues and organs of the larva as well as into imaginal disks. The fate of these cell lines in a *D. melanogaster* embryo, according to Poulson [71], is shown in Fig. 7.4.

The determination and segregation of presumptive cells producing imaginal disks apparently takes place during blastoderm formation, approximately 2.5 h after

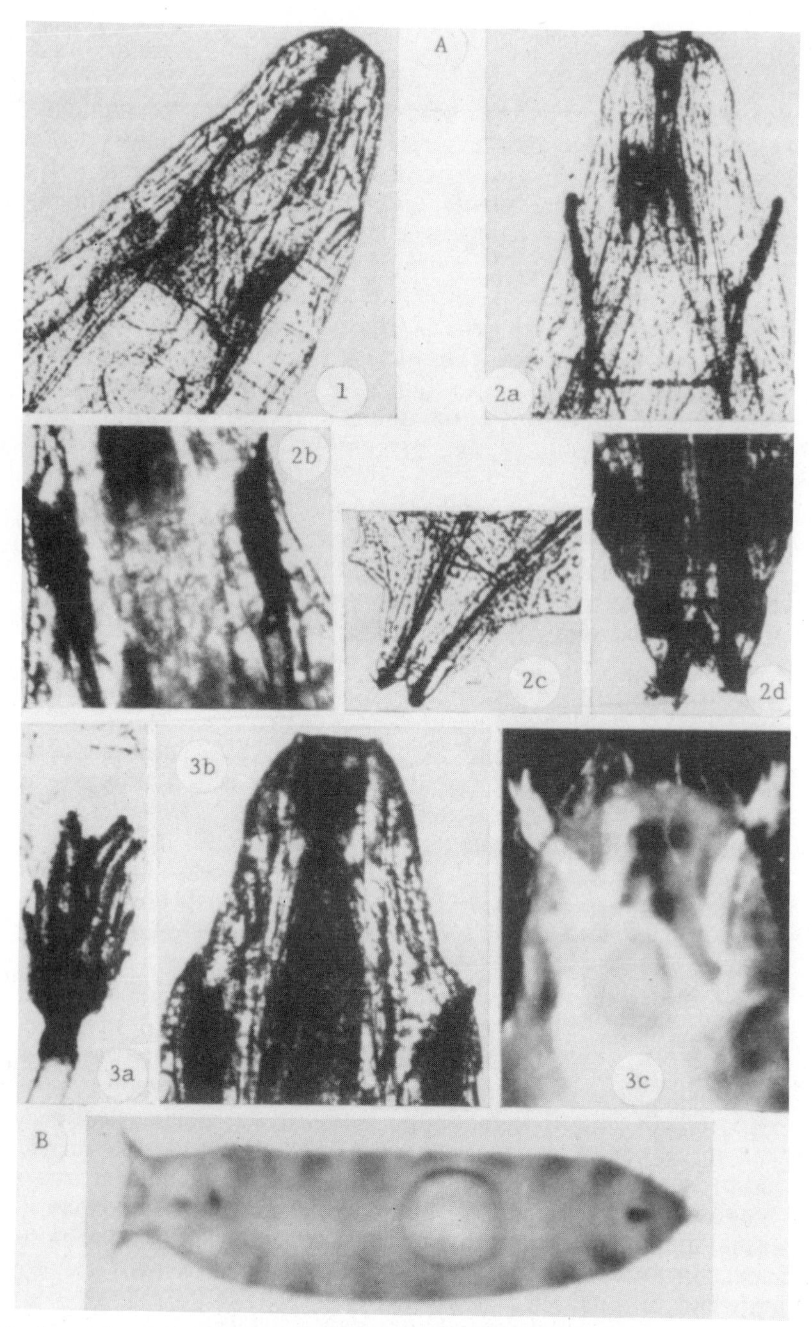

Fig. 7.5. Some diagnostic features of larvae of different ages in *Drosophila*. A. Structure of anterior and posterior spiracles. 1) 1st Larval instar; 2) 2nd larval instar; anterior spiracles before molt (a), immediately before molt (b); posterior spiracle before molt (c), immediately before molt (d); 3) 3rd larval instar: excised anterior spiracle (a), anterior spiracle *in situ* (b), in the late 3rd larval instar (c). B. Prepupa (gas bubble in body cavity), ventral view.

TABLE 7.2. Stages of *D. melanogaster* Oregon-R Embryonic Development at 23-25°C (from [80])

Stage	Time after oviposition (hours)	Characteristic features
1	2	Syncytial blastoderm. Completion of synchronous divisions
2	2.5	Formation of cell walls around blastodermal nuclei. Formation of cleavage furrows
3 4	3 3.5	Cellular blastoderm pregastrulation cell movements. Early gastrula. Ventral groove; plate thickened under polar cells
5	3.75	Head groove present, ventral groove becomes deeper. Development of the gonad rudiment begins
6	4	Invagination of the anterior and posterior rudiments of the midgut. Enlargement of the gonad rudiment. Formation of the embryonic envelopes
7	4.5	Dorsal folds. Mesoderm flattens as the gonad rudiment increases
8	5	Enlargement of the gonad rudiment completed. Single large neuroblasts; yolk mass enclosed in the primary gut
9	5.5	Invagination of the anterior and posterior gut begins. Mitoses in neuroblasts
10	6	Segmentation of the mesoderm rudiments of the anterior and posterior gut becomes deeper
11	7	Tracheal invaginations. Salivary gland plate appears.
12	8	Segmentation of the head and body, attachment of muscles
13	9	Beginning of the embryo shortening. Salivary glands go deeper inside
14	10	Head folding. Dorsal closure
15	11	Compact gonad rudiments
16	12	Primary gut looks like a sac; it is not compressed. Deep anterior sac
17	13	Midgut contraction. Beginning of chitinization
18	14	Ventral nervous system becomes more dense; first muscular contractions
	16	Nervous system continues to become more dense; regular muscular contractions
	18	Larval differentiation almost completed. Active movements
19	20	First air in tracheae; active movements continue
20	22-24	Larva hatches from egg membranes

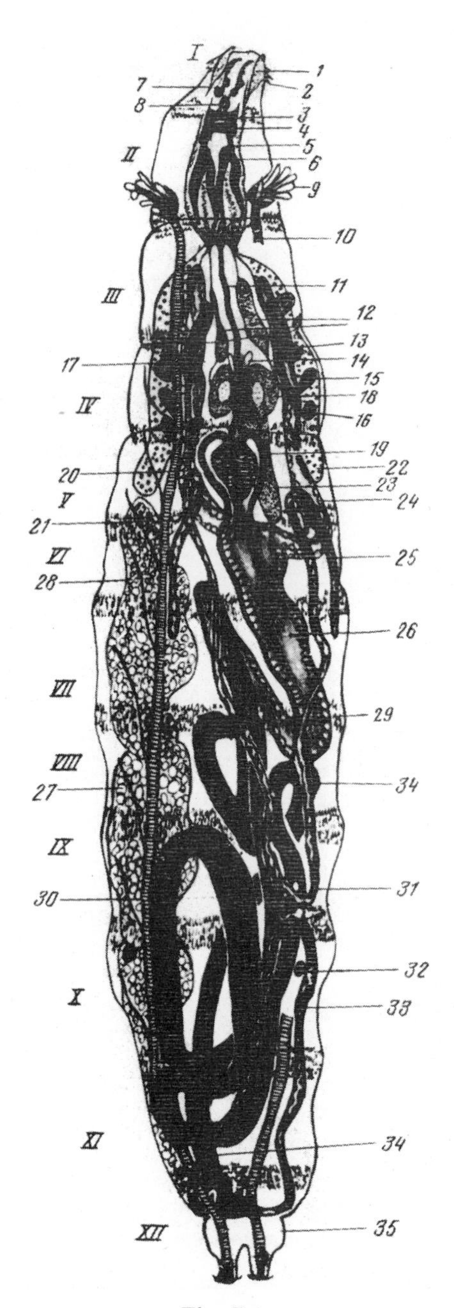
I
1
2
7
8
3
4
5
6
II
9
10
III
11
12
13
14
15
18
16
17
IV
19
22
23
24
20
V
21
VI
25
28
26
VII
29
VIII
34
27
IX
31
30
32
X
33
XI
34
XII
35

Fig. 7.6.

fertilization at 25°C [4, 15, 19, 25, 66]. Ten to twenty hours after fertilization the imaginal disks become histologically distinguishable; at the end of the larval period they are 50-500 μm in diameter [66].

There are relatively few studies of early embryogenesis. According to some authors [57, 76], protein and DNA synthesis in the embryo begins during the division of the pre-blastoderm nuclei. Synthesis of RNA, at least of the ribosomal type, apparently begins later, since the pre-blastoderm nuclei do not contain nucleoli [58]. The nucleoli become visible when the nuclei migrate to the surface of the egg before the cell boundaries are completed [25]. The duration of the later stages of embryonic development and brief details about them are given in Table 7.2.

7.8.2. Postembryonic Development

After hatching from the egg, *Drosophila* larvae increase in size considerably. Immediately after hatching, the larva slightly exceeds 0.5 mm; if the cultivation conditions are good, the size increases to 5 mm by the end of this stage. The body of the larva consists of 12 segments, comprising 1 head, 3 thoracic, and 8 abdominal segments. The cuticle of the larva is transparent, and the inner organs are clearly visible (Figs. 7.5 and 7.6). The sex of the larvae may be determined easily by inspecting the gonads located in the Vth abdominal segment. The testes are markedly larger than the gonads of the females. They appear as light-colored vesicles surrounded by fat bodies. The ovaries are small and they are difficult to see clearly in live larvae [16].

Larvae possess two systems of organs: the larval and the imaginal. In *D. melanogaster* the determination of individual regions of egg cytoplasm to develop into definite organs of the larval system ends by the time of oviposition [30]; by hatching, the organs of this system are completely differentiated [71]. The number of cells in the larval organs is constant. As the larvae grow the cell volume in the organs increases. The chromosomes of most larval organs are polytene. During larval developmental stages the organs functional in adults are represented by the imaginal disks, groups of cells determined to develop into specialized structures (Fig. 7.7). Fate maps have been constructed for all the major disks of *D. melanogaster* [27, 29, 77, 79]. In contrast to the larval organs with their constant number of cells, the imaginal disks increase in size as a result of cell divisions. For example, the number of cells in the wing disk during late larval development is

Fig. 7.6. *Drosophila melanogaster*, internal organs of late 3rd larval instar [82]. Body segments: I) head; II–IV) thorax; V–XII) abdomen. 1) Right antenno-maxillar complex; 2) chitin edge; 3) H-shaped part of mouth skeleton; 4) dorsal part of the cephalo-pharyngeal plate; 5) external pharyngeal epithelium; 6) right cephalo-pharyngeal plate; 7) left mouth hook; 8) mouth angular plates; 9) papillae of the anterior spiracle [8, 9]; 10) trachea; 11) esophagus; 12) antenna/eye disks; 13) right wing disc; 14) right disk of foreleg; 15) right haltere disk; 16) disk of hind leg; 17) anterior transverse tracheae; 18) right hemisphere of suprapharyngeal ganglion; 19) outer proventricular wall; 20) light cells of the proventriculus; 21) posterior part of the foregut; 22) right salivary gland; 23) ventral ganglion; 24) paired right appendages of the stomach; 25) anterior Malpighian tubule; 26) stomach; 27) fat bodies; 28) tracheoles; 29) bristle ring, situated on the anterior edge of abdominal segment IV; 30) midgut; 31) beginning of hindgut; 32) ovary (length, 40 μm), normally positioned at the end of the Vth ventral segment; 33) posterior Malpighian tubule; 34) hind gut; 35) posterior spiracle (spinal vessels and muscles not shown).

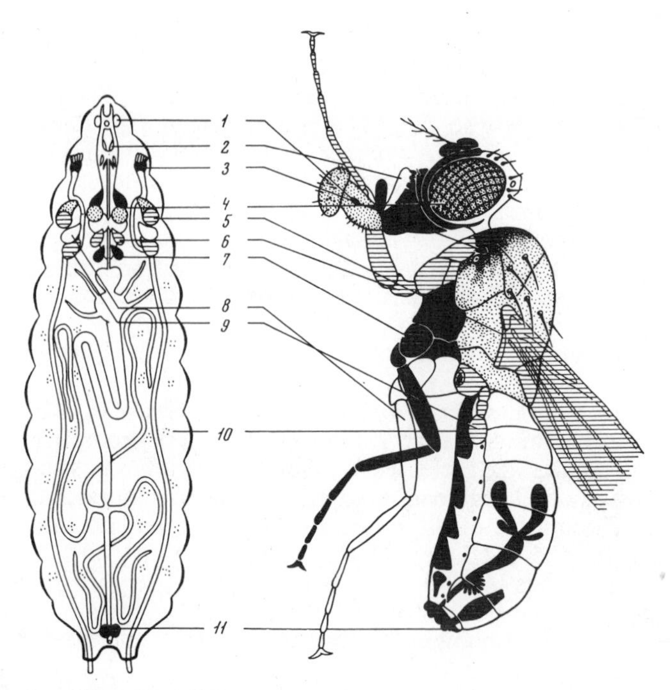

Fig. 7.7. Location of imaginal disks in the larva and the corresponding structures in the adult [66]. 1) Labium; 2) clypeus-labium; 3) humer; 4) eye/antenna; 5) wing-thorax; 6, 7, 8) 1st, 2nd, and 3rd pairs of legs, respectively; 9) haltere; 10) ventrum; 11) genitalia (gonad disks not shown).

16,000-17,000 [12], and 18 h after puparium formation this number reaches 52,000 [26]. Differentiation of imaginal disks into adult structures takes place at the pupa stage and is controlled by ecdysone [25, 31].

The post-embryonic development of *Drosophila* may be divided into ten morphologically distinguishable stages [9] (Table 7.3), the total duration of which for *D. melanogaster* at 25°C is approximately 192 h.

The duration of stages presented in Tables 7.2 and 7.3 is approximate, the exact length depending on the insect genotype and the action of a number of environmental factors [47, 50, 60]. Thus, when beginning to work with various *Drosophila* stocks one should specifically determine the duration of those stages which are of interest to the investigator under given particular culture conditions. A fuller characterization of individual developmental stages is presented below.

Drosophila larvae undergo two molts which divide this period into three larval instars. A few minutes before each molt the larva stops feeding and lies motionless on the medium surface. During the molt the old cuticle and its specialized parts (mouth structures and spiracles) are replaced by the new structures characteristic of the next larval instar [9]. The molt begins with the larval mouth hooks becoming active and biting through the old cuticle; this is followed by active muscle contractions. After the cuticle is ruptured, the larva rapidly leaves it and moves around in the medium.

TABLE 7.3. Main Stages of *D. melanogaster* Postembryonic Development [9]

Stage	Time (hours)	Characteristic features
1	0	Hatching from egg membranes. I larval instar. Head ends of the tracheal stems are branched*
2	25	First molting, II larval instar. Head ends of the tracheal stems have a clublike shape
3†	48	Second molting, III larval instar. Formation of papillae
4	96	Appearance of puparium, which is white
5	98	Pigmented puparium
6	100	Molting of the prepupa
7	108	Pupation; the cephalic complex, wings and legs protrude outside
8	145	Beginning of eye pigmentation
9	165	Beginning of bristle pigmentation
10	192	Hatching of the adult

*The structure of the head end of tracheal stems is the most convenient for determining the larval instar and is additionally introduced into the table as a diagnostic feature.
†In contrast to *D. melanogaster*, for *D. virilis* stage 3 may be conveniently subdivided into three phases (see text).

The larval instar can be easily determined by the body size, and the structure of the mouth parts and of the spiracles [9]. We believe that the structure of the anterior spiracles is the most suitable trait for determining the larval instar, since the body size varies depending on the cultivation conditions, and determining the mouth structure requires fixed preparations, thus killing the larvae. By contrast, the structure of spiracles is easily visible in live larvae. The larval tracheae consist of two big lateral stems (see Fig. 7.6), which end in the anterior and posterior spiracles. The anterior spiracles in the two first larval instars do not function, while the posterior ones are functional throughout larval life. Prior to molts the posterior spiracle of each larval instar becomes thicker and looks like a yellow bulge inside which one can see the old spiracle with its lighter color; this is lost together with the old cuticle (Fig. 7.5, A, 2c, d). The formation of the double spiracle is evidence of the approaching molt.

Stage 1: First larval instar. The anterior ends of the tracheae in larvae of this instar do not thicken, and fine branching in the thoracic segments may be seen (Fig. 7.5A, 1).

Stage 2: Second larval instar. During this stage, the anterior spiracle is distinctly marked, a thickening appears on the anterior end of the tracheal stem, and its end acquires the shape of a pin head (Fig. 7.5A, 2a). By the end of the second larval instar a plate consisting of tightly compressed papillae may be seen at the anterior end of the tracheal stem (Fig. 7.5A, 2b).

Stage 3: Third larval instar. During this stage the papillae move somewhat away from each other and the ends of the tracheal stems acquire a brush shape with 7.9 open-ended papillae (Fig. 7.5A, 3a, b). In species of *Drosophila* group *virilis*, in contrast to *D. melanogaster*, the third larval instar may be divided into early, middle, and late phases on the basis of the anterior spiracle pigmentation. During the early phase the spiracle lacks pigmentation. During the middle phase, approximately 24 h after the second molt, a brown ring appears near the base of the *D. virilis* spiracle. The pigmentation of this ring increases with larval age. During the late phase, approximately 6 h prior to puparium formation, an orange spot appears near the base of the anterior spiracle [69]. Complete pigmentation of the anterior spiracle takes place a few hours prior to puparium formation.

Stage 4: Prepupae with nonpigmented cuticle. By the end of the third larval instar, shortly before pupation, the larvae leave the medium, larval motility decreases, the anterior spiracle protrudes outside (Fig. 7.5), and the larvae then completely lose mobility, become shorter, and acquire the shape of a pupa (Fig. 7.5B). Puparium formation takes place during this period. By the time it is complete, the larval segmentation disappears, while the cuticle is still free of pigment. This corresponds to the prepupa stage [9], which lasts several minutes.

Stage 5: Pigmentation of the prepupa puparium. Puparium undergoes hardening and pigmentation. Puparium pigmentation begins from the external cuticle surface and gradually proceeds to the deeper layers. The formation of the puparium in *D. melanogaster* at 25°C continues for approximately 3.5 h.

Stage 6: Prepupa molting. Thirty minutes after puparium formation the epidermis separates from the puparium and the larva transforms into a headless prepupa with no external wings and legs, surrounded by a thin cuticle.

Stage 7: Pupation. Pupation takes place 12 h after puparium formation. During this process the structures corresponding to the head, wings, spiracles, and legs turn inside out. A typical pupa with a head, thorax, and abdomen is formed. It is surrounded by three membranes: the external puparium, the intermediate prepupal cuticle, and the inner pupal cuticle.

The larval organs and tissues such as salivary glands, fat bodies, and the intestinal tract undergo histolysis during the pupal stage. Malpighian tubules and nervous tissues undergo a relatively small change during metamorphosis. This period involves histogenesis of the organs functional in the adult insect.

Stage 8: Beginning of eye pigmentation.

Stage 9: Beginning of bristle pigmentation. (No additions to stages 8 and 9.)

Stage 10. Several hours prior to eclosion, wings and bright red eyes can be distinctly seen through the semitransluscent pupal membranes. Newly emerged flies, after leaving the puparium, have an elongated, spindlelike, weakly pigmented body, and short wings which are not yet well spread. By these criteria they can be easily distinguished from adult flies.

The duration of adult life is genetically determined and may differ not just for the various species of *Drosophila* but even for individual stocks within a given species. At 25°C the mean lifespan of *D. melanogaster* is 36.5 days, while for *D. virilis* it is 54 days [24]. The duration of adult life also depends on environmental conditions, primarily those which directly or indirectly affect the intensity of the metabolism. Such factors as temperature, moisture, feeding conditions, culture density, and illumination affect the duration of adult life and the rate of physiological aging [54].

REFERENCES

1. J. Agrell, "Physiological and biochemical changes during insect development," in: *The Physiology of Insecta*, Vol. 1, Academic Press, New York (1964).
2. D. T. Anderson, "The epigenetics of the larva in Diptera," *Acta Zool.* **43**, 221 (1962).
3. D. T. Anderson, "The comparative embryology of the Diptera," *Annu. Rev. Entomol.* **11**, 22 (1966).
4. D. T. Anderson, "The development of holometabolous insects," in: *Developmental Systems of Insects*, C. H. Waddington and C. H. Counce (eds.), Academic Press, New York (1972).
5. M. Ashburner and E. Novitski (eds.), *Genetics and Biology of Drosophila*, Vols. 1a-c, Academic Press, New York (1976).
6. M. Ashburner and T. R. F. Wright (eds.), *Genetics and Biology of Drosophila*, Vols. 2a-d, Academic Press, New York (1978–1980).
7. S. P. Bainbridge and M. Bownes, "Studying the metamorphosis of *Drosophila melanogaster*," *J. Embryol. Exp. Morphol.* **66**, 57–80 (1981).
8. W. K. Baker, "Position-effect variegation," *Adv. Genet.* **14**, 133–169 (1968).
9. D. Bodenstein, "The postembryonic development of *Drosophila*," in: *Biology of Drosophila*, Wiley, New York (1950).
10. W. Bownes, "A photographic study of development in the living embryo of *Drosophila melanogaster*," *J. Embryol. Exp. Morphol.* **33**, 789–801 (1975).
11. C. B. Bridges and K. S. Brehme, *The Mutants of Drosophila melanogaster*, Carnegie Inst. Wash. Publ. (1944).
12. P. S. Bryant, "Cell lineage relationships in the imaginal wing disk of *Drosophila melanogaster*," *Dev. Biol.* **22**, 389–411 (1970).
13. H. Burla, "Systematik, Verbreitung, und Oekologie der *Drosophila* Arten der Schweiz," *Rev. Suisse Zool.* **58**, 23–175 (1951).
14. A. Butz and P. Hayden, "The effects of age of male and female parents on the life cycle of *Drosophila melanogaster*," *Ann. Entomol. Soc. Am.* **55**, 617–618 (1962).
15. L. N. Chan and W. Gehring, "Determination of blastoderm cells in *Drosophila melanogaster*," *Proc. Natl. Acad. Sci. USA* **68**, 2217–2221 (1971).
16. K. W. Cooper, "Normal spermatogenesis in *Drosophila*," in: *Biology of Drosophila*, Wiley, New York (1950).
17. S. S. Counce, "The analysis of insect embryogenesis," *Annu. Rev. Entomol.* **6**, 295 (1961).
18. S. S. Counce, "Developmental morphology of polar granules in *Drosophila*," *J. Morphol.* **112**, 129–145 (1963).
19. C. H. Counce, "The causal analysis of insect embryogenesis," in: *Developmental Systems of Insects*, C. H. Waddington and C. H. Counce (eds.), Academic Press, New York (1972).
20. M. S. David and M. F. Clavel, "Essai de définition d'une température optimale pour le dévéloppement de la *Drosophila*," *C. R. Acad. Sci., Paris* **262**, 2159 (1966).
21. M. Demerec (ed.), *Biology of Drosophila*, Wiley, New York (1950).
22. M. Demerec and B. P. Kaufmann, *Drosophila Guide. Introduction to the Genetics and Cytology of Drosophila melanogaster*, Carnegie Inst. Wash. Publ. (1961).

23. W. W. Doane, "*Drosophila*," in: *Methods in Developmental Biology*, F. H. Wilt and N. K. Wessels (eds.), T. Y. Crowell, New York (1967).
24. E. R. Felix and F. J. Ramirez, "Differential life shortening induced by irradiation with electrons in species of *Drosophila*," *Ann. Inst. Biol., Univ. Nac. Auton. Mex.* **38**, *Ser. Biol. Exp.* (1), 5–10 (1967).
25. I. W. Fristrom, "The developmental biology of *Drosophila*," *Annu. Rev. Genet.* **4**, 325–346 (1970).
26. A. Garcia-Bellido and J. K. Merriam, "Perimeters of the wing imaginal disk development of *Drosophila melanogaster*," *Dev. Biol.* **24**61–87 (1971).
27. W. J. Gehring, "Imaginal disks: determination," in: *The Genetics and Biology of Drosophila*, Vol. 2c, M. Ashburner and T. R. F. Wright (eds.), Academic Press, New York (1978).
28. W. J. Gehring (ed.), "Genetic mosaics and cell differentiation," in: *Results and Problems in Cell Differentiation*, Vol. 9, Springer-Verlag, Berlin (1978).
29. W. Gehring and R. Nothiger, "The imaginal disks of *Drosophila*," in: *Developmental Systems of Insects*, C. W. Waddington and C. H. Counce (eds.), Academic Press, New York (1972).
30. K. Geigy, "Erzeugung reinimaginaler Defekte durch ultraviolette Eibestrahlung bei *Drosophila melanogaster*," *Wilhelm Roux's Arch. Entwicklungsmech. Org.* **125**, 406–447 (1931).
31. L. S. Gilbert and H. A. Schneiderman, "The content of juvenile hormone and lipid in Lepidoptera. Sexual differences and developmental changes," *Gen. Comp. Endocrinol.* **1**, 453–472 (1961).
32. E. K. Ginter and V. E. Bulyzhenkov, "Determination of *Drosophila* imaginal disks and its genetic control," in: *Drosophila in Experimental Genetics*, V. V. Khvostova, M. D. Golubovski, and L. I. Korochkin (eds.) [in Russian], Nauka, Novosibirsk (1978).
33. E. K. Ginter and B. A. Kuzin, "The insects," in: *Methods of Developmental Biology*, T. A. Dettlaff, V. Ya. Brodsky, and G. G. Gause (eds.) [in Russian], Nauka, Moscow (1974).
34. J. R. Girton and P. J. Bryant, "The use of cell lethal mutations in the study of *Drosophila* development," *Dev. Biol.* **77**, 233–243 (1980).
35. M. D. Golubovsky, "Certain general characteristics of genome organization and phenotypic expression of mutations in *Drosophila*," *Genetika* **8**, 143–157 (1972).
36. E. Hadorn, "Transdetermination," in: *The Genetics and Biology of Drosophila*, Vol. 2c, M. Ashburner and T. R. F. Wright (eds.), Academic Press, New York (1978).
37. E. Hadorn, R. Hurliman, G. Minder, G. Schubiger, and M. Staub, "Entwicklungsleistungen embryonaler Blasteme von *Drosophila* nach Kultur im Adultwirt," *Rev. Suisse Zool.* **75**, 557–569 (1968).
38. O. Hess and G. F. Meyer, "Genetic activities of the Y-chromosome in *Drosophila* during spermatogenesis," *Adv. Genet.* **14**, 171–223 (1968).
39. P. E. Hildreth and J. C. Lucchesi, "Fertilization in *Drosophila*. I. Evidence for the regular occurrence of monospermy," *Dev. Biol.* **6**, 262 (1968).
40. I. H. Herskowitz, *Bibliography of the Genetics of Drosophila, II*, Alden Press, Oxford (1953).
41. I. H. Herskowitz, *Bibliography of the Genetics of Drosophila, III*, Indiana University Press, Bloomington (1958).
42. I. H. Herskowitz, *Bibliography of the Genetics of Drosophila, IV*, McGraw-Hill, New York (1963).

43. M. J. Hollingsworth and J. V. Burcombe, "The nutritional requirements for longevity in *Drosophila*," *J. Insect. Physiol.* **16**, 1017–1025 (1970).
44. K. Illmensee, "Nuclear and cytoplasmic transplantation in *Drosophila*," in: *Insect Development*, P. A. Lawrence (ed.), Halsted Press, Oxford (1976).
45. S. B. Joon and A. S. For, "Permeability of premature eggs from *Drosophila* collected with the 'ovitron'," *Nature (London)* **206**, 910 (1965).
46. V. T. Kakpakov, "Nutrient media for *Drosophila*," in: *Genetic Problems in Studies Using Drosophila*, V. V. Khostova, L. I. Korochkin, and M. D. Golubovsky (eds.) [in Russian], Nauka, Novosibirsk (1977).
47. G. A. Kerkut and L. I. Gilbert (eds.), *Comprehensive Insect Physiology, Biochemistry, and Pharmacology*, Vol. 1, Embryogenesis and Reproduction, Pergamon Press, Oxford–New York (1985).
48. R. B. Khesin, "Physiological differences between two populations of *Drosophila melanogaster*," *Dokl. Akad. Nauk SSSR* **59**, 167–170 (1948).
49. A. F. Khishin, "The response of the immature testis of *Drosophila* to the mutagenic action of x rays," *Z. Indukt. Abstamungs- und Vererbungslehre* **87**, 57 (1955).
50. J. C. King, "Differences between populations in embryonic developmental rates," *Am. Nat.* **93**, 171 (1959).
51. R. C. King, *Ovarian Development in Drosophila melanogaster*, Academic Press, New York (1970).
52. E. A. Koch and R. C. King, "The origin and early differentiation of the egg chamber of *Drosophila melanogaster*," *J. Morphol.* **119**, 283–304 (1966).
53. B. A. Kuzin, "The imaginal disks of *Drosophila* and methods of working with them," in: *Genetic Problems in Studies Using Drosophila*, V. V. Khvostova, L. I. Korochkin, and M. D. Golubovsky (eds.) [in Russian], Nauka, Novosibirsk (1977).
54. M. J. Lamb, "Ageing," in: *The Genetics and Biology of Drosophila*, Vol. 2c, M. Ashburner and T. R. F. Wright (eds.), Academic Press, New York (1978).
55. D. L. Lindsley and E. H. Grell, *Genetic Variations of Drosophila*, Carnegie Inst. Wash. Publ. (1967).
56. D. L. Lindsley and K. T. Tokuyasu, "Spermatogenesis," in: *Genetics and Biology of Drosophila*, Vol. 2d, M. Ashburner and T. R. F. Wright (eds.), Academic Press, New York (1980).
57. R. A. Lockshin, "Insect embryogenesis and macromolecular synthesis during early development," *Science* **134**, 775 (1966).
58. A. P. Mahowald, "Ultrastructural differentiations during formation of the blastoderm in the *Drosophila melanogaster* embryo," *Dev. Biol.* **8**, 186–204 (1963).
59. A. P. Mahowald, "Polar granules of *Drosophila*. II. Ultrastructural changes during early embryogenesis," *J. Exp. Zool.* **167**, 237–262 (1968).
60. N. N. Medvedev, *Practical Genetics* [in Russian], Nauka, Moscow (1966).
61. V. A. Mglinets, "Homeostatic mutations in *Drosophila* and problems of developmental genetics," in: *Drosophila in Experimental Genetics*, V. V. Khvostova, M. D. Golubovsky, and L. I. Korochkin (eds.) [in Russian], Nauka, Novosibirsk (1978).
62. H. K. Mitchell, "Biochemical aspects of *Drosophila*," *Annu. Rev. Genet.* **1**, 185–199 (1967).
63. H. J. Muller, *Bibliography on the Genetics of Drosophila*, Oliver and Boyd, Edinburgh (1939).

64. S. M. Newman and G. Shubiger, "A morphological and developmental study of *Drosophila* embryo ligated during nuclear multiplication," *Dev. Biol.* **79**, 128–131 (1980).
65. J. F. Nohidez, "The internal phenomena of reproduction in *Drosophila*," *Biol. Bull.* **39**, 207–230 (1920).
66. R. Nöthiger, "The larval development of imaginal disks," in: *The Biology of Imaginal Disks*, Vol. 5, H. Ursprung and R. Nöthiger (eds.), Springer-Verlag, Berlin (1972).
67. W. Ouweneel, "Normal and abnormal determination in the imaginal disks of *Drosophila* with special reference to the eye disks," *Acta Embryol. Exp.* **1**, 95–119 (1970).
68. S. T. Patterson and G. B. Mainland, "The Drosophilidae of Mexico," *Univ. Texas Publ.* **4445**, 9–101 (1944).
69. S. T. Patterson and W. S. Stone, *Evolution in the Genus Drosophila*, Macmillan, New York (1952).
70. B. S. Pettit and R. W. Rasch, "Tritiated-histidine incorporation into *Drosophila* salivary chromosomes," *J. Cell Physiol.* **68**, 325–333 (1966).
71. D. F. Poulson, "Histogenesis, organogenesis, and differentiation in the embryo of *Drosophila melanogaster* Meigen," in: *Biology of Drosophila*, Wiley, New York (1950).
72. M. Rabinowitz, "Studies on the cytology and early embryology of the egg of *Drosophila melanogaster*," *J. Morphol.* **69**, 1–36 (1941).
73. R. Ransom (ed.), *A Handbook of Drosophila Development*, Elsevier Biomedical Press, Amsterdam–New York–Oxford (1982).
74. D. B. Roberts (ed.), *Drosophila. A Practical Approach*, IRL Press, Oxford–Washington (1986).
75. A. Rotschild, W. V. Eisenberg, and A. W. Varquez, "A simple rearing container for *Drosophila*," *J. Hered.* **59**, 98 (1968).
76. M. Sabour, "Nucleic acid and protein synthesis during early embryogenesis of mealybug *Pseudococcus obscurus* Essig.," Doctoral Dissertation, University of California at Berkeley (1969).
77. G. Schubiger, "Anlageplan, Determinations-Zustand und Transdeterminations-Leistungen der mannlichen Vorderbeinscheibe von *Drosophila melanogaster*," *Wilhelm Roux's Arch. Entwicklungsmech. Org.* **160**, 9–40 (1968).
78. A. Shearn, "Mutational dissection of imaginal disk development in *Drosophila melanogaster*," *Am. Zool.* **17**, 585 (1977).
79. S. Shearn, "Mutational dissection of imaginal disks in *Drosophila melanogaster*," in: *Genetics and Biology of Drosophila*, Vol. 2c, M. Ashburner and T. F. R. Wright (eds.), Academic Press, London (1978).
80. B. P. Sonnenblick, "The early embryology of *Drosophila melanogaster*," in: *Biology of Drosophila*, Wiley, New York (1950).
81. W. P. Spencer, "Collection and laboratory culture," in: *Biology of Drosophila*, Wiley, New York (1950).
82. E. H. Strasburger, *Drosophila melanogaster Meig. Eine Einfuhrung in den Bau und die Entwicklung*, Berlin (1935).
83. M. W. Strickberg, *Experiments in Genetics with Drosophila*, Wiley, New York (1962).
84. A. H. Sturtevant, "Culture method for *Drosophila*," in: *Culture Methods for Invertebrate Animals*, New York (1937).

85. A. H. Sturtevant, "The classification of the genus *Drosophila* with descriptions of nine new species," *Univ. Texas Publ.* **4213**, 5–51 (1942).
86. P. G. Svetlov and G. F. Korsakova, "Critical periods of development of macrochaetae in the life cycle of *Drosophila melanogaster*," *Dokl. Akad. Nauk SSSR* **176**, 226–228 (1967).
87. H. D. Swanson and G. A. Swanson, "A new technique for manipulation and study of *Drosophila melanogaster*," *J. Hered.* **62**, 203 (1971).
88. C. H. Waddington, *Principles of Embryology*, Allen and Unwin, Ltd., London (1957).
89. M. R. Wheeler, "Taxonomic studies on the Drosophilidae," *Univ. Texas Publ.* **4920**, 157–195 (1949).
90. E. Wieschaus and J. Szabad, "The development and function of the female germline in *Drosophila melanogaster* and cell lineage study," *Dev. Biol.* **68**, 29–46 (1979).

Chapter 8

THE HONEYBEE *Apis mellifera* L.

D. V. Shaskolsky

8.1. INTRODUCTION

The biological [1, 2, 6, 11-18, 21, 22, 28, 29, 31, 38] and apicultural aspects [5, 25, 27, 42, 55] of the honeybee have been discussed in many publications.

As an experimental animal the honeybee has many advantages. The female, or queen, lays a large number of eggs over a long period extending from spring to autumn. The embryos of one fertilized queen, that is, a queen inseminated by several males, can be obtained for many months and even up to 2-3 years. Although these embryos develop from eggs fertilized by the sperm of several males, the number of males whose sperm is stored in the queen's spermatheca for years is not necessarily large. The spermatozoa of each haploid male are genetically identical. The abundance of larvae and pupae of a similar age in the honeycomb enables sufficient amounts of material to be collected for biochemical studies. Material can be obtained even from a beehive in a town, on a roof, or in an attic, for example.

Normal honeybee males or haploid drones are extremely suitable for investigating the morphological and biochemical manifestations of recessive mutations. The availability of body pigmentation and eye pigmentation mutations [69] enables genetically marked material to be used.

Honeybee embryos have been used in experiments to separate parts of the embryo [50] and for nuclear transplantation [11, 12]. They have been used for studying the nucleus and chromosomes by electron microscopy [16, 17], and determining the amount of DNA in different nuclei [34]. The development of a female larva into a queen, worker bee, or intermediate forms was studied in detail some time ago [23, 24, 26, 29]; a more sophisticated and up-to-date analysis now seems appropriate.

The discovery of a series of multiple alleles of the gene controlling the sex in the honeybee [33, 67] suggests that this species could be particularly interesting for studying the mechanisms of sex determination. The opportunities which the honeybee offers for studying problems of developmental biology remain to be fully realized.

Investigations into honeybee development have an immediate practical aspect since apiculture is a major aspect of the economy. The pollination of field and garden plants by honeybees necessary for crop production is much more important in economic terms than the direct revenue from honey and wax.

8.2. TAXONOMY AND THE NUMBER OF CHROMOSOMES

The honeybee *Apis mellifera* Linnaeus, 1758 (sometimes called *Apis mellifica* Linnaeus, 1761) is one of the most highly organized insects. Domestication has changed the honeybee only slightly. A swarm lost from an apiary may settle in the hollow of a tree and carry on living normally under natural conditions.

Two out of the four species belonging to the genus *Apis* L. have been domesticated. The honeybee *A. mellifera*, which has a number of subspecies initially found in Europe, Africa, and Western Asia, is the most important species economically and can now be found throughout all the continents. As well as this species, the Middle Indian honeybee *A. indica* F. (synonym *A. cerana*) is bred in India, eastern Asia, and Oceania; this species also builds its nest out of several honeycombs. The gigantic Indian honeybee *A. dorsata* F. and the midget Indian honeybee *A. florea* F., which live in India and Indonesia, are not used for breeding. Their colonies build just one open honeycomb.

The honeybee chromosomal complex in early development corresponds to the Dzierzon rule [19], according to which females develop from fertilized eggs and males from unfertilized eggs; the embryonic cells of *A. mellifera* females have a diploid chromosomal number $2n = 32$, while haploid male cells have $n = 16$ chromosomes. These chromosomal numbers apply to embryonic cells, but some cells may be polyploidized from the larval stage onward. The number of chromosomes is identical in the three remaining species of the genus *Apis* [20]. The sixteen chromosomes of the *A. indica* haploid male may be divided according to size into eight similar pairs, while the 32 female chromosomes may be grouped into eight quartets [7]. This indicates that the chromosomal complement of the honeybee underwent duplication relatively recently in its evolution.

8.3. CASTES AND COLONY DEVELOPMENT

Honeybees live in colonies which consist of four castes: the queen, worker bees, common drones, and the embryos of diploid drones. Females develop from fertilized eggs heterozygous in the gene of sex. A female larva develops into either a queen or a worker bee, depending on nutrition.

8.3.1. The Queen

The queen lives for several years. She lays eggs, but cannot collect food, feed larvae, or build the nest. The queen's long abdomen contains two large ovaries, each consisting of 110-210 ovarioles. Her genital apparatus contains a seminal receptacle or spermatheca; in a young fertilized female this organ immediately after fertilization contains between 5 million and nearly 8 million spermatozoa of several drones. It is generally believed that after the beginning of oviposition the queen does not mate again. The average queen in the high season may lay 800-1500 eggs a day for weeks. A highly productive queen may lay up to 3000 eggs a day, possibly more.

In the warm nest located in the center of the beehive, the queen lays one egg on the bottom of each empty cell previously cleaned by bees. Hardly ever making any mistakes, she lays fertilized eggs into the worker cells and the queen cell cups, and deposits unfertilized eggs into the drone cells. The ovipositing queen, who has either lost her store of spermatozoa in the spermatheca or did not have any from the beginning, lays only unfertilized eggs into all the cells. Such a queen is usually referred to as a "drone queen."

As a rule, an individual beehive contains the eggs of just one queen. Two queens meeting in a beehive fight until one (usually the older) dies.

8.3.2. Worker Bees

The worker bees belonging to the summer brood live for about a month,while autumn bees live until spring. The number of them in a colony varies from a few thousand to 50,000-70,000, depending on the season and strength of the colony. They collect nectar and pollen from flowers, propolis from resinous buds, and sweet secretions from plant-sucking insects. They bring water, transform nectar into honey, feed all the larvae, queen, and drones, secrete wax from the abdomen, and use it to build the honeycomb. They also clean, heat, ventilate, and guard the beehive; specifically, they prevent worker bees of other colonies from entering their beehive. Worker bees alter the form of their activity according to their age (with marked deviations) and the requirements of the colony. Initially they do a variety of work inside the beehive (household bees) and at the age of 2-3 weeks they become active outside the beehive (field bees).

The two ovaries in a worker bee contain 1-12 ovarioles, but no eggs are formed in them. In special circumstances, for example if the queen is absent from the colony for a long time, the ovaries of young bees may begin to function. These worker bees are called laying workers; they are capable of laying approximately 20 unfertilized eggs during their life. They lay eggs into any cell, whether or not there are any previously laid eggs already in them; eggs are usually laid onto the cell walls, not on the bottom. These eggs result in the usual drones.

8.3.3. Common (Haploid) Drones

The drones develop from unfertilized eggs in normal colonies only during spring and summer. They vary in number from several hundred to thousands; they are bigger than worker bees, and have no sting. Drones do not participate in the activities of worker bees. They are sexually mature by the twelfth day after leaving the pupae. The amount of spermatozoa in one drone averages 11 million, and the volume of the sperm ejaculate is 1-1.33 mm^3. Drones reared in worker cells (the hunchback brood) are smaller, but their spermatozoa are normal although there are fewer. The haploid drone is hemizygous for all genes. The genotype of all its spermatozoa, with the exception of mutated ones, is identical.

On days when the weather is good the drones fly out of the beehives and gather in special places, sometimes far from the apiaries. Here they find young queens on their wedding flight. The drones mate with the queens in mid-air, and die immediately afterward. During the spring and summer, each colony admits drones into its

TABLE 8.1. Dynamics of Development (in days) of Four Honeybee Castes in a Beehive at a Temperature of about 35°C (from [60], with modifications)

Brood	Developmental phase	Worker bee	Queen	Drone	
				usual (haploid)	diploid
Open (cells are open)	Egg (embryo)	3	3	3	3
	Larvae	6-7	5	7-9	0-1
Sealed (cells are sealed with cups)	Prepupa	2-3	2	4-6	–
	Pupa	9	6	10-14	–
Total		20-22	16-17	24-32	3-4

Note. Variation in time of development is due predominantly to temperature fluctuations. This variation most affects the drone brood, which is frequently located at the border of the heated nest.

beehive and feeds them. At the end of the swarming season, particularly when the period of nectar and pollen collection is over, the worker bees no longer feed the drones and do not admit them into the beehive (queenless colonies excepted).

8.3.4. Diploid Drones

These drones develop from fertilized eggs, homozygous for any of the multiple alleles of the sex gene. This is the fourth caste of the honeybee discovered between 1951-1963 by Mackensen and Woyke [33, 67]. It was previously unknown. Members of this caste do not usually reach maturity because nurse bees have an instinct to kill the larvae of diploid drones. Initially these larvae are fed, but in the first hours after hatching they are killed and eaten by the nurse bees.

A proportion of the queens in any apiary lays eggs which result in the birth of individuals belonging to this caste. The percentage of such eggs among the fertilized ones varies from queen to queen and is usually low, only in extremely rare cases reaching 50% (however, by inbreeding one can easily get a queen to produce 50% of diploid drones). On average the proportion of such eggs is in inverse proportion to the number of alleles of the sex gene in the population [52]. Laying eggs which produce diploid drones can be detected owing to the presence of a patchy brood pattern. This results from a certain percentage of empty cells in a good honeycomb that contains pupae or larvae of approximately the same age. The presence of individuals belonging to this caste represents a rudimental mode of sex determination that existed previously [53, 54].

8.3.5. Living Conditions in the Beehive: Cell and Honeycomb

Before the adult forms emerge from the pupae, development proceeds in the wax cells of the three main castes, which vary somewhat in size for different subspecies and races. The duration of development for the four castes in honeycomb cells is given in Table 8.1. A mass of adjacent worker cells is arranged in strict vertical layers in the honeycomb; the hexahedral cells are located horizontally in the honeycomb. The drone honeycomb consists of large cells with a configuration similar to that of the workers' cells.

The queen cells are oriented vertically with the holes directed downward. They are located individually, usually near the edge of the honeycomb. They do not touch each other, so stay rounded. In contrast to the worker and drone cells, the queen cells are built during the larval growth stage and used only once. The construction of a swarm queen cell begins with a queen cup being made, which has a hole with a diameter close to that of a worker cell, but somewhat narrower than the largest inner diameter of the cup. The empty cup can be left for a long period of time. After the egg is laid into it, the "sown cup" is rapidly transformed into an open queen cell containing a larva, and then into a sealed queen cell with a pupa. If the queen is lost, the colony lays down emergency queen cells for the available young larvae. The number of queen cells varies from 2 or 3 to 100 or more, depending on the quality of the colony, including hereditary factors.

Cells which have been used several times for brood rearing become darker and reduced in volume, owing to the presence of cocoon remnants; consequentially the bees developing inside them are somewhat smaller.

The spaces between the ledges and walls (or the ceiling) in the beehive measure 8-9 mm in width or height. They are left free by honeybees, larger spaces are taken up by the honeycomb or connections, while smaller ones are glued by propolis ("bee-glue," derived from plant resins) and wax (the space between the honeycombs is wider). The discovery of this "bee space" in the 19th century enabled beehives to be developed which could be disassembled with the honeycomb in replaceable frames.

As well as for breeding, the honeycomb is used by the bees for processing or storing food such as nectar, honey, and stored pollen. The worker cells as well as the drone cells store honey. These cells are often longer than normal. Cells with mature honey are sealed by a continuous waxcomb and cells with the pupating brood are covered with porous cups consisting of wax (60% for worker cells and 10-15% for drone cells) mixed with plant fibers and pollen grains.

Cells containing brood are located in the center of the beehive, where they form a nest of varying volume but with a more or less round shape. The temperature in the nest is maintained by the bees at 34-35.5°C. The space under the honeycomb may be free. The aperture is normally located laterally at the lowest level of the beehive, whereas the store of honey is usually placed by the bees in the upper part far from the aperture. In winter the bees collect in a round cluster occupying several adjacent honeycombs close to the stored honey. They maintain the temperature from 6°C at the cluster periphery to 28-32°C in its center.

8.3.6. Colony and Individual

Not only an individual, but the colony as a whole, comprises a biological unit of the honeybee species. The behavior of an individual worker bee or a queen depends not only on its phenotype and external conditions, but also on the status of the colony in general. The status depends on the phase of the annual and circadian cycle of the colony, heredity, the number of adult household and field bees, the amount of open and sealed brood in the honeycomb, the quantity of food, the free space for building honeycombs, the design of the beehive, weather conditions, availability of nectar and pollen, presence or absence of artificial food in the beehive, and possible irritating factors.

The number of eggs deposited by the queen depends on the strength of the colony (the number of bees in the colony) and the stage of colony development. For example, while preparing to swarm, the colony enters the "swarming mood," during which the queen temporarily discontinues oviposition. The queen's maximal potential may only be realized in a strong colony.

8.3.7. The Annual Cycle of Colony Development; Basics of Care

By the end of winter (February-March) the queen begins oviposition by forming a nest, which is small at first. Initially the queen only deposits eggs into the worker cells. Worker bees spend the first warm spring day "flying around," emptying their colons, the content of which by the end of winter weights up to a third of the body weight (healthy bees do not defecate in the beehive). In spring the queen gradually increases the intensity of oviposition and the volume of the nest, and the appearance of young bees increases correspondingly; therefore, the volume of the beehive should be increased. The old autumn bees are gradually replaced by young ones. Except in winter, bees usually die outside the beehive. In the presence of nectar and pollen or artificial feeding, the bees begin to build the honeycomb. When the colony is growing, the beehive should always have free space that may be used for building honeycombs. Soon thereafter, particularly in strong colonies, the bees begin to build drone cells which the queen uses for laying eggs; this is an early stage of swarming preparations. The colony continues to grow, and the household bees are completely occupied nursing the brood; growth continues until the queen reaches the limit of oviposition. If there is no large nectar and pollen yield, and if the bees are not busy (building the honeycomb, for example), the colony begins immediate preparations to swarm. It may not be easy to overcome the "swarming mood" of a colony. Preparing to swarm depends on many factors and may proceed at quite different rates. The cups are built well in advance, but not necessarily used. The queen does not deposit eggs in these cups immediately after they are built, and not simultaneously either. She reduces the rate of oviposition and then stops it altogether. When the queens in the queen cells are ready to emerge, the bees, on a clear day, will fly out of the beehive *en masse* along with the old fertilized queen and settle nearby, for example on a tree branch. Reconnaissance bees inform the swarm about suitable sites and, after a while, the swarm rises up again and flies slowly to the selected site, sometimes several kilometers away, and settles there. The apiculturist should collect the settled swarm without letting it escape, and place it into a free beehive in the evening. After the first swarm the colony may release a second, and sometimes even subsequent swarms, all of which include virgin queens. The colony which released the swarm also remains with a virgin queen. The virgin queen flies out to mate at the age of 6-8 days. In 1 or 2 days she mates with roughly a dozen drones. If the mating does not take place for a month, the queen begins laying unfertilized eggs and becomes a "drone queen." Oviposition may be induced artificially by CO_2 treatment [45].

Swarming is a form of reproduction in the honeybee colony. The proportion of swarming colonies at the apiary depends on heredity and seasonal conditions, but mainly on the system of apiculture, in other words, the apiculturist. Modern apiculture uses artificial swarming rather than natural swarming, which results in losses.

During the swarming season 2-year-old, or sometimes even 1-year-old, queens are replaced by young ones. When necessary, the colonies themselves conduct the replacement of the old or unsatisfactory queen (she is classed as such if, for example, one of her legs is damaged) by rearing the young queen and replacing the old one independently on swarming. In many regions of the USSR, the end of the swarming season coincides with the main period of nectar and pollen collection. This mobilizes the colony for nectar collection and sometimes results in the arrest of swarming preparations. During nectar collection the honeycomb is needed not only to hold ripe honey, but also to store nectar and to process it. The end of the honey-collection period, particularly if it has been done quickly, enables the colonies to prepare for winter. The workers drive the drones out of the beehive and may even throw out the drone brood. If some nectar collection or artificial feeding carries on, only workers cells continue to be built. The rate of oviposition by the queen slows down. The artificial feeding of colonies, which is carried out before the winter season to increase the stores or replace honey with sugar, should not be started too late. When the bees transfer food into the honeycomb, the queen's egg laying is stimulated. It is important that the young bees resulting from this stimulation reach field age and have time to fly around before winter. Optimally, before winter, the main proportion of the colony should comprise young bees which have not participated in the nectar and pollen collection or even in the transfer of artificial food into the honeycomb. These bees should, however, winter after flying around for a time during which they are able to empty their intestines. Honeybee colonies may be left to winter in the open after making additional thermal insulation from the outside; another possibility is to leave them in a dry quiet place at a temperature ranging from 0-4°C. If the temperature rises, the bees become excited and a cluster will disintegrate. For wintering, each colony should have 18-30 kg of honey or sugar transferred by the bees into the honeycomb; it is also recommended that a frame with stored pollen be left. This is used as protein food for growing larvae at the end of the winter. The bees winter well on flower honey and better still on sugar, but they may perish during the winter if they use honey from the sweet secretions of plant-sucking insects; in the summer this honey is quite harmless. The increased content of dextrins in this honey, or in the flower honey with its admixture, results in the colon of the wintering bees being overfilled. This leads to the need to drink and causes diarrhea and the death of the colony.

Keeping bees is facilitated by making sure that the normal colony always maintains the necessary conditions in the beehive, such as high temperature and sufficient humidity for the brood. One should supply colonies with enough good honeycomb, space for them to build, and food; they should be protected from irritants. Colonies with only household bees should, in addition, be supplied with water. Details of apiculture are not presented here; they can be found in more specialized treatises (see, for example, [27]).

8.4. GAMETES

8.4.1. Structure of Eggs

The honeybee egg is shaped like a slightly bent cylinder rounded on its poles, narrowing a little at the posterior part near the future ventral side of the embryo.

The egg moves in the ovariole and oviduct, posterior pole forward. The laid egg is attached to the cell bottom at the posterior pole. The convex side of the cylinder corresponds to the future ventral side of the embryo. The size of the egg varies for different subspecies of the honeybee and differs with each season, too. During spring and autumn the size of the egg in European bees is 1.6 × 0.33 mm (for some queens, the length may attain 1.8 mm), while in summer, during intensive oviposition, the size is 1.4 × 0.32 mm. The weight of the deposited egg averages 0.13 mg, dropping by the end of its development to 0.1 mg.

The egg of the honeybee is surrounded by two membranes. The chorion, an elastic but tough external membrane, gives the egg its main protection. The chorion is produced by follicular cells, the boundaries of which are imprinted on its surface. The inner vitelline membrane is not so thick. The anterior pole contains the micropyle or "the micropylar region," a definition suggested by Snodgrass [57].

The nucleus is located under the egg surface on the convex side at a distance that is somewhat less than the width of one egg from the anterior pole, in a small conelike thickening of the periplasm; both maturation divisions, which are not accompanied by the extrusion of polar bodies, proceed while the nucleus is in this position. During oviposition the mature unfertilized and fertilized eggs are at the stage of the first maturation division (Tryasko, personal communication).

Unfertilized eggs can be obtained in the way described by Tryasko [63]. The ovipositing queen is removed from the colony for a few hours in a glass jar containing a drop of honey on the side, or placed into a microversion of the beehive which has mesh instead of a bottom, and contains 40-60 bees. The queen deposits several dozen, sometimes up to 100, eggs. They are not placed into cells and, according to available observations, are always unfertilized.

8.4.2. Structure of Spermatozoa and Method of Collection

Microscopic examination of the sperm, diluted with a small quantity of water, reveals the movement of spermatozoa [56]. Each sperm consists of a long head, neck, and tail (total 240 μm) [56].

The sperm is best obtained from mature drones just returned from a flight. A drone of known origin can be grown in a beehive equipped with a diaphragm through which only worker bees can pass; this prevents drones from flying around. Before extracting the sperm, the drones should have emptied their colons so as to prevent sperm contamination; furthermore, they should have filled their abdominal air sacs with air, which helps to turn the endophallus inside out. To achieve this they have to fly around against a window or in a special setup [45]; alternatively, holding the drone by the thorax and head, one may sweep it in the air, allowing his wings to work.

To extract sperm, the drone should be carefully rotated with two fingers and either his thorax should be slightly pressed or his head should be torn off. As a result, his genitals turn inside out, depending on the state of drone readiness; the process occurs either at once or in two steps, the horns being initially everted, followed by the endophallus which is everted upon subsequent pressing. The surface of the endophallus may contain a creamy drop of sperm and a white droplet of mucus. The sperm is taken by using a thin pipette or capillary. In the latter case it stays viable longer without drying up. The sperm can be kept for a long time under artificial conditions [47, 59].

Obtaining drones in early summer presents no problem. By the end of summer and during autumn normal colonies do not keep drones and live only in queenless colonies. During this period the drones may be reared from laying worker bees, as described by Schasskolsky and Alber [45]. Using this technique, drones can be obtained until October in the central regions of Russia [51]; the last batches may be reared from a sealed brood in an incubator.

8.5. FERTILIZATION, ARTIFICIAL INSEMINATION, AND MOSAICISM

Egg activation in the honeybee is not associated with fertilization and all eggs, whether fertilized or unfertilized, develop immediately after they are laid.

8.5.1. Fertilization

The egg is inseminated immediately prior to oviposition (0.1-0.5 min prior to laying). As it passes through the oviduct, the egg comes into contact with a very small portion of a secretion which is produced by the accessory gland and contains spermatozoa; this secretion is delivered by a special apparatus consisting of ducts and specially controlled valves. The whole process of laying the egg, including the preceding inspection of the cell, which requires the anterior part of the body to be placed into it, takes the queen less than a minute to perform. There are reasons for believing that egg insemination takes place no earlier than 0.5-0.1 min prior to laying.

When the queen is young and only beginning to lay eggs, the number of spermatozoa entering the egg is greater than in the case of old queens. According to Woyke [66], 60% of the eggs receive only one spermatozoon, while the remaining 40% receive more. Petrunkevich [39] has found up to 7 sperms in fertilized eggs. Nachtsheim [37] believed that 3-7 sperms per egg is a normal number, and even 10 spermatozoa per egg occurs quite frequently. The first meiotic division ends immediately after the egg is laid. This is followed by the second maturation division and, simultaneously, by the division of the first polar nucleus. The female pronucleus is located deeper, and moves further inside the egg, than the other nuclei. The remaining three polar nuclei stay near the surface, and two of them fuse. In the fertilized egg the female pronucleus meets the nucleus of one of the sperm cells which have entered the egg. The female pronucleus moves further into the unfertilized egg, almost up to the future dorsal side of the embryo. Cleavage divisions begin in the unfertilized eggs approximately 45 min after they occur in the fertilized egg [41].

Using sperm to wet the anterior pole of the freshly laid unfertilized egg enabled Barrat [3] to achieve egg fertilization outside the queen's body. Queens could be reared from such eggs [41].

8.5.2. Artificial Insemination

The artificial insemination of queens has been developed in detail in recent years. It is usually performed by extracting sperm from several drones in accordance with the natural polyandry of the honeybee [61]; this polyandry involves mating each young queen with several drones within one or a few days.

Insemination with the sperm from 8-12 drones yields a normal fertilized queen. The sperm of 3-4 drones is sufficient for a good but rather short-term oviposition.

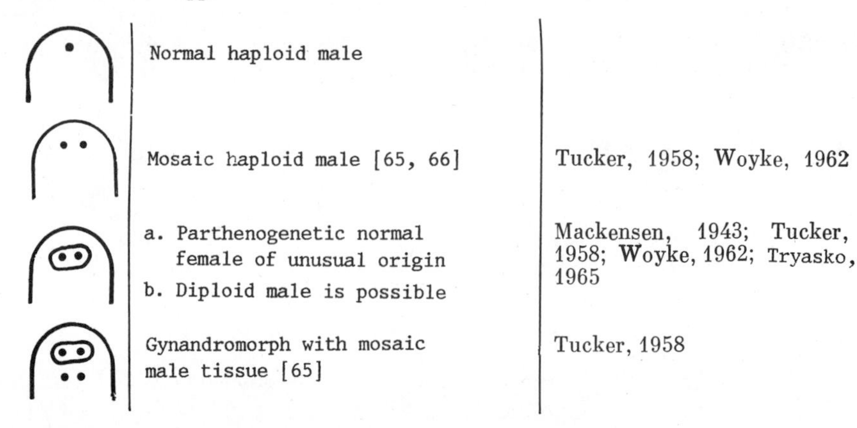

From unfertilized eggs

Description	References
Normal haploid male	
Mosaic haploid male [65, 66]	Tucker, 1958; Woyke, 1962
a. Parthenogenetic normal female of unusual origin b. Diploid male is possible	Mackensen, 1943; Tucker, 1958; Woyke, 1962; Tryasko, 1965
Gynandromorph with mosaic male tissue [65]	Tucker, 1958

From eggs fertilized by one spermatozoon

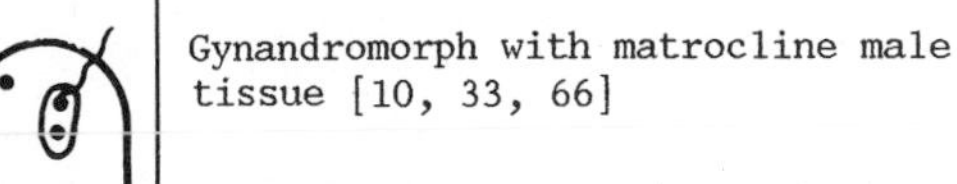

Description	References
a. Heterozygote in the sex gene, normal female [33, 67, 68] b. Homozygote in the sex gene, diploid male	Mackensen, 1951; Woyke, 1963, 1965
Gynandromorph with matrocline male tissue [10, 33, 66]	Mackensen, 1951; Woyke, 1962; Drescher et. al., 1963

From eggs fertilized by several spermatozoa

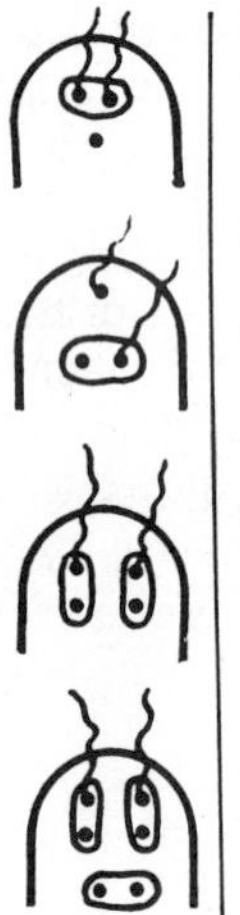

Description	References
Gynandromorph with diploid female tissue of patrocline origin [30]	Laidlaw., et al., 1964
a. Gynandromorph with male tissue of patrocline origin [10, 44] b. Mosaic male with inclusions of diploid tissues [10, 43]	Rothenbuhler et al., 1952; Drescher et al., 1963 Rothenbuhler, 1957; Drescher et al., 1964
Mosaic female, diploid [58, 56]	Taber, 1955; Woyke, 1962
Mosaic female with diploid parthenogenetic tissue [66]	Woyke, 1962

Fig. 8.1. Origins of normal and abnormal bees (modified from [69]). 1) Spermatozoon; 2) female pronucleus; 3) zygote.

Inseminating the queen living in a small colony using sperm from just one drone is sometimes successful even during the second season, but quite often the queen lays an increasing number of unfertilized eggs, which eventually completely replace the fertilized ones.

Insemination may be performed on young virgin queens which have not yet begun to lay eggs. The queen is attached to a rack under the dissecting microscope and anesthetized with CO_2. About 0.01 ml of undiluted sperm is drawn into a special syringe. The sting chamber is extended by two hooks, the vaginal valve is then deflected using a special probe, a syringe is inserted through the vagina into the unpaired oviduct, and the sperm carefully supplied. It is better to inseminate twice using two portions of sperm; the second insemination should be performed 2 days after the first. With sufficient experience one can obtain more than 90% of inseminated queens.

In experimental studies it is often very interesting to inseminate the queen using the sperm from only one male. Modern versions of artificial insemination techniques have been described in detail by Ruttner [45].

The genetic origin of the offspring can be traced by using the "cordovan" gene which controls body pigmentation or other genes (for a list of genes see [69, 22]).

In order to obtain progeny from one individual queen and not her daughter, after the unnoticed release of swarm or replacement, the queen should be marked. This marking may involve placing one or more spots on the thorax using shellac dissolved in acetone with an added aniline dye.

In particularly demanding experimental situations, when only the descendants of one particular queen are desired, one should get rid of individuals derived from the progeny of laying workers. This is particularly important because a small percentage of unfertilized eggs develops in the females [62]. To achieve this, one has to redistribute the frames with brood in order to establish a colony consisting of bees which differ in color from the daughters of a particular queen; for example, a colony consisting of bees belonging to another race may be formed.

The experimental rearing of adult bees is performed in an incubator or beehive, containing a honeycomb with the sealed brood placed in an insulated box made of fine mesh. The drones may be reared in a box formed by a diaphragm. The queens may be reared in the queen cells.

8.5.3. Mosaics and Gynandromorphs

These have been repeatedly described in the honeybee. They are produced when more than one active sperm is found in the egg, or when meiotic divisions are impaired, or when the nuclei of polar bodies show abnormal behavior (facilitated by a temporary drop in temperature) [10]. The origin of various abnormal bees, as well as that of normal ones, is shown in Fig. 8.1, taken from Woyke [69] with some modifications.

8.6. OBTAINING AND INCUBATING DEVELOPING EGGS

Eggs at a known stage of development can best be obtained in the laboratory by using a special observation beehive [48]; the queen can be observed laying eggs, the cells used noted, and time of egg deposition recorded. Subsequently, the cells are taken out, and the eggs are withdrawn and placed into an incubator.

TABLE 8.2. Normal Development of the Honeybee Embryo at 35.5°C (compiled by the author on the basis of data of DuPraw [18])

Stage	Time after oviposition (hours*)	Cleavage	External features	Internal features
1 – beginning	0	–	Deposition of eggs into the cell	First maturation division usually at anaphase
1 – middle	2-3	I	Appearance of a clear space between the membrane and embryo near the anterior pole	Division of the zygote nucleus yielding two energids in the "cleavage center," near the anterior egg pole
2 – beginning	4.5 ± 0.2	IV	Clear space appears near the posterior pole	Beginning of energid movement toward the posterior pole
2 – middle	–	VII-VIII	No changes	The energids spread throughout the length of the embryo change the direction of movement and begin to move to the surface
3 – beginning	8.1 ± 0.1	X	The embryo surface becomes spotted	Emergence of nuclei into the periplasm on the embryo surface begins from the differentiation center (presumptive thorax region)
3 – middle	–	XI-XII	No changes	Superficial cleavage and the formation of blastoderm
4 – beginning	10.6 ± 0.5	2-3 asynchronous divisions	Disappearance of clear space near both poles	
4 – middle	16	Mitoses stop	No changes	Appearance of the inner periplasm between the blastoderm and yolk, initially in the presumptive thorax region (see below). The cells of blastoderm form one layer of columnar cells
4 – end	–	–	Appearance of clear space near the anterior pole; the cells of females are twice as large	Inner periplasm surround the embryo poles

5 – beginning	24.5 ± 1.4	Mitoses are absent	Appearance of clear space near the posterior end	Periplasm covers all surfaces by one cell layer
5 – end	–	–	–	Columnar blastoderm cells are located in one layer
6 – beginning	32.3 ± 1.4	–	Migration of cells to the anterior pole in ventral direction. Thickening in the anteroventral part corresponding to the rudiment of the midgut	–
6 – first half	–	–	The embryo looks like an hourglass. The amnion-serosa cap is formed on the anterior pole	Beginning from the thoracic region, dorsolateral ectodermal strips migrate over the mesoderm
6 – middle	–	Mitoses reappear	Ectodermal folds on both sides are connected ventrally in the anterior third of the embryo	Bilaminar structure appears for the first time in the differentiation center presumptive thorax region
6 – end	–	–	Closing of lateral folds of the amnion-serosa is complete	The embryo acquires the shape of "gothic arch"
7 – beginning	38.3 ± 1.4	–	Ectodermal folds merge up to the posterior end of the embryo. When viewed laterally the embryo resembles a canoe, its hollow formed by thin mesoderm and dense rudiment of the midgut anteriorly and posteriorly. Yolk on the dorsal side is covered only by thin amnion-serosa. The amnion-serosa completely covers the whole anterior part of the embryo up to the transverse groove of the ventral side.	
7 – middle	–	–	–	Groove of the ventral side moves in the posterior direction; the amnion-serosa closes on the ventral side
8 – beginning	44.0 ± 1.1	–	Posterior pole becomes covered with amnion-serosa. The embryo is greatly contracted	–
8 – middle	–	–	An ectodermal thickening, the rudiment of the labrum, is formed dorsally at the anterior pole	Formation of the midgut from the anterior and posterior rudiments

TABLE 8.2 (continued)

Stage	Time after oviposition (hours*)	Cleavage	External features	Internal features
9 – beginning	46.9 ± 2.0	–	Segregation of head and beginning of segmentation	–
9 – middle	–	–	The labrum projects as a horn. The embryo elongates and stretches the amnion-serosa	Appearance of the nervous system, oral appendages and posterior gut
10 – beginning	66.6 ± 2.6	–	–	–
10 – middle	–	–	The larva turns through 180° along its longitudinal axis: the ventral side is now turned to the concave side of the chorion. The amnion-serosa stretches by larval movement and breaks up, air enters tracheal tubes	–
10 – end	70.0 ± 2.0	–	Chorion ruptures. The larva creeps out through the hole	

*Time of onset of some of the substages is not given in the paper by DuPraw [18].

Another technique, suggested by Petrunkevich as long ago as 1903, can be used [40]. A good honeycomb devoid of eggs is placed for one or more hours into the center of the normal (preferably strong) colony; the queen uses the comb to deposit eggs. The comb may be placed into a chamber formed by a diaphragm which does not restrict the worker bees but limits the queen's movements. Egg-laying is more successful if the adjacent frames (combs) contain abundant unsealed brood and stored pollen as well as numerous nurse bees, while the cells in the comb used by the queen are cleaned (polished) by the bees themselves. Such a frame can sometimes be taken from another colony whose queen has not yet started oviposition.

The comb with the eggs taken out of the beehive should be protected from direct sunlight, wind, and prolonged chilling. If the eggs are not used immediately, they should be prevented from drying up by wrapping the comb in a wet towel. It is more convenient to take eggs out of dark cells where they are attached to cocoon remnants rather than wax. Using a sharpened dental scalpel, part of the cell base containing the egg can be excised and picked up using small watchmaker forceps. Care should be taken not to touch the egg. The excised piece is transferred to a Petri dish covered with a layer of wax. The cell base with the attached egg is glued to the wax and a piece of wet cotton wool is placed nearby to maintain high humidity. Eggs may also be incubated in paraffin oil. For this purpose, the region of the cell base containing the egg is held in watchmaker forceps while the egg is carefully pushed by the rounded tip of a glass microneedle into an individual cup containing the warmed oil. It is better to use relatively light paraffin oil, since heavier types cause developmental abnormalities. When the eggs are submerged in oil, they are prevented from drying up. Incubation is conducted at 35.5°C. Embryos placed into oil become translucent. When the lens of a microscope is placed into oil, one can see the cells and nuclei on the surface of the embryo [18].

8.7. NORMAL EMBRYONIC DEVELOPMENT

The life cycle of the honeybee consists of an embryonic period as well as larval, pupal, and adult stages.

Embryonic development has been studied in great detail using fixed preparations by Nelson [38] and Schnetter [48]. Schnetter distinguished four periods subdivided into 21 stages, and described both the outer and inner structure of the embryo. DuPraw [18] traced sequential changes in individual embryos at 35.5°C using time-lapse cinematography and distinguished 10 stages of development; each was divided into early, middle, and late substages. Below the main structural characteristics of embryos, as well as their changes during each stage, are given according to DuPraw [18]. Figure 8.2 illustrates embryos at different stages, also taken from DuPraw (see also Table 8.2).

Stage 1 (Fig. 8.2, 1). Duration about 4.5 h. The stage begins with oviposition and ends when the cleavage nuclei and the surrounding cytoplasm (the energids) begin to migrate rapidly from the anterior part of the egg toward the posterior pole. Two maturation divisions, karyogamy and the first three synchronous cleavage divisions, take place during this period in the anterior part of the egg (cleavage center). There is a 30-35 min interval between divisions. In the middle of the stage the protoplasm becomes slightly compressed in the anterior direction, leaving a clear space full of liquid under the membrane near the anterior pole.

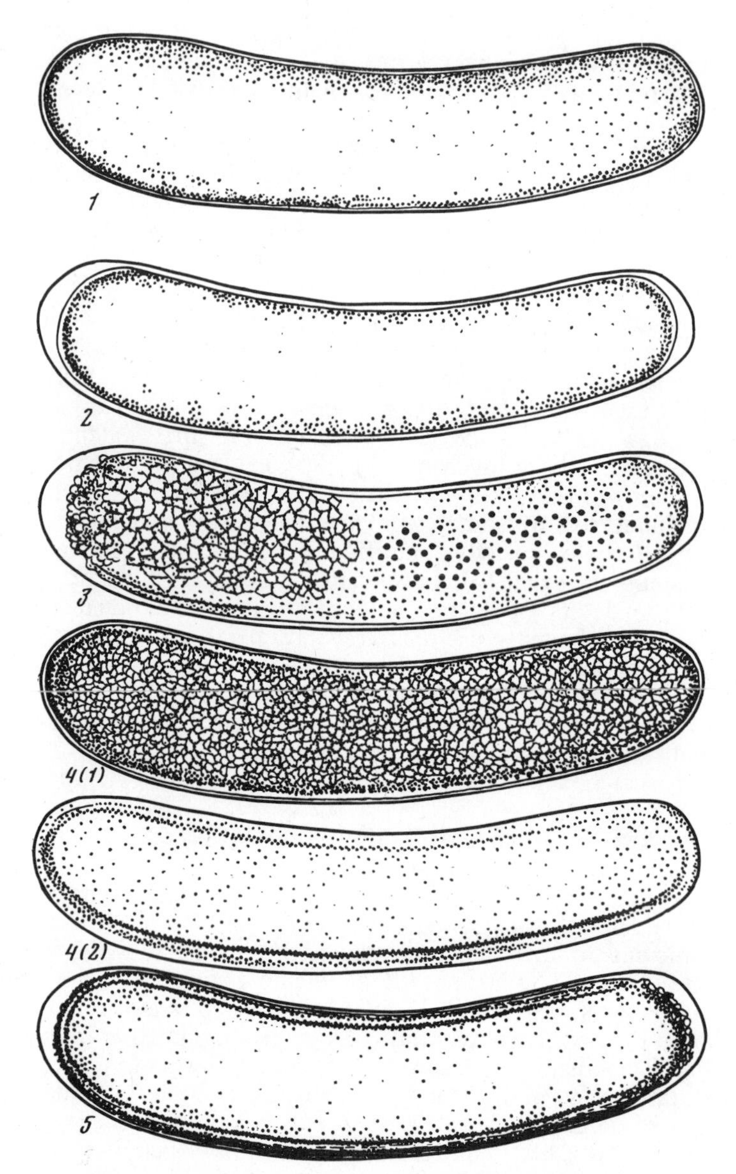

Fig. 8.2. Normal developmental stages of the honeybee, *Apis mellifera* L. [18]. Sketch numbers correspond to stage numbers. Lateral views are given of all stages; 6(2) is viewed from above. Anterior pole is on the left; posterior pole, attached to cell, is on the right. The convex side of the embryo is ventral. 1) Beginning of stage 1; 2) middle of stage 2; 3) early phase of stage 3; 4(1)) early phase of stage 4; 4(2)) late phase of stage 4; 5) middle of stage 5; 6(1)) early phase of stage 6; 6(2)) middle of stage 6; 7, 8, 9) middle of stages 7, 8, 9, respectively; 10) late phase of stage 10.

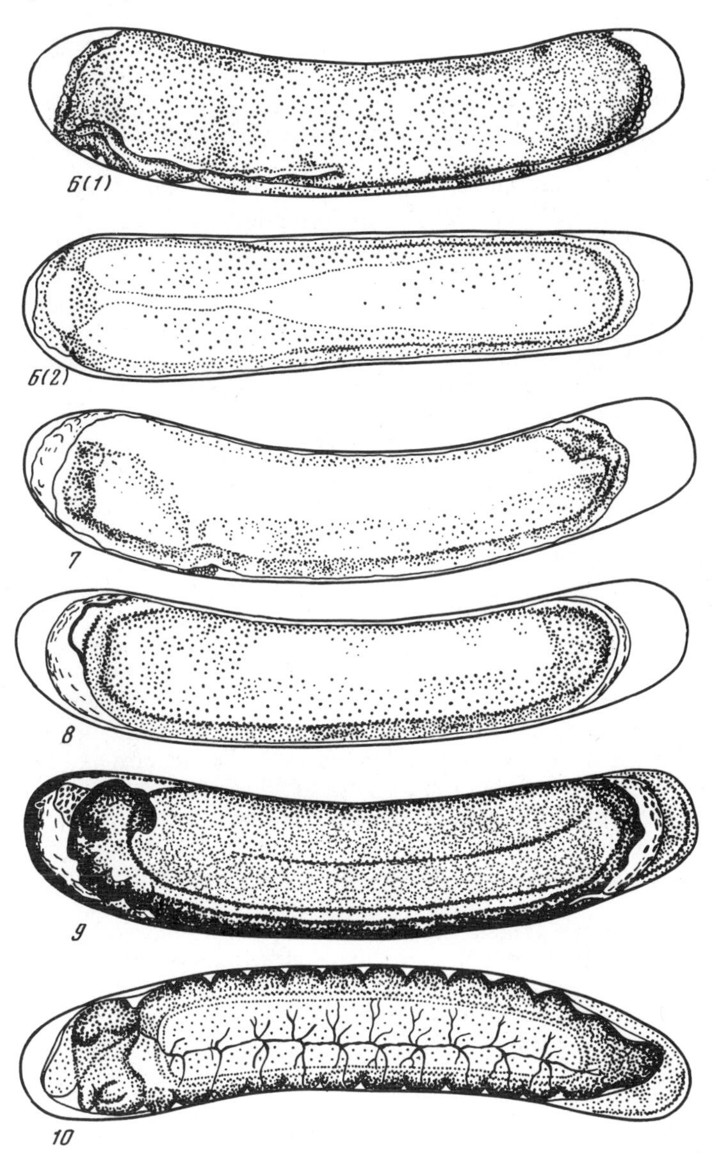

Stage 2 (Fig. 8.2, 2). Duration about 3.6 h. This stage begins during cleavage division IV as a sudden rapid migration of energids from the anterior position toward the posterior pole of the egg. Synchronous cleavage divisions V-VII continue during this migration and the nuclei disperse uniformly throughout the egg. This process takes 110-115 min. At the beginning of this stage a clear space appears at the posterior pole owing to a certain compression of yolk (Fig. 8.2, 2). At the 128-nuclei stage, nuclear migration toward the posterior pole stops and the nuclei begin to move radially from the protoplasm toward the surface, while the

synchronous cleavage divisions continue (VIII-X). At the 1024-nuclei stage the energids enter the periplasm at the surface of the embryo. The radial migration of nuclei takes 100-115 min. At first the energids reach the surface in the presumptive thorax region, which is the differentiation center. During this stage the embryo has a characteristic spotted appearance.

Stage 3 (Fig. 8.2, 3). Duration about 2.5 h. This stage represents the period of "superficial cleavage" or blastoderm formation. It begins with the surface around each cleavage nucleus being indented; protuberances, particularly distinct in the anterior part, are formed. Just as in the later cleavage stages, this process is not quite synchronous and takes place earlier in the presumptive thoracic region. This region, the differentiation center, stains intensely with thionine [48]. The spindles are located tangentially to the egg surface. These divisions are not completely synchronous, but the interval between them (35 min) is similar to that of the observed preceding divisions. Numerous furrows are formed. The protoplasm begins to expand, and the egg again occupies the whole volume inside the membrane.

Some energids (about 10%) do not enter the periplasm in the course of cleavage, but remain in the protoplasm forming the yolk nuclei, subsequently forming vitellophages. A similar process of cleavage may take place in enucleated eggs as well; this is referred to as the pseudocleavage.

Stage 4 [Fig. 8.2, 4(1) and 4(2)]. Duration about 14 h. The transition from the end of stage 3 to the beginning of stage 4 is marked by the resorption and disappearance of the clear fluid-filled spaces at the two poles of the egg. Externally, the embryo closely resembles a newly laid egg but differs from it only because blastoderm cells are present along the boundary profile. The embryo shows no major changes during this period. Approximately 5.5 h after the beginning of stage 4 (16-h embryo), a thick layer appears between the blastoderm and the yolk mass, which initially can be seen better in the anterior ventral part of the protoplast at the differentiation center. This layer comprises the inner embryonic blastema or inner periplasm. It gradually spreads to the spinal side of the embryo (Fig. 8.2). It is absent during the early phases of stage 4, but in later phases it surrounds both poles of the embryo.

By the end of stage 4, the protoplast again contracts anteriorly. This is accompanied by the appearance of a clear region near the anterior pole (Fig. 8.2, 5). The blastoderm cells begin to rearrange into a layer of columnar cells. They soon undergo a series of asynchronous mitoses, the spindles of which are located obliquely to the surface; this yields two cell layers. The only region of the blastoderm which does not change is the narrow longitudinal streak along the center line of the spinal side. It does not form any part of the embryo, but may contribute to the extra-embryonic membrane (amnion-serosa); in this region only one layer of spheroid or cuboid cells can be found. The phase of the two-layer embryo lasts about 4 h. During this period the cells are located in a very narrow space between the external chorion and the inner basal membrane. The total number of cells reaches 19,000 in the diploid worker bee and 34,000 in the haploid drone [41]. This difference is equivalent to 2-3 additional mitotic divisions per cell. However, one cannot be sure that all cells participate similarly in proliferation.

Beginning at about 16 h, two major changes develop more or less simultaneously in the blastoderm structure: first, the inner periplasm makes its appearance beneath the basement membrane in the presumptive thorax region; second, in the same region, the blastoderm cells begin to rearrange themselves as a single layer of almost identical columnar cells with uniformly apical nuclei. Both of these changes progress slowly until the two layers converge, resulting in the disappearance of the basement membrane followed by the fusion between the inner periplasm and the cytoplasm of the blastoderm cells. At this stage, the blastoderm cells, which previously were discrete and typically uninucleate, are once more in continuity with each other through the inner blastoderm layer. The exception is the narrow longitudinal mid-dorsal strip. Even in late embryos the inner periplasm tapers to an end on either side of this strip. The other parts of the blastoderm at any given moment can be found in widely different transition stages until the inner periplasm formation is completed. By that time the plasma layer has a rather grainy appearance and a regular profile: it is thickened in the region of the differentiation center and tapers in thickness toward the anterior and posterior poles of the embryo and from the ventral midline toward the dorsolateral areas. This tapering may be regarded as a manifestation of anterior–posterior, medial–lateral, and ventral–dorsal gradients in the embryo. The mapping of rudiments has been performed by Schnetter [48] and Reinhardt [41]. Little is known about the formation of the inner periplasm and the interrelationships between its appearance and its transition to the columnar blastoderm. At the end of stage 4, the honeybee blastoderm is similar in structure to the *Drosophila* periplasm, which has incomplete columnar cells arising from the inner periplasm during superficial cleavage and are separated from each other by flask-shaped furrows. After the appearance of the inner periplasm until the latter half of stage 6, no further mitoses can be observed in the embryo. At the end of stage 4 or the beginning of stage 5 the drones can be distinguished from worker bees by the size of their cells: in the worker bee they are about twice as large as those of the drone.

Stage 5 (Fig. 8.2, 5). Duration about 8.3 h. When the embryo is approximately 24-h old, a new yolk contraction begins. This is accompanied by the slow reappearance of clear fluid-filled spaces at the anterior and posterior poles. By the beginning of stage 5, the periplasm covers the whole egg with a layer of cells. No significant outer changes take place at this stage, but many important morphogenetic movements are being prepared. They take place at the next stage (stage 6). This period signals the re-establishment of basal cytoplasmic membranes which complete the separation of blastoderm cells from the ventral yolk mass; the inner periplasm is distributed more or less uniformly between the surface cells. In the middle of stage 5, traces of periplasm are still conspicuous in the distinctly granular basal cytoplasm of the neighboring blastoderm cells. By the end of this stage, the granular plasm can no longer be distinguished and is apparently replaced in many or all of the cells by a single large vacuole located basally to the nucleus. This vacuole persists during stages 6 and 7 and survives longer in the ectoderm than it does in the mesoderm. DuPraw assumes that the inner periplasm is not absorbed by the blastodermal cells, but incorporated as such.

Stage 6 [Fig. 8.2, 6(1) and 6(2)]. Duration about 6.2 h. No significant changes take place at the beginning of stage 6. Longitudinal sections show an ellipse of several hundred virtually identical columnar cells enclosing the central yolk mass. The nuclei in these cells are located apically. The most differentiated cells of the embryo, which possess translucent vacuolized cytoplasm, form a cap over the anterior pole.

The map of rudiments in the blastoderm at the beginning of stage 6 was described by Schnetter [48]. It consists of six longitudinal strips made up as follows. A mid-ventral strip occupying about one-sixth of the egg circumference; its anterior and posterior tips form the endoderm, while the middle part forms the mesoderm.

Two dorsolateral ectodermal strips, each with a width of about one-third of the egg circumference, move during stage 6 in the ventral direction and cover the middle part of the ventral mesoderm; these two strips form the spinal parts of the extra-embryonic membrane (amnion-serosa) which are stretched into squamous epithelium as the ectoderm migrates ventrally. A narrow mid-dorsal strip, according to Nelson [38], forms a syncytium on the surface of the yolk mass.

The beginning of stage 6 is marked by a sudden ventrally directed migration of blastoderm cells at the anterior pole. The anterior midgut rudiment [Fig. 8.2, 6(1)] appears as a conspicuous thickening in the anteroventral blastoderm. The similar but less-conspicuous posterior midgut rudiment soon appears near the posterior embryo pole. After the appearance of the anterior midgut rudiment the cap of vacuolized cells lying over the anterior pole separates from the yolk, which is covered by the basal membrane and forms a hoodlike cell layer between the chorion and the protoplast [Fig. 8.2, 6(2) and 7]; this is the anterior part of the amnion-

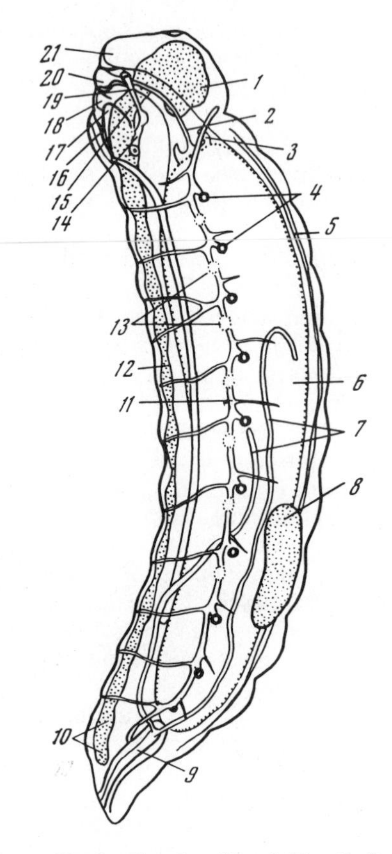

Fig. 8.3. Honeybee larva immediately after hatching [48]. 1) Suprapharyngeal ganglion; 2) foregut; 3) mushroomlike body; 4) stigmata; 5) heart; 6) midgut; 7) Malphigian tubules; 8) ovary; 9) hindgut; 10) fused ganglia 12 and 13; 11) trachea; 12) ventral nervous chain; 13) endocytes; 14) tectorium 2; 15) subpharyngeal ganglion; 16) weaving gland; 17) maxilla; 18) labium; 19) tectorium 1; 20) mandible; 21) labium.

serosa which is continuous with the two dorsolateral membrane rudiments. More or less simultaneously with the separation of the anterior serosa the two ventrolateral sheaths of ectoderm begin to migrate ventrally over the mesodermal plate. Early in stage 6 these ectodermal folds can be seen as short longitudinal borders lying ventrolaterally just behind the anterior midgut rudiment [Fig. 8.2, 6(1)]; thereafter the folds gradually become longer in a posterior direction while they move slowly toward the ventral midline.

From the ventral side the embryo, at this stage, exhibits a conspicuous hourglass configuration which is clearly visible under oblique lighting. By the middle of stage 6 the ectodermal folds from each side meet ventrally in the anterior third of the embryo (differentiation center) [Fig. 8.2, 6(2)], where a full two-layer condition

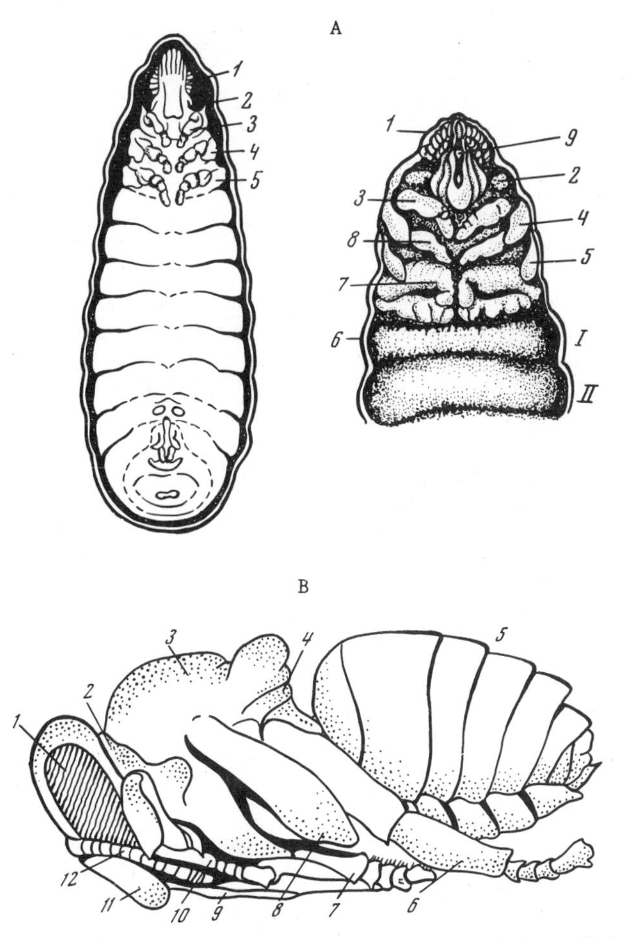

Fig. 8.4. Prepupa and pupa of the honeybee. A. Prepupa, ventral view. 1) Antenna; 2) compound eye; 3) foreleg; 4) forewing; 5) hind wing; 6) larval integument; 7) hind leg; 8) mid-leg; 9) oral appendages; I, II) abdominal segments [31]. B. Pupa, lateral view. 1) Compound eye; 2–4) thorax – anterior, middle, and posterior regions, respectively; 5) abdomen; 6) hind leg; 7) mid-leg; 8) wing; 9) proboscis; 10) antenna; 11) mandibles; 12) antennae.

is first established, with the ectoderm overlying the mesoderm. Now the embryo acquires the shape of a "gothic arch." The fusion of the lateral folds spreads from the thoracic region in a posterior direction. The protoplasm contraction, which began at the end of stage 5, gradually progresses and the "gothic arch" becomes shorter and narrower. By the end of stage 6 the lateral folds completely meet each other and the arch disappears altogether. Time-lapse films of whole embryos in oblique light also show clearly that the ectodermal sheaths move ventrally, thereby resulting in the concentration of the original blastoderm cells in the ventral half of the embryo; at the spinal side, the cells of the serosa stretch out, abandoning their original columnar configuration, and cover the free protoplast surface with squamous epithelium. By the end of stage 6 the yolk mass of the embryo's spinal half is only covered by this thin amnion-serosa. As the result of this displacement, true embryonic tissues become located on the ventral surface of the protoplast where the midplate, consisting of endoderm and mesoderm, is shaped rather like a canoe [48]; the bow and stern correspond to the dense midgut rudiment, while the concave hole between them comprises a thin inner layer of mesoderm. The columnar ectoderm is located externally to the mesoderm, but does not as yet spread over the rudiment of the midgut. Finally, the amnion-serosa forms a kind of dorsal canopy whose ventral margins fit exactly into the corresponding dorsal margins of the ectoderm (Fig. 8.2, 7).

Initially uniform blastoderm cells at stage 6 also undergo differentiation: the nuclear diameter after fixation reaches 10 μm in the columnar ectoderm, 7.5 μm in the anterior and posterior endodermal aggregates, but only about 6 μm in the mesoderm. DuPraw described the changes in the ectodermal and mesodermal cells during this period, changes which may relate to inductive interactions between them (see also Sauer [46] and Bertzbach [4]).

Stage 7 (Fig. 8.2, 7). Duration about 7.5 h. During this period the ventral amnion-serosa is gradually detached from the embryonic tissues. In profile, embryos of this stage show a conspicuous "dent" with a transverse groove behind it. The groove is formed at the posterior margin of the amnion-serosa. The amnion-serosa completely closes the whole anterior part of the embryo up to the transverse groove on the ventral side (Fig. 8.2, 7); it is located only dorsally behind this groove, where it is still attached to the lateral margin of the ectoderm. As the amnion-serosa continues to grow, the groove slowly moves toward the posterior pole. The early, middle, and late phases of stage 7 can be distinguished by their positions. By the end of stage 7, the amnion-serosa covers both poles of the embryo and is located in the fluid-filled spaces between the protoplast and the chorion (Fig. 8.2, 8).

The protoplast continues to shorten throughout stage 7. Correspondingly, the canoe-shaped embryo curves dorsally around the anterior and posterior poles to such an extent that both midgut rudiments come to lie on the dorsal surface of the yolk mass.

Stage 8 (Fig. 8.2, 8). This is a short transition period lasting about 2.4 h. It begins with the amnion-serosa covering the embryo's posterior pole. The protoplast contracts to a minimum during this stage. An ectodermal thickening corresponding to an unpaired rudiment of the labium forms near the anterior pole. The midgut forms from the anterior and posterior midgut rudiments. Neither the head nor the paired structures have appeared yet.

Stage 9 (Fig. 8.2, 9). Duration about 19.5 h. Stage 9 includes periods in which the larva has already developed, but is not yet capable of active muscular contractions. The stage begins from the appearance of a well-defined head and 13 body segments, namely 3 thoracic and 10 abdominal. Organogenesis takes place during this period. The labrum juts out like a unicorn's horn. Three pairs of head appendages, the mandibles and two maxillae, become prominent as knobs of tissue along the ventral sides of the head. Two large ganglia are laid down as well as the ventral nervous chain. The two ganglia are connected with the circumesophageal ring. A proctodeal invagination occurs near the posterior tip of the abdomen; it does not open into the midgut until the end of the larval feeding period which precedes pupation. The tracheal system is laid down. From the middle of this stage, the embryo starts to elongate, stretching the amnion-serosa by means of the head and posterior tip of the abdomen. The posterior end of the body becomes more pointed (Fig. 8.2, 9).

Stage 10 (Fig. 8.2, 10). Duration about 3 h. This stage corresponds to hatching. It begins with active muscular movements in the head region resembling faint "nodding"; gradually this movement becomes more vigorous and extends to more anterior segments until the whole larva shows a kind of rhythmic pulsing. As soon as the larva has enough freedom of movement, it begins a 180° rotation around its long axis, bringing its ventral side to the concave surface of the chorion.

After rotation has been accomplished, the peristalsis-like movements become more vigorous and the larva expands somewhat in volume. The amnion-serosa becomes visibly stretched each time the larva moves its tail. Finally, the serosa breaks up into fragments. Soon after this fragmentation the tracheal tubes rapidly fill up with air. This makes the tracheal system opaque and the tubes become clearly visible from the outside. The larva continues its active movements; several minutes later the chorion suddenly breaks up in one or several places, and the larva creeps out from the chorion through the gradually enlarging hole.

It has been shown [13, 14] that in the course of hatching the chorion becomes partially dissolved by the hatching hormone which is secreted no earlier than 10 min prior to normal larva hatching.

The normal table of honeybee embryonic development (Table 8.2) has been compiled using the data of DuPraw [18], who studied the embryonic development in paraffin oil at a constant temperature by time-lapse cinematography.

A relative dimensionless unit of development τ_0 (which has been suggested by Dettlaff and Dettlaff [8]) according to DuPraw is 30-35 min at 35.5°C. It has not been used in Table 8.2. Since a constant temperature is maintained when the honeybees are rearing their young, τ_0 may be used to compare honeybee development with the embryonic development of other insect species.

8.8. LARVAL DEVELOPMENT

The larvae of worker bees and drones after hatching are fed with the nurse bees' "milk," secreted by their salivary glands. From the third day onward they are fed with a mixture of honey and pollen. The queens' larvae are fed only on the "milk," but its composition is somewhat different.

A hatched larva weighs 0.08-0.1 mg. The larval weight of worker bees belonging to two races (in milligrams) is given on the next page as a function of time of development (Smaragdova, cited from [60]).

Age of larva (days)	1	2	3	4	5	6
Grey mountain Caucasian race	1.3	3.8	12.4	46.1	78.7	133.9
Central Russian race	1.6	6.7	14.5	48.0	91.0	160.5

The developing larvae of the worker bee pass through four molts, each of which lasts about 30 min. These molts take place after hatching at 12-18, 36, 60, and 78-89 h (Bertold, cited in [60]). The larvae may be cultured outside the beehive [35].

The structure of the larva immediately after hatching is shown in Fig. 8.3. The midgut and Malpighian tubules are separated from the posterior gut and only become connected with it prior to pupation, thereby making defecation possible. The development of larvae and pupae has been described by Myser [36].

8.9. THE PREPUPA AND PUPA OF THE WORKER BEE

After the cell is sealed by worker bees, the larva weaves a primitive cocoon. If, for some reason, the cell remains unsealed, the larva usually falls out. After weaving the cocoon the larva becomes immobile. This is the prepupal stage (Fig. 8.4A). Segregation of the definitive head, thorax, and abdomen proceeds under the larval integument. Eyes, antennae, and oral apparatus are formed on the head; rudiments of both pairs of wings and of three pairs of legs are formed on the thorax. The pupal organs appear, while the larval inner organs undergo histolysis. Instead of the 4 large Malpighian tubules, up to 100 thin ones are formed. The 7 ganglia of the ventral nervous chain of the pupa and the adult insect are derived by the partial fusion of 11 larval ganglia.

The fifth molt takes place at the end of the prepupal stage, the larval integument being discarded. The pupa looks like the adult honeybee but is devoid of pigmentation. The antennae, legs, and wings are pressed to the body and are in a non-stretched state (see Fig. 8.4B). The pupa gradually darkens, beginning with the eyes.

After the sixth molt, the adult bee perforates the cell cap with its mandibulae and emerges from the honeycomb. Its first actions are to clean the cells and feed the older larvae with honey and stored pollen. The young bees feed themselves, and their salivary glands develop, thereby making it possible for them to feed the younger larvae.

REFERENCES

1. W. W. Alpatov, "Biometrical studies on variation and race of the honeybee (*Apis mellifera* L.)," *Quart. Rev. Biol.* **4**, 1–58 (1929).
2. W. W. Alpatov, *Varieties of the Honeybee and Their Use in Agriculture* [in Russian], MOIP, Moscow (1948).

3. G. Barrat, "Fertilizing drone eggs – an experiment," *Am. Bee J.* **59**, 12, 415–416 (1919).
4. R. Bertzbach, "Experimentelle Untersuchungen über den Einfluss von Röntgenstrahlen auf die Embryonalentwicklung der Honigbiene," *Wilhelm Roux's Arch. Entwicklungsmech. Org.* **152**, 524 (1960).
5. A. Bryukhanenko, *Apiculture* [in Russian], Moscow (1926).
6. R. Chauvin, *Traité de Biologie de l'Abeille*, Vols. I-V, Paris (1968).
7. G. B. Deodikar, C. V. Thakar, and P. N. Shah, "Cytogenetic studies in Indian honeybees. I. Somatic chromosome complements in *Apis indica* and its bearing on evolution and phylogeny," *Proc. Ind. Acad. Sci.* **49**, 3, 194–206 (1959).
8. T. A. Dettlaff and A. A. Dettlaff, "Concerning the dimensionless characteristics of developmental duration in embryology," *Dokl. Akad. Nauk SSSR* **134**, 199–202 (1960).
9. W. Drescher and W. C. Rothenbuhler, "Gynandromorph production by egg chilling," *J. Hered.* **54**, 5, 195–201 (1963).
10. W. Drescher and W. C. Rothenbuhler, "Sex determination in the honeybee," *J. Hered.* **55**, 3, 90–96 (1964).
11. E. J. DuPraw, "Supporting evidence for nuclear transplant using eggs of the honeybee," *Anat. Rec.* **132**, 429 (1958).
12. E. J. DuPraw, "Further development in research on the honeybee egg," *Glean. Bee Cult.* **88**, 2, 105–111 (1960).
13. E. J. DuPraw, "A unique hatching process in the honeybee," *Trans. Am. Microsc. Soc.* **80**, 185 (1961).
14. E. J. DuPraw, "Techniques for the analysis of cell function and differentiation using eggs of the honeybee," *Proc. 16th Int. Congr. Zool., Washington*, Vol. 2 (1963), p. 238.
15. E. J. DuPraw, "The organization of honeybee embryonic cells. 1. Microtubules and amoeboid activity," *Dev. Biol.* **12**, 53 (1965).
16. E. J. DuPraw, "The organization of nuclei and chromosomes in honeybee embryonic cells," *Proc. Natl. Acad. Sci. USA* **53**, 1, 161–168 (1965).
17. E. J. DuPraw, "Macromolecular organization of nuclei and chromosomes: a folded fiber model based on whole mount electron microscopy," *Nature (London)* **206**(4982), 338–343 (1965).
18. E. J. DuPraw, "The honeybee embryo," *in: Methods in Developmental Biology*, F. H. Wilt and N. K. Wessels (eds.), T. Y. Crowell, New York (1967).
19. J. Dzierzon, "Über die Fortpflanzung der Biene," *Bienenzeitung* **1**, 131 (1845).
20. H. Fahrenhorst, "Nachweis ubereinstimmender Chromosomen-Zahlen (n = 16) bei allen 4 Apis-Arten," *Apidologie* **8**(1), 89–100 (1977).
21. K. Frisch, *Aus dem Leben der Biene*, Springer-Verlag, Berlin–New York (1977).
22. G. K. Goetze, *Die Honigbiene in natürlichen und künstlicher Zuchtauslese*, Vols. 1–2, Verlag Paul Parey, Hamburg (1964).
23. H. Gontarskie, "Mikrochemische Futtersaftuntersuchungen und die Frage der Koniginnentstehung," *Leipzig Bienenzeitung* **63**, 157 (1949).
24. H. Gontarski, "Über physiologische Unterschiede bei Bienen verschiedener Abstammung," *Z. Bienenforsch.* **2**, 98–108 (1953).
25. K. Grout, *The Hive and the Honeybee*, Hamilton (1946).
26. P. M. Komarov, "Types of honeybee intermediate forms," *Zool. Zh.* **14**, 1, 171–192 (1935).

27. A. M. Kovalev, A. S. Nuzhdin, V. I. Poltev, and G. F. Taranov, *Textbook of the Apiculturist* [in Russian], Kolos, Moscow (1973).
28. G. A. Kozhevnikov, "Materials concerning natural history of the honeybee," *Izv. Obshestva Lyubitelei Estestvoznaniya, Antropologii i Etnografii, 49, Tr. Zool. Otd.* **14**, Parts 1–2, 325 (1900–1905).
29. G. A. Kozhevnikov, "Transition forms between the queen and worker bee," *Biol. Izv.* **1** (1922).
30. H. H. Laidlaw and K. W. Tucker, "Diploid tissue derived from accessory sperm in the honeybee," *Genetics* **40**, 6, 1439–1442 (1964).
31. F. A. Lavrekhin and S. V. Pankova, *Biology of the Honeybee Colony* [in Russian], Kolos, Moscow (1969).
32. O. Mackensen, "The occurrence of parthenogenetic females in some strains of honeybee," *J. Econ. Entomol.* **36**, 465–467 (1943).
33. O. Mackensen, "Viability and sex determination in the honeybee (*Apis mellifera* L.)," *Genetics* **36**, 5, 500–509 (1951).
34. R. W. Merriam and H. Ris, "Size and DNA contents of nuclei in various tissues of male, female, and worker honeybee," *Chromosoma* **6**, 522–538 (1954).
35. A. S. Michael and M. Abramovitz, "A method of rearing honeybee larvae *in vitro*," *J. Econ. Entomol.* **48**, 1, 43–44 (1955).
36. W. C. Myser, "The larval and pupal development of the honeybee *Apis mellifera* Linnaeus," *Ann. Entomol. Soc. Am.* **47**, 683 (1954).
37. H. Nachtsheim, "Cystologische Studien über die Geschlechtsbestimmung bei der Honigbiene (*Apis mellifica* L.)," *Arch. Zellforsch.* **11**, 169–241 (1913).
38. J. A. Nelson, *The Embryology of the Honeybee*, Princeton University Press, Princeton (1915).
39. A. Petrunkewitsch, "Die Richtungskörper und ihr Schicksal im befruchteten und unbefruchteten Bienenei," *Zool. Jahrb. Abt.* 1, **14**, 573–608 (1901).
40. A. Petrukewitsch, "Das Schicksal der Richtungskörper im Drohnenei," *Zool. Jahrb.* Abt. 1, **17**, 481 (1903).
41. E. Reinhardt, "Kernverhältnisse Eisystem und Entwicklungsweise von Drohnen und Arbeiterineneiern der Honigbiene (*Apis mellifera*)," *Zool. Jahrb.*, Abt. 1, **78**, 167–234 (1960).
42. A. Root and E. R. Root, *The ABC and XYZ of Bee Culture*, Medina, Ohio (1959).
43. W. C. Rothenbuhler, "Diploid male tissue as new evidence on sex determination in honeybees," *J. Hered.* **48**, 160–168 (1957).
44. W. C. Rothenbuhler, J. W. Gowen, and O. W. Park, "Androgenesis with zygogenesis in gynandromorphic honeybees (*Apis mellifera* L.)," *Science* **115**, 637–638 (1952).
45. F. Ruttner (ed.), *The Instrumental Insemination of the Queen Bee*, Apimondia, Bucharest (1976).
46. E. Sauer, "Keimblatterbildung und Differenzierungsleistungen in isolierten Eiteilen der Honigbiene," *Wilhelm Roux's Arch. Entwicklungsmech. Org.* **147**, 302–354 (1954).
47. F. Savada and M. C. Chang, "Tolerance of honeybee sperm in deep freezing," *J. Econ. Entomol.* **57**, 891–892 (1964).
48. M. Schnetter, "Morphologische Untersuchungen über das Differenzierungszentrum in der Embryonalentwicklung der Honigbiene," *Z. Morphol. Okol. Tiere* **29**, 114–195 (1934).

49. M. Schnetter, "Physiologische Untersuchungen über das Differenzierungszentrum der Embryonalentwicklung der Honigbiene," *Wilhelm Roux's Arch. Entwicklungsmech. Org.* **131**, 285–323 (1934).
50. M. Schnetter, "Die Entwicklung von Zerglarven in geschnurten Bieneneiern," *Zool. Anz. Suppl.* **9**, 82 (1936).
51. D. W. Schasskolsky, "Genetische Analyse der Biene nach der Nachkommenschaft der Arbeitsbienen," *Arch. Bienenkunde* **16**, 1–8 (1935).
52. D. V. Shaskolsky, "Distribution of a series of multiple alleles in theoretical populations in relation to the biology of reproduction in the honeybee," *Genetika* **4**, 41–55 (1968).
53. D. V. Shaskolsky, "A rudimental sex determination manner by a multiple allele series in Hymenoptera and lethal egg distribution," *13th International Congress of Entomology (Proceedings)*, Vol. 1 (1971), p. 349.
54. D. V. Shaskolsky, "A means for sex determination by a series of multiple alleles reproducing unequal sex ratios in panmictic populations of ancestors of Hymenoptera," *14th International Congress of Genetics (Abstracts, Part 1)* (1978), p. 482.
55. V. Yu. Shimanovskiy, *Methods of Apiculture* [in Russian], Leningrad (1926).
56. I. V. Smirnov, "New data about drone sperm," *Pchelovodstvo*, No. 2 (1953).
57. R. E. Snodgrass, *Anatomy of the Honeybee*, Comstock Publishing Association, Ithaca (1956).
58. W. S. Taber, "Evidence of binucleate eggs in the honeybee," *J. Hered.* **46**, No. 4, 156 (1955).
59. W. S. Taber and M. S. Blum, "Preservation of honeybee semen," *Science* **131**, 1734–1735 (1960).
60. G. F. Taranov, *Anatomy and Physiology of the Honeybee* [in Russian], Kolos, Moscow (1968).
61. V. V. Tryasko, "Criteria of the honeybee queen insemination," *Pchelovodstvo*, No. 11, 25–34 (1951).
62. V. V. Tryasko, "Natural parthenogenesis in the honeybee," in: *20th International Apiculture Congress* [in Russian], Kolos, Moscow (1965).
63. V. V. Tryasko, "Obtaining eggs from queens kept outside the beehive," in: *21st International Apiculture Congress* [in Russian], Kolos, Moscow (1967).
64. V. V. Tryasko, "Female parthenogenesis in the honeybee," in: *Problems of Apomixis in Plants and Animals* [in Russian], Nauka, Novosibirsk (1973).
65. K. W. Tucker, "Automictic parthenogenesis in the honeybee," *Genetics* **43**, 299–316 (1958).
66. J. Woyke, "Geneza powstawania niezwykych pszczol.," *Pszeel. Zesz. Nauk* **6**(2), 49–64 (1962).
67. J. Woyke, "Drone larvae from fertilized eggs in the honeybee," *J. Apicult. Res.* **2**, 19–24 (1963).
68. J. Woyke, "Genetic proof of the origin of drones from fertilized eggs of the honeybee," *J. Apicult. Res.* **4**, 7–11 (1965).
69. J. Woyke, "Genetic aspects of artificial insemination," in: *The Instrumental Insemination of the Queen Bee*, F. Ruttner (ed.), Apimondia, Bucharest (1976).

Chapter 9

THE SILKWORM *Bombyx mori*

V. V. Klimenko

9.1. INTRODUCTION

The scientific biography of the silkworm contains a wealth of observations of tremendous biological significance; therefore it seems inevitable that this economically important insect will become an established laboratory species in developmental biology. The advantages of using the silkworm in this area include primarily the abundance of available genetic information relating to development [20, 52, 53]. However, cytogenetic investigations of silkworm chromosomes, despite the progress in this field [35], present considerable difficulties because of the large chromosomal number ($2n = 56$) and their morphological uniformity. The high polyploidy of the silk-gland is as yet insufficiently studied at the chromosomal level; the supposedly polytenic chromosomes of this organ do not appear to possess disk organization.

One unquestionable advantage of the silkworm is associated with the possibility of artificial parthenogenetic and androgenetic cloning which, when combined with the ease of biological synchronization, provides the required amounts of isogenic material at any developmental stage from the mature egg to the adult form. Professor Astaurov's fundamental studies in the field of artificial reproduction of the silkworm were reprinted a few years ago [13], and will only be discussed briefly in this chapter.

The cytogenetic mechanism of cloning achieved in the silkworm by the heat-activation procedure is based on the lack of one meiotic reductional division following high-temperature treatment. In unfertilized eggs activated by heating, the mitotic division proceeds more or less successfully, and the maternal genotype, whether diploid or polyploid, passes unchanged to the parthenogenetic offspring (Figs. 9.9 and 9.10).

Heating the eggs immediately after spermatozoa have penetrated them induces gynogenetic development of the ameiotic type, the development being activated by spermatozoa which introduce the acrosome, centriole, extrachromosomal chromatin, and cytoplasmic self-reproducing structures into the egg [50].

Various treatments of the unfertilized egg at doses lower than those needed for ameiotic parthenogenesis enable meiotic parthenogenesis to be induced. In this case the reduction division reaches completion. The haploid pronucleus divides to yield two genetically identical daughter nuclei which then fuse to form the diploid cleavage nucleus. Such a cytogenetic mechanism leads to the appearance of completely homozygous males which are of great value in various studies [56]. In the case of meiotic parthenogenesis of tetraploids, the role of cleavage nucleus is played by the diploid pronucleus formed in the course of meiosis. Because of this, the resulting descendants (almost always females) are heterogeneous, which enables the isogenic clone to be transformed into a large number of new genetically nonidentical clones [49].

The silkworm was the first animal species in which androgenetic development up to sexual maturity was obtained [3, 4, 23]. Discussions on the general significance of androgenesis in the study of nucleocytoplasmic relations have been given in the Astaurov critical review [8], while its practical application has been described in studies about the relative significance of radiation damage for the nucleus and the cytoplasm [6, 7, 9-11, 58]; other papers in this series deal with the interaction of the nucleus and cytoplasm of various species (see [13]).

A technique allowing bipaternal androgenetic hybrids to be obtained has been developed for the silkworm [45]; this opens up new opportunities for studying nucleocytoplasmic relations.

Methods allowing one to prevent the activation of eggs following the penetration of spermatozoa, or various physical and chemical treatments, are of crucial significance in developmental studies. The eggs with blocked activation are viable and can be successfully activated later. This technique is particularly important for studying the nature of factors which unblock meiosis or lead to activation [49].

The intimate relationship between parthenogenesis and polyploidy, which can be observed in nature, is also seen in artificial temperature-induced parthenogenesis. The latter occurs best with the hybrid forms. Furthermore, the first parthenogenetic generation consists to a large extent (up to 30%) of female $2n/4n$ mixoploids. The tetraploid oocytes of such animals may be subjected to the second heat activation step in order to yield tetraploid clones. Crossing tetraploid females with diploid males results in triploid females. In these artificial temperature-induced activation is the only possible way of reproduction since bisexual reproduction results in aneuploidy and the death of the embryos. Triploid parthenogenetic clones of the silkworm have been obtained in this way.

The silkworm is the first animal species to yield to the efforts of experimenters combining bisexual reproduction with a constant level of polyploidy; this resulted in the construction of the new tetraploid race of the silkworm. Up to now this race has passed through roughly two dozen generations. Bisexual tetraploids have been obtained as follows. Initially, triploid hybrid females were obtained in crosses between *Bombyx mori* ($4n$) and *B. mandarina* ($2n$) and parthenogenetic offspring were obtained. The appearance of the mixoploid females $3n/6n$ could be expected among parthenogenetic triploid females just as at the diploid level. Crosses between the mixoploid $3n/6n$ females and diploid males of *B. mandarina* should, after the fertilization of hexaploid oocytes, yield both male and female tetraploids. Such allotetraploid males have indeed been found; in contrast to autotetraploids they showed partial fertility [12]. Further studies of these experimentally constructed parthenogenetic and bisexual forms may provide interesting opportunities [27].

The relative ease with which polyploid parthenogenetic forms are obtained and reproduced comprises yet another important biological advantage of the silkworm.

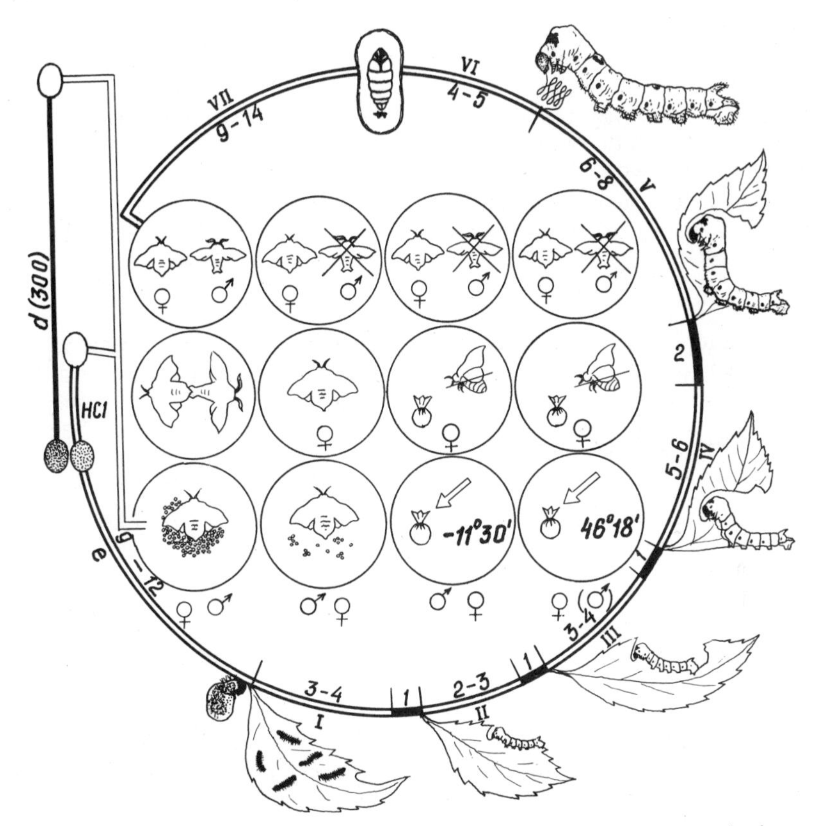

Fig. 9.1. Life cycle of the silkworm *Bombyx mori*. I–V) Larval instars; VI) spinning and prepupa; VII) pupa. Arabic numerals refer to the duration (in days) of corresponding instars, as well as to intermediate molts, etc. If the eggs obtained using one of the four techniques described (vertical columns, see text) are treated with HCl prior to diapause (d), embryogenesis (e) proceeds to the hatching of the 1st larval instar.

It could well be that the silkworm polyploids derived by other techniques, such as centrifugation, colchicine treatment, or heating of freshly inseminated eggs, may be successfully cloned using the temperature-activation procedure [30].

Cooling eggs which begin parthenogenetic development results in the appearance of autotetraploid forms [40].

The activation of eggs in the bodies of live moths of different ploidy, followed by insemination at various stages of parthenogenetic ameiotic development, is of considerable methodological significance. This provides ample opportunities for obtaining various types of polyploids combining several genomes of genetically nonidentical forms [51].

The modification of cytogenetic mechanisms participating in egg maturation and control of early developmental stages resulted in the elaboration of five experimental techniques enabling the sex to be controlled in the silkworm.

Using methods of genetic engineering at the gene and chromosomal level, Strunnikov [47] has solved the industrially important problem of sex ratio control in *Bombyx mori*.

Temperature-induced parthenogenesis and androgenesis allows artificial sex control in the silkworm; this has industrially important implications [48].

Combining artificial reproduction techniques opened up wide perspectives for solving such problems as genome analysis, obtaining genetically identical copies of completely homozygous males, and eliminating deleterious recessive genes from the gene pool of silkworm populations. The methods used provide new approaches to other equally important goals in experimental biology [47, 48].

One advantage of using the silkworm in developmental biology relates to the possibility of obtaining mosaics when two haploid derivatives of the female nucleus are fertilized by two different spermatozoa. Other types of mosaicism have also been described [26, 52].

Normal and double fertilization, androgenesis, meiotic and ameiotic types of parthenogenesis combined with different levels of ploidy provide opportunities for the construction of a wide variety of experimentally modified cytogenetic structures of developing silkworm embryos; quite a number of these opportunities have already been fulfilled. The initial discovery of artificial parthenogenesis in animals was made by Tikhomirov in 1886 [57], using the silkworm. This stimulated further studies of the phenomenon, initially using marine invertebrates and then other animal species as well. The reversibility of activation-induced changes in the course of artificial and normal activation of the egg, recently described in *Bombyx mori*, may again prove this species to be an extremely suitable model for analyzing the problem of egg activation at the molecular level [28].

Studies on the physiology and endocrinology of metamorphosis, synthesis of silk proteins, radiation-induced and chemical mutagenesis, the use of the silkworm for testing environmental pollution, its application in virology, and other subjects have been described in a recent monograph [54]. The greatest interest has always centered around the function of the silk gland. The interests of practical silk farming focused on this system which, owing to its simplicity, attracted the attention of biochemists and molecular biologists, including genetic engineers. The silk fibroin gene has been isolated and cloned by Morrow et al. [34]. Important reviews dealing with the synthesis of silk proteins include special issues of *Biochimie* [63], *Experientia* [64], and several others [1, 24].

9.2. TAXONOMY AND DISTRIBUTION

The silkworm *Bombyx mori* L. is a holometabolous insect belonging to the subclass Pterygota, order Lepidoptera, suborder Heteroneura, family Bombycidae (true silkworms).

The continental form of *Bombyx mandarina* Moore, the haploid chromosomal number of which is identical to that of *Bombyx mori* L. (28 chromosomes), appears to comprise the wild ancestor of the domesticated silkworm [14]. The island (Japanese) race of *Bombyx mandarina* has 27 pairs of chromosomes [35].

All varieties of silkworm, existing in the form of one or several stocks, may be divided into several regional groups: Japanese, Chinese, or Southeast Asian. The widest collection of silkworms, amounting to 154 stocks, is available in Japan. Two hundred improved stocks have been obtained from this collection [22, 62].

Each of these groups has a characteristic life cycle duration, constitution, resistance to disease, fertility, size, shape, color, and quality of the cocoons. One important characteristic of the variety is its voltinism, that is, its capacity to give one

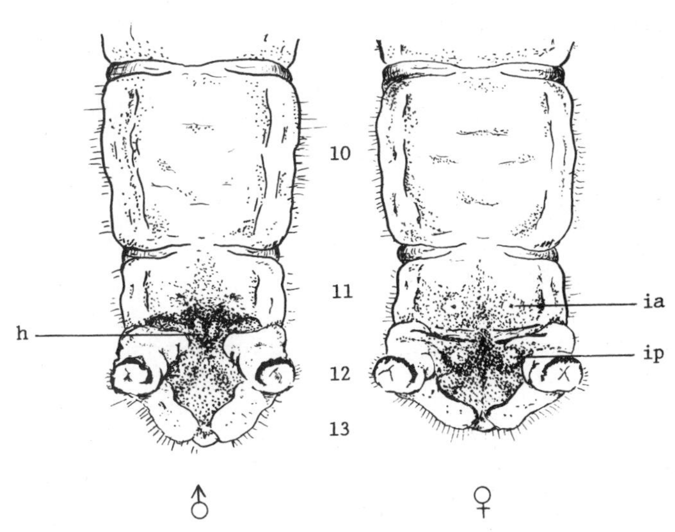

Fig. 9.2. Sexual dimorphism in *Bombyx mori* larvae. 10–13) Segments from the ventral side. h, Herold imaginal bud; ia and ip, anterior and posterior Ishiwata imaginal buds.

(monovoltinic), two (bivoltinic), or more (polyvoltinic) generations per year, under conditions simulating natural ones and including lower temperatures during spring incubation and higher temperatures in summer. With most of the varieties the larval stage is divided into five instars by four molts, but varieties with three molts are also known.

Silk is produced in more than twenty countries. The leading three are Japan, China, and Korea. The area of silkworm distribution is limited to regions where mulberry trees grow, and also depends on economic factors. Artificial media for *Bombyx mori* have been described in experimental studies [25, 53].

9.3. LIFE CYCLE AND THE GENE POOL

Typically, the life cycle of *Bombyx mori* (Fig. 9.1) includes the egg stage, which may last at least 300 days in monovoltinic varieties but may be reduced to 9-12 days by artificially shortening the diapause using hydrochloric acid treatment. It is followed by the larval stage, which is divided into five instars by four molts; the next stage is that of prepupa, which lasts about 2-3 days after the final spinning of the cocoon; the pupal stage then lasts 9-14 days, and is followed by the adult stage.

Genetic analysis has been performed for 328 traits of *Bombyx mori*. The majority of these traits belong to the egg and larval stages; this is a unique aspect of this animal. Mutations affect the following characteristics of the egg: size, shape, pigmentation of the chorion, pigmentation of the yolk, pigmentation of the serosa. Newly hatched larvae are usually dark brown with rather long bristles on all segments of the body. Mutations yielding larvae with reddish-brown pigmentation and short bristles have been described. A marked number of hereditary traits at later larval stages are associated with the pigmentation and spotted pattern of the larvae, which become particularly clearly manifested under the fourth molt (Fig. 9.2).

Other traits include body shape and hemolymph pigmentation. A number of traits are associated with the pupal stage; specifically they include the pigmentation, shape, and structure of the cocoon. At the adult stage there is a hereditary underdevelopment of the glueing glands in females, which in this case cannot glue eggs to the substrate, so the eggs become scattered. Detailed information about these and other hereditary traits may be found in a series of monographs and reviews [19, 20, 52, 53, 61, 62].

9.4. CULTURE

There is a vast amount of literature on silkworm rearing; only the leading and authoritative works are indicated here. These books, however, frequently either omit details which are important to developmental biology, or mention them only briefly [2, 16, 25, 32, 33, 37, 53]. Therefore, omitting the description of techniques used for feeding and rearing larvae, a more detailed description is given below of manipulations commonly used in laboratory practice.

9.4.1. Obtaining Synchronously Developing Eggs

Unfertilized mature oocytes at metaphase I are extremely easy to obtain from the ovarioles by either rubbing or grating them through a suitably sized capron mesh. Eggs obtained in this way are collected in a beaker containing some water, and then washed with water on a piece of cheesecloth stretch over a funnel until the residues of the ovarioles are completely removed from the surface of the eggs.

Freshly inseminated eggs are obtained as follows. Silkworm moths should first be prepared for oviposition. At the day of flight they mate several times with males which emerged on the preceding days and were kept in a cool place at about 15°C. It is not difficult to continue the mating until evening, when even virgin moths begin laying individual unfertilized eggs. At this time, mated females begin intensive oviposition at 25°C, and the moment the eggs are laid may be taken as the time of insemination. Two sheets of thick smooth paper are placed on a table in a dark place. Between 50 and 100 laying females are rapidly placed onto one sheet and allowed to continue oviposition for a certain length of time, depending on the required accuracy of synchronization (e.g., from 30 sec to 5-10 min). Thirty seconds before the end of the planned period the females are quickly transferred to the adjacent sheet of paper, and egg collection begins again. If only freshly inseminated eggs are required, then the eggs deposited on the first sheet are placed in a refrigerator, or are put into an envelope which is then stored in an incubator during the set period. At the required time the eggs are fixed or placed into a refrigerator at 0°C, where the necessary amount of material is gradually accumulated. Later this material is either fixed or subjected to other treatments. It should be borne in mind that cold treatment may lead to very serious artifacts in cytogenetic and other studies, if the eggs are allowed to develop further after chilling.

Large quantities of synchronously developing embryos may be very easily obtained by activating the eggs extracted from moths using the artificial parthenogenesis method [13]. For this purpose the moths should be taken from a parthenoclone with a high capacity to parthenogenesis [13]. The techniques used for obtaining parthenogenetic and androgenetic eggs have been described in another book of this series [38, 39].

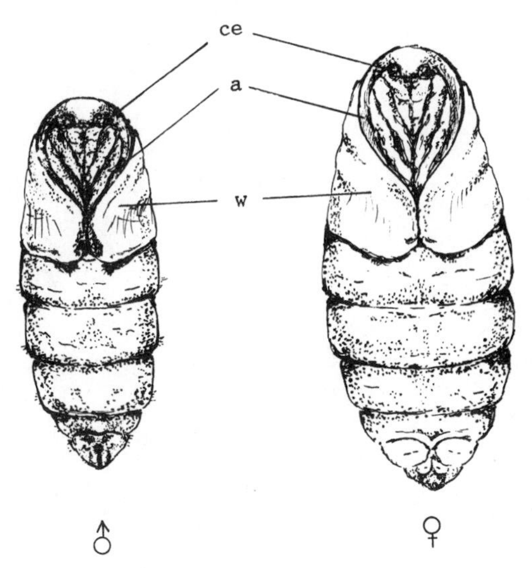

Fig. 9.3. Sexual dimorphism in *Bombyx mori* pupae. a, antenna; ce, compound eye; w, wings.

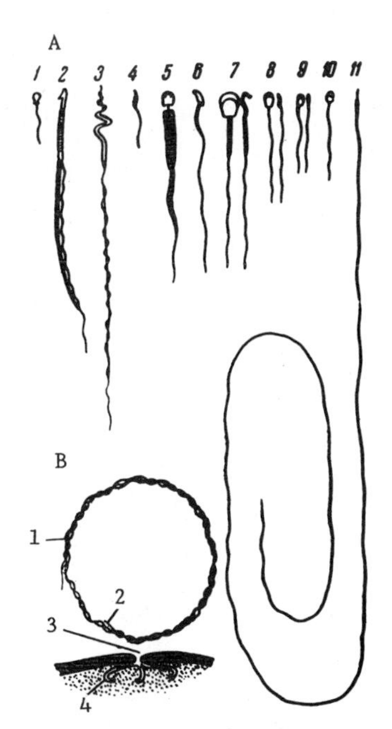

Fig. 9.4. A) Spermatozoa of silkworms (11) (from [46]) compared with those of vertebrates (1–10). B) Rotating spermatozoon of silkworm near the micropylar pole of the egg. 1) Spermatozoon; 2) sperm head; 3) micropyle; 4) micropylar channels.

9.4.2. Larvae and Pupae

Larvae just hatched from the eggs are used. To synchronize a large number of such larvae there must be correct conditions for hatching them from the eggs within a short period of time. For this purpose the hatching occurring in the morning is delayed for a day by a complete exclusion of light followed by bright illumination. The larvae hatched during a certain interval are either fixed or fed to allow development to the necessary stage.

The larvae of subsequent instars are collected immediately after the molts. The procedure consists of placing the required number of larvae on a sheet of paper during their sleeping period. Thereafter, the larvae which undergo molting within a certain period of time are collected. This technique is inapplicable in silkworm farming, but is useful in laboratory practice.

The pupae are similarly synchronized, but in order to see the prepupae one should cut the cocoons with a razor blade immediately after spinning has been completed.

9.4.3. Adults

The adults fly out during the early hours of the morning, and the material may be easily synchronized by the above technique, during the *en masse* emergence from the cocoons or from pupae, if the outer part of the cocoon has been removed.

The flight of moths may be delayed by keeping the cocoons at lower temperatures (12-18°C), but one should bear in mind that this may subsequently affect the quality of the silkworm eggs that the moths will lay.

Mating takes place during the morning. Insemination requires about 1 h of mating time, but is frequently prolonged to 2-4 h. A given male may mate several times within one day. Intervals between matings should be 2-3 h, and during this period the males are kept in a paper envelope at 10-12°C. At this temperature the fertilizing capacity of males remains unchanged for 2-3 weeks. The males may also be kept at 4-5°C.

The life cycle as well as the normal and experimental methods used for obtaining silkworm eggs of a conventional monovoltinic variety are shown in Fig. 9.1. The left vertical column in the figure refers to fertilized silkworm eggs obtained by the conventional procedure described above. The second column corresponds to spontaneous or accidental parthenogenesis. The third column corresponds to artificial meiotic parthenogenesis discovered by Sato [42] and Koltzov [31]; it allows completely homozygous males to be obtained, while females are rather rare. The most effective method of obtaining homozygous males in the course of meiotic parthenogenesis has been suggested by Terskaya and Strunnikov [55]. The fourth column corresponds to artificial ameiotic parthenogenesis, discovered by Astaurov [5]; using this procedure, one may clone the female genotypes, while the males are virtually absent.

Irrespective of the technique used for obtaining eggs of the monovoltinic silkworm varieties, several days later the egg enters diapause and the next cycle does not start until 10 months later. In the practice of silkworm breeding, the eggs are stored for this long period of time under strictly controlled temperature and humidity conditions [16, 41]. Similar rules should be followed in the laboratory. However, if there is a need to obtain the next generation quickly, the onset of diapause is prevented by treating the eggs with hydrochloric acid [53], as follows.

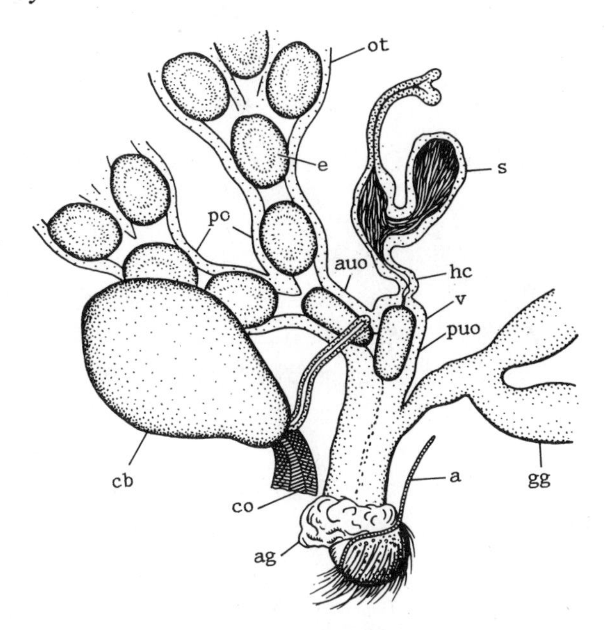

Fig. 9.5. Female genitalia of the silkworm moth [46]. ot, ovarian tubes; e, egg; po, paired oviduct; auo, puo, anterior and posterior part, respectively, of unpaired oviduct; ag, aromatic glands; a, apophysis; v, vestibulum; s, seminal receptacle with helical channel (hc); cb, copulatory bursa; co, copulatory orifice; gg, glueing glands.

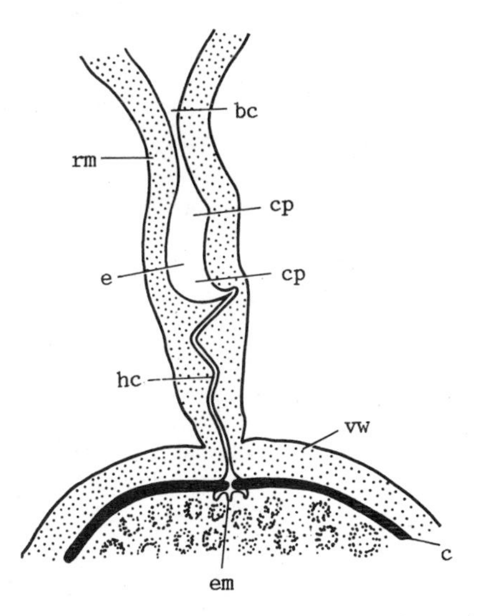

Fig. 9.6. Path of silkworm spermatozoon from the spermatheca to the egg [46]. bc, beginning of channel; rm, ring muscle; e, expansion with cylindrical part (cp); hc, helical channel; vw, vestibulum wall; em, micropyle; c, chorion.

TABLE 9.1. Developmental Stages of a Monovoltinic Variety of *Bombyx mori* L. at 25°C [17, 18]

Stage	Time after fertilization (hours)	Diagnostic features
		Prediapause development
0	0	Metaphase I of meiosis (see Fig. 10.8)
1	1-1.5	Metaphase II of meiosis (Fig. 10.12)
2	2-3	Fusion of pronuclei (Fig. 10.13)
3	3-4	Beginning of cleavage
4	10-12	Blastoderm formation
5	15-16	Formation of the embryonic streak
6	18-22	Formation of the amniotic fold, gradually closing over the embryonic streak which sinks inside; amnion and serosa formed. The yolk gradually segregates to yield spheres from periphery to the center. Formation of amnion complete. The primary streak appears on the outer side of the embryonic streak; 18 bulges are produced on its inner side from yolk cells; mesoderm differentiation proceeds inside the bulges

Stage	Time after beginning of incubation (days)	Diagnostic features
		Postdiapause development
A	0	Diapause embryonic streak. The embryo looks like an elongated plate, which is wider near the anterior micropylar end than at the posterior end. Its borders are somewhat bent inside and the entire embryo is curved along its long axis. The primary groove passes along the middle line of the concave side of the embryo (when the embryo is greatly contracted the groove may remain unseen). There are 18 mesodermal metameres
B	1	Elongated embryonic streak with distinct metamerism. The embryo undergoes elongation and thinning. There is a distinct metamerism of the ectoderm corresponding to that of the mesoderm. The widened anterior and of the embryo forms two symmetric lateral protuberances (head laminae)
C	3-4	Elongated embryo with undifferentiated thoracic appendages. Metameric thickenings of the ectoderm grow more prominently in the anterior region of the embryo than in the posterior one. Seven first pairs of ectodermal protuberances located symmetrically with respect to the midline of the body comprise the rudiments of future head and thorax appendages: antenna, mandibles, two pairs of maxillae, and three pairs of thoracic legs. Appendages develop in the

TABLE 9.1 (continued)

Stage	Time after beginning of incubation (days)	Diagnostic features
		anteroposterior direction. Ectodermal protuberances are poorly marked on the ventral segments. The two depressions of the stomodeum and proctodeum are marked between the head lamina and the posterior segment. Thoracic legs are not differentiated
D	5	Weakly contracted embryos with marks of ventral appendages. False legs will develop from these appendages at segments III-IV and IX(X). The embryo is somewhat contracted; the head segments are brought into proximity
E	6-7	Shortening of the embryo and elongation of the stomodeum and proctodeum. The ventral and head neural ganglia are formed, as are the lateral walls of the larva. The anterior head border is somewhat bent inside the egg because of fusion and displacement of the head segments. A pair of ectodermal depressions appear near the base of the labrum; they form the silk-gland duct
F	8	Beginning of blastokinesis. Embryo reaches maximal contraction. Preparation for blastokinesis is marked by the separation of the embryo posterior end from the membranes. Initial differentiation of primary germinal cells begins
G	9	S-shaped stage of blastokinesis; the embryo is bent in the shape of a letter S
H	9-10	End of blastokinesis. The lateral walls of the larva are completely formed. The embryo contracted in the course of blastokinesis undergoes quick extension
I	10-11	Stage of caudal elongation. Ventral tip reaches the middle of the egg ventral curvature. Differentiation of organs
K	11	Stage where the embryo bends. Ventral end almost touches the head. The embryo is S-shaped
L	11-12	Helically coiled embryo. The posterior ventral end glides over the head and bends to the egg center, forming a helix. Eggs become lighter in color
M	12-14	The first instar larvae hatch

The first of the two techniques is used if the next generation must be reared after 11-12 days. A hydrochloric acid solution with a specific gravity of 1.075 at 15°C is prepared. Silkworm eggs aged 20-25 h (at 25°C) are treated with this solution at 46°C for 4-8 min, depending on the exact age of the eggs and the silkworm variety. After treatment the acid is washed away with water and the eggs are left to incubate at 25°C. On the following day, the eggs acquire a rose-brownish pigmentation. At this stage they may be placed into a refrigerator (5°C) if it is necessary to

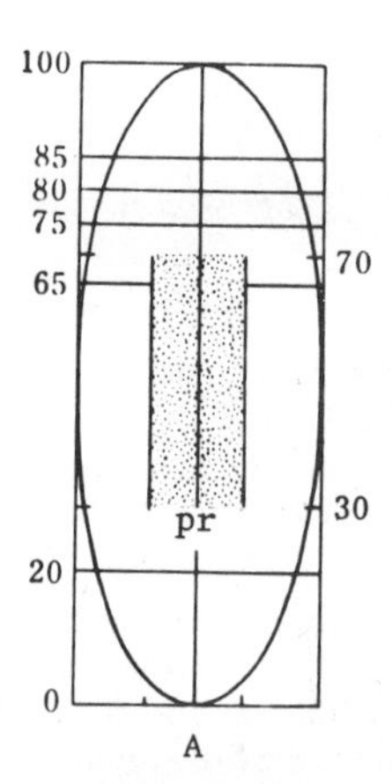

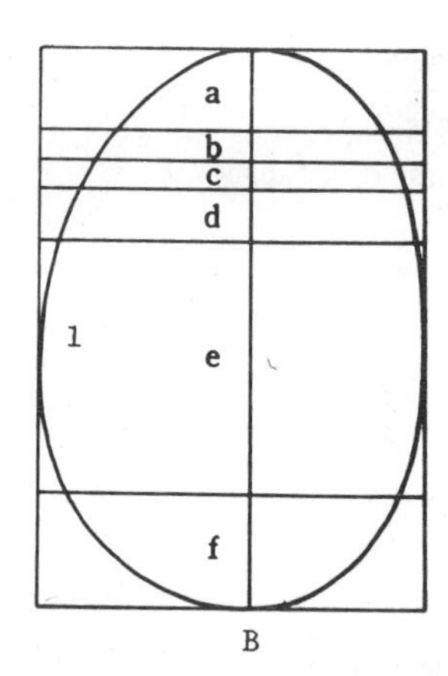

Fig. 9.7. Fate map of the silkworm egg (modified from [52]). A) Ventral view; B) lateral view. Hatched area indicates the determined area of the mesoderm. a, f) Extraembryonic parts; b) head segment; c) jaw segment; d) thoracic region; e) abdominal region; pr) presumptive region of germ cell differentiation; 1) ventral side. Relative size of a region is given a number proportional to the maximum diameter of the egg (taken as 100).

delay hatching. This is normally expected after 10 days, but not longer than 3 weeks.

The second technique uses hydrochloric acid with a specific gravity of 1.10 at 15°C. The rose-brownish eggs, aged 42-50 h at 25°C, are placed into a refrigerator at 5°C for 40-60 days. Prior to treatment with hydrochloric acid (48°C for 5-10 min), the eggs are kept for about 3 h at room temperature.

9.4.4. Sexual Dimorphism

Planned crosses frequently necessitate the separation of the males from the females. This can be done very easily using genital disks as criteria at the beginning of the fourth and fifth instars. With a good magnifying glass the male and female larvae can be separated during the third and even the second instar. The paired Herold organs are present in males between the eleventh segment, which carries the caudal spike, and the twelfth segment; they are located on the ventral side under the hypodermis. In the females these segments carry Ishivata disks (Fig. 9.2). Sex identification at the pupal stage presents no difficulties (Fig. 9.3); the sexual dimorphism of adult moths is illustrated in Fig. 9.3.

9.5. GAMETES, INSEMINATION, FERTILIZATION, AND PARTHENOGENESIS

References to studies on gametogenesis can be found in a number of reviews [41, 52, 53].

The silkworm shows internal insemination. The insemination of eggs outside the body has not been successful because the forward movement of the spermatozoa

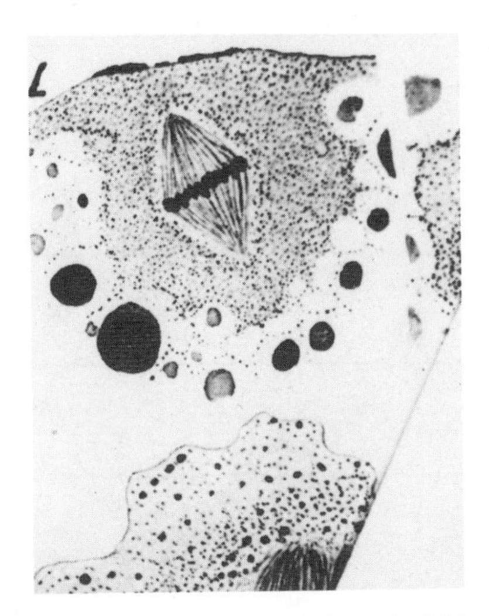

Fig. 9.8. Metaphase I in the newly laid egg [18].

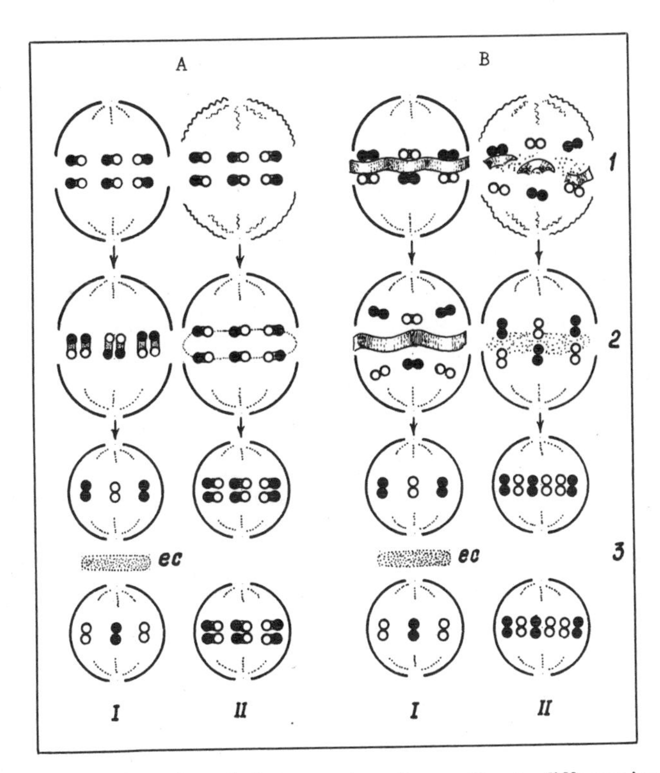

Fig. 9.9. Identification of homologs during metaphase I according to different interpretations of the elimination of segregated chromatin (ec) during the first meiotic division compared to the cytogenetic mechanism of heat-induced inactivation. A) According to Frolova [21]. B) New scheme. I) Normal; II) following heat inactivation. 1) Metaphase; 2) anaphase; 3) metaphase II.

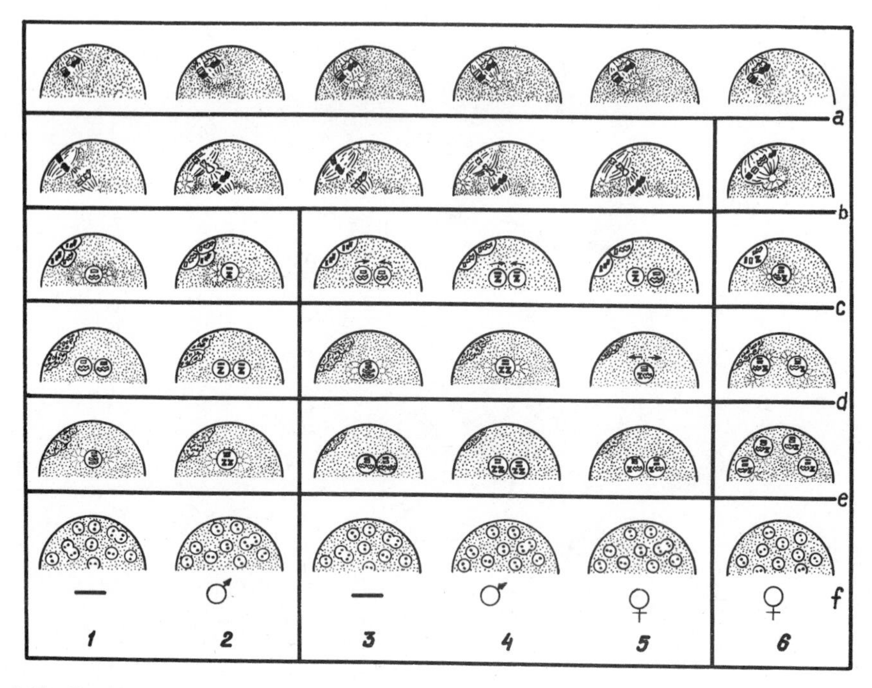

Fig. 9.10. Parthenogenetic variants of *Bombyx mori*. See text for explanations (modified from [12]).

required to make contact with, and penetrate, the egg does not depend on the spermatozoon itself, but is supported by the female's complex genital system [46]. The spermatozoa resemble very long filaments up to 600-800 μm in length with a diameter of about 1 μm; the head is symmetrical and shaped like an awl. In the female genital tract the spermatozoon rotates, forming a helix with 2-2.5 turns (Figs. 9.4 and 9.5).

Artificial insemination in the silkworm has been described [36, 44]. For this purpose the sperm is taken from the freshly inseminated females by opening the dissected copulative bursae containing spermatophores introduced by a male. The spermatophores are then introduced into the genital tract of another female.

Physiological polyspermy is a characteristic feature of the silkworm. The polyspermy for 230 newly inseminated eggs is given below (from Kawaguchi [52]):

No. of spermatozoa in the egg	0	1	2	3	4	5	6	8	11
No. of eggs	13	30	84	80	17	3	1	1	1

Freshly laid eggs of *Bombyx mori* (see Fig. 9.7) have a somewhat irregular ellipsoid shape; the long axis has a length of 1.3-1.5 mm, the middle axis 1.1-1.25 mm, and the short axis 0.5-0.7 mm. The weight of the egg slightly exceeds 0.5

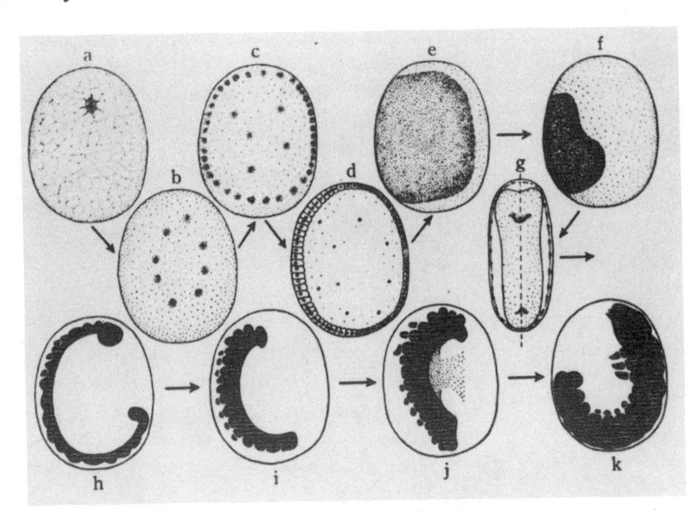

Fig. 9.11. Developmental stages of the *Bombyx mori* embryo [52].

mg. The egg is covered on the outside by a translucent rigid membrane or chorion which protects it from mechanical damage, drying up, and chemical agents. The chorion is wetted by the material secreted by the glueing gland, and the egg becomes tightly glued to the substrate the moment it is laid. The outer surface of the chorion shows the imprints of follicle and nutrimentary cells. The micropyle has only one external orifice. The channel passes through the chorion and the vitelline membrane and branches to yield 3-4 canaliculi (see Fig. 9.4). This system of micropyle channels follows the spermatozoa to penetrate the egg (Fig. 9.6).

The eggs of *Bombyx mori* are typically centrolecithal. The moment the egg is laid, the nucleus is at metaphase I. The spindle is located in the cytoplasmic islet

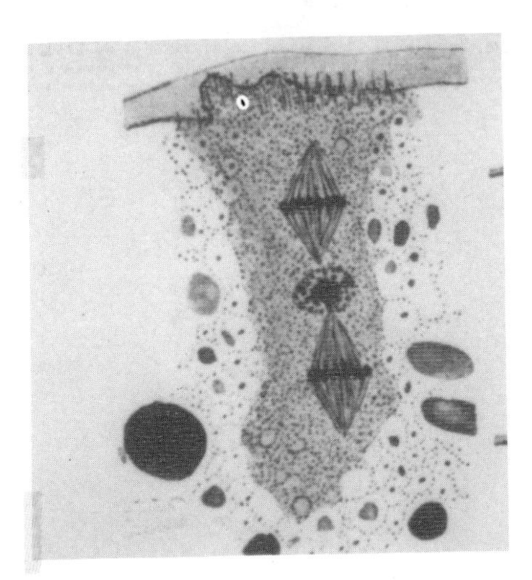

Fig. 9.12. Metaphase II of the fertilized egg [18].

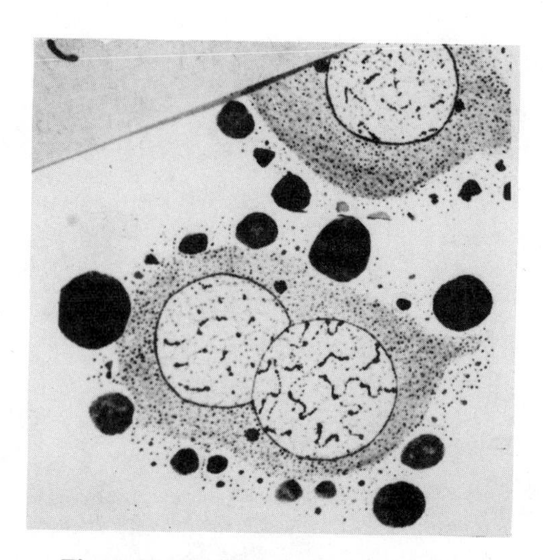

Fig. 9.13. Fusion of pronuclei [18].

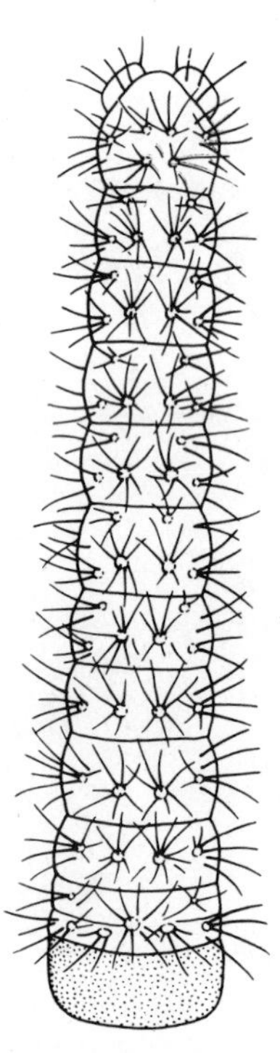

Fig. 9.14. First larval instar, enlarged [52].

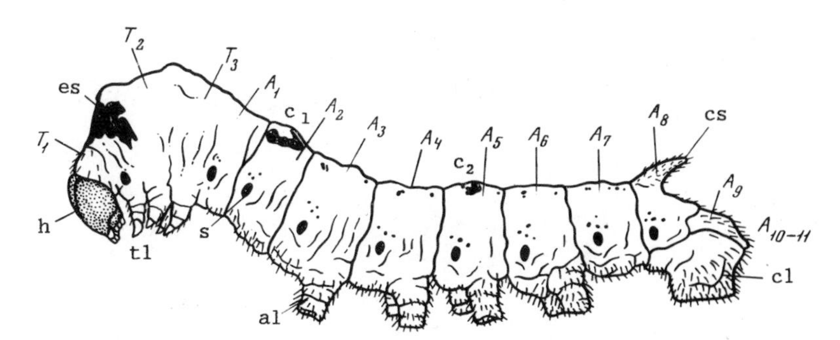

Fig. 9.15. Fifth larval instar [52]. h, head; T_1–T_3, thoracic segments; A_1–A_{11}, abdominal segments; es, eye spots (mask); c_1, c_2, crescents; tl, thoracic legs; al, abdominal legs; cl, caudal legs; s, spiracle; cs, caudal spike.

associated with the egg cortex, but lies far from the micropyle (Fig. 9.8).

The so-called elimination chromatin characteristic of Lepidoptera and some other insects begins to segregate at anaphase I soon after the egg has been inseminated [21]. A reexamination of this process has allowed the old schemes of homologous alignment of chromosomes at metaphase I to be revised and a new mechanism for the artificial ameiotic parthenogenesis, discovered by Astaurov [29, 59, 60], to be suggested. Both mechanisms are presented here (Fig. 9.9).

It has already been mentioned that one important advantage of *Bombyx mori* is that several different cytogenetic variants of parthenogenesis may be obtained. A brief description of them is given below (Fig. 9.10); in this figure, columns 1-5 correspond to the variants of meiotic parthenogenesis when the egg either spontaneously (which is extremely rare), or after treatment with special agents, leaves metaphase I (column a), passes through metaphase II (column b), and after completing maturation divisions yields four haploid derivates (column c). In cases 1 and 2, when the polar bodies do not participate in forming the diploid cleavage nucleus, only the rare absolute male homozygotes carrying ZZ sex chromosomes are viable. The WW constitution is nonviable. In cases 3-5 polar bodies participate in forming the diploid nucleus of the parthenogenetic embryo. In such a case (5) females with maternal genotype may appear. Case 6 refers to ameiotic parthenogenesis when the reduction division is inhibited by heating for 18 min at 46°C (5), while the remaining equational division leaves the maternal genotype unchanged, and this serves as the basis for cloning.

9.6. STAGES OF NORMAL EMBRYONIC DEVELOPMENT

A map of the unfertilized *Bombyx mori* egg presumptive regions has been established using the cauterization technique [52] (see Fig. 9.7).

The table of normal development presented here (Table 9.1) does not pretend to be complete or highly accurate, particularly with regard to time parameters. Essentially, it provides mean data. Fig. 9.11 presents schematic drawings of several developmental stages of *Bombyx mori* which, however, give only an approximate picture of the material (see also Figs. 9.12 and 9.13).

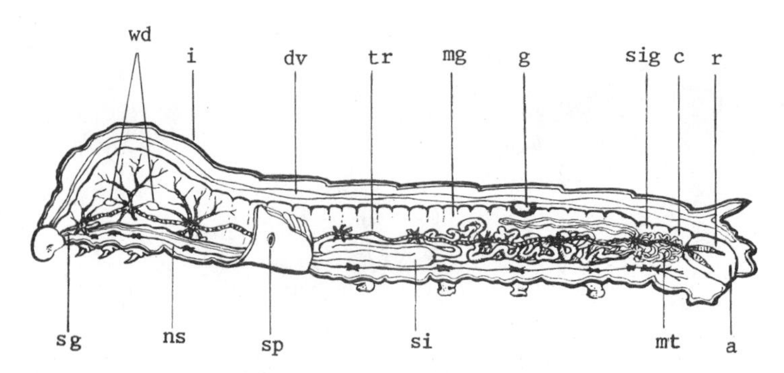

Fig. 9.16. Internal organs of the larva [52]. a, anus; c, colon; dv, dorsal vessel; g, gonad; i, integument; mg, midgut; mt, Malpighian tubule; ns, nervous system; r, rectum; si, small intestine; sg, salivary gland; sig, silk gland; sp, spiracle; tr, trachea; wd, wing disk.

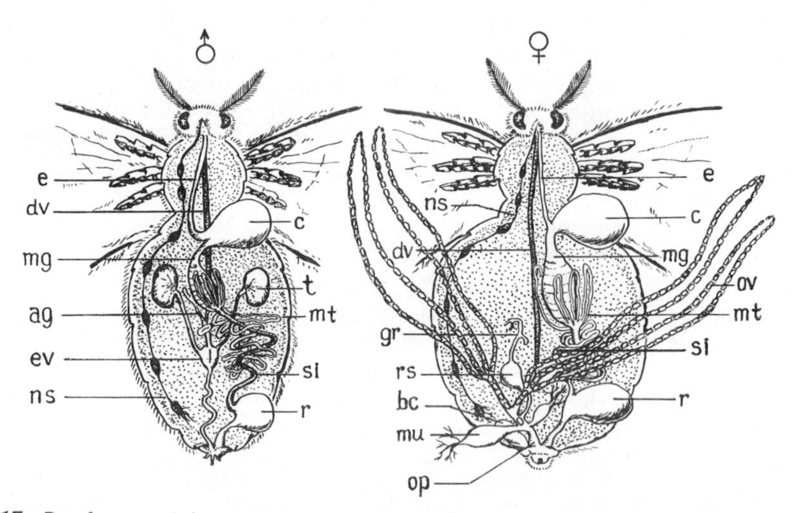

Fig. 9.17. *Bombyx mori*, internal organs of the adult male and female (modified from [54]). ag, accessory gland; bc, bursa copulatrix; c, crop; dv, dorsal vessel; ev, ejaculatory vesicle; gr, glandula receptaculi; mg, midgut; mu, mucous gland; mt, Malpighian tubules; ns, nervous system; e, esophagus; op, oviposter; ov, ovarioles (note that the relative length of these is 4-5 times greater than shown); r, rectum; rs, receptaculum seminis; si, small intestine; t, testis.

9.7. STAGES OF NORMAL POST-EMBRYONIC DEVELOPMENT

The life-cycle diagram (see Fig. 9.1) shows that the larval growth of *Bombyx mori* is not associated with any essential changes of the larval shape, with the exception of newly hatched larvae which have relatively longer bristles (Fig. 9.14). The outer and inner structure of the larva is shown in Figs. 9.15 and 9.16. Metamorphosis takes place at the pupal stage (see Fig. 9.3). The inner structure of the adult male and female *Bombyx mori* is shown in Fig. 9.17.

REFERENCES

1. H. Akai, "Hormonal control of silk production in silkworm *Bombyx mori*," *Japan Agric. Res. Q.* **13**, 116–122 (1979).
2. B. L. Astaurov, *Silkworm Breeding in Japan and the Tasks of Silkworm Breeding in the USSR* [in Russian], Selkhozgiz, Moscow (1933).
3. B. L. Astaurov, "Artificial parthogenesis and androgenesis in the silkworm," *Byull. Vses. Akad. Sel'skokhoz. Nauk im. V. I. Lenina* **12**, 47–52 (1936).
4. B. L. Astaurov, "Experiments on experimental androgenesis and gynogenesis in the silkworm," *Biol. Zh.* **6**, 1–50 (1937).
5. B. L. Astaurov, *Artificial Parthogenesis in the Silkworm Bombyx mori (Experimental Investigation)* [in Russian], Izd. Akad. Nauk SSSR, Moscow (1940).
6. B. L. Astaurov, "Experimental proof for the absence of direct damaging action of x rays on the living cell cytoplasm," *Dokl. Akad. Nauk SSSR* **58**, 887–890 (1947).
7. B. L. Astaurov, "Direct evidence for the nuclear nature of biological action of x rays and independence of the final results of x-irradiation on primary changes in the cytoplasm," *Zh. Obshch. Biol.* **8**, 421–441 (1947).
8. B. L. Astaurov, "Significance of experiments of merogony and androgenesis for the theory of development and heredity (review and perspectives)," *Usp. Sovrem. Biol.* **25**, 49–88 (1948).
9. B. L. Astaurov, "Differential effect of radiation damage on the nucleus and cytoplasm as a consequence of their functional specificity," *Byull. Mosk. Ova. Ispyt. Prir., Sec. Biol.* **63**, 35–48 (1953).
10. B. L. Astaurov, "Ionizing radiation and heredity (the genetic theory of radiation damage)," *Priroda* **4**, 55–67 (1962).
11. B. L. Astaurov, "Functional principle in the evaluation of relative significance of the radiation damage of the nucleus and cytoplasm (genetic theory of radiation disease)," in: *Primary Mechanisms of Biological Action of Ionizing Radiation* [in Russian], *Tr. Mosk. Ova. Ispyt. Prir.* **7**, 140–161 (1963).
12. B. L. Astaurov, *Developmental Cytogenetics of the Silkworm Bombyx mori and Its Experimental Control* [in Russian], Nauka, Moscow (1968).
13. B. L. Astaurov, *Parthenogenesis, Androgenesis, and Polyploidy* [in Russian], Nauka, Moscow (1977).
14. B. L. Astaurov, M. D. Golysheva, and I. S. Roginskaya, "The chromosomal complex of the *ussuri* geographic race of the wild silkworm *Bombyx mandarina* M. in connection with the problems of origin of domesticated silkworm *Bombyx mori* L.," *Tsitologiya* **1**, 327–331 (1959).
15. B. L. Astaurov and V. P. Ostryakova-Varshaver, "Complete heterospermic androgenesis in interspecific hybrids of the silkworm (experimental analysis of the relative role of the nucleus and the cytoplasm in development and heredity)," *Izv. Akad. Nauk SSSR, Ser. Biol.* **2**, 154–175 (1957).
16. C. Ayuzawa, I. Sekido, K. Yamakawa, U. Sakurai, W. Kurata, Y. Yaginuma, and Y. Tokoro, *Handbook of Silkworm Rearing*, Fuji Publishing Co., Tokyo (1972).
17. P. Bartoloni, "Osservazioni sullo sviluppo embrionale in *Bombyx mori* L.," *Ann. Sper. Agrar., N. S.* **11**, 591–622 (1957).
18. E. Bataillon and Tschou-Su, "Les processus cinétiques dans l'oeuf de *Bombyx mori*," *Arch. Anat. Microsc.* **29**, 285–372 (1933).

19. H. Chikushi et al., *Genes and Genetic Stocks of the Silkworm*, Kegaku Publ. Co., Tokyo (1972).
20. H. Doira, "Genetic stocks of the silkworm," in: *The Silkworm*, Y. Tazima (ed.), Kodansha, Tokyo (1978).
21. S. L. Frolova, "Peculiar features of the maturation of unfertilized eggs of the silkworm *Bombyx mori* activated by high temperature," *Dokl. Akad. Nauk SSSR* **27**, 601–603 (1940).
22. T. Gamo, "Recent concepts and trends in silkworm breeding," *Farming Jpn.* **10**, 11–22 (1976).
23. H. Hasimoto, "Formation of an individual by the union of two sperm nuclei in the silkworm," *Bull. Sericult. Exp. Sta. Jpn.* **8**, 455–464 (1934).
24. Y. Horie and H. Watanabe, "Recent advances in sericulture," *Ann. Rev. Entomol.* **25**, 49–71 (1980).
25. T. Ito and M. Kobayashi, "Rearing of the silkworm," in: *The Silkworm*, Y. Tazima (ed.), Kodansha, Tokyo (1978).
26. K. Katsuku, "Weitere Versuche über erbliche Mosaikbildung und Gynandromorphismus bei *Bombyx mori* L.," *Biol. Zbl.* **55**, 361–383 (1935).
27. V. V. Klimenko, "On the number of cells in diapausing embryos of the silkworm having different ploidy," *Ontogenez* **5**, 357–362 (1974).
28. V. V. Klimenko, "Mechanism of artificial parthenogenesis in the silkworm. III. Reversibility of secondary activation changes after thermal and cryogenetic activation," *Genetica* **18**, 64–72 (1982).
29. V. V. Klimenko and T. L. Spiridonova, "Elimination of chromatin and artificial parthenogenesis in the silkworm," *Tsitologiya* **21**, 793–797 (1979).
30. V. V. Klimenko and T. L. Spiridonova, "Polyploidy and parthenogenesis in the silkworm *Bombyx mori*," *Izv. Akad. Nauk Mold. SSR, Ser. Biol. Khim. Nauk* **4**, 32–36 (1982).
31. N. K. Koltzov, "Artificial parthenogenesis in the silkworm," *Probl. zhivotnovodstva* **4**, 55 (1932).
32. S. Krishnaswami, N. Narasimhanna, K. Suryanarayan, and S. Kumaraj, "Sericulture manual 2: Silkworm rearing," *FAO Agric. Bull.* **15**, 1–131 (1973).
33. E. N. Mikhailov, *Silkworm Breeding* [in Russian], Selkhozgiz (1950).
34. J. Morrow, J. W. Wozney, and A. Efstradiadis, "A study of the silk fibroin gene," in: *Recombinant Molecules: Impact on Science and Society*, Miles International Symposium, No. 10, Raven Press, New York (1977).
35. I. Murakami, "Cytological evidence for holocentric chromosomes of the silkworms *Bombyx mori* and *B. mandarina* (Bombycidae, Lepidoptera)," *Chromosoma* **47**, 167–178 (1974).
36. S. Omura, "Artificial insemination of *Bombyx mori*," *J. Fac. Agric. Hokkaido Imper. Univ., Sapporo* **38**, Pt. 2 (1936).
37. E. F. Poyarkov, *The Silkworm Bombyx mori* [in Russian], Izd. Sredneaziat. Inst. Shelkovodstva (1929).
38. N. N. Rott and V. N. Vereiskaya, "Diploid parthenogenesis," in: *Methods of Developmental Biology* [in Russian], Nauka, Moscow (1974).
39. N. N. Rott and E. P. Terskaya, "Diploid androgenesis," in: *Methods of Developmental Biology* [in Russian], Nauka, Moscow (1974).
40. V. Z. Ruban, "Experimental construction of autotetraploids in the silkworm," *Genetica* **19**, 115–120 (1983).

41. B. Sakaguchi, "Postembryonic development of the silkworm," in: *The Silkworm*, Y. Tazima (ed.), Kodansha, Tokyo (1978).
42. H. Sato, "Untersuchungen über die kunstliche Parthenogenese des Seidenspinners," *Biol. Zbl.* **51**, 389–394 (1931).
43. I. A. Shcherbakov, *Technology of Silkworm Egg Production* [in Russian], Selkhozgiz, Moscow (1952).
44. M. I. Shevyakova, "Effectiveness of artificial insemination in the silkworm *Bombyx mori*," *Shelk* **1**, 16–19 (1960).
45. V. A. Strunnikov, "Construction of bipaternal androgenetic hybrids in the silkworm *Bombyx mori*," *Dokl. Akad. Nauk SSSR* **122**, 516–519 (1958).
46. V. A. Strunnikov, "The process of egg insemination in the silkworm *Bombyx mori* L., *Zh. Obshch. Biol.* **20**, 35–42 (1959).
47. V. A. Strunnikov, "Sex control in practical silkworm breeding," *Priroda* **7**, 36–47 (1972).
48. V. A. Strunnikov, "Sex control in the silkworm," *Nature (London)* **255**, 111–113 (1975).
49. V. A. Strunnikov and V. M. Maresin, "Characteristic features of silkworm egg activation," *Dokl. Akad. Nauk SSSR* **251**, 720–724 (1980).
50. V. A. Strunnikov, L. V. Strunnikova, and Yu. M. Pavlov, "Artificial ameiotic gynogenesis in the silkworm," *Dokl. Akad. Nauk SSSR* **258**, 491–494 (1981).
51. L. V. Strunnikova, "New method to construct polyploids in the silkworm," *Dokl. Akad. Nauk SSSR* **248**, 1456–1460 (1979).
52. J. Tazima, *The Genetics of the Silkworm*, Academic Press, London (1964).
53. Y. Tazima, H. Doira, and H. Akai, "The domesticated silkmoth *Bombyx mori*," in: *Handbook of Genetics*, Vol. 3, R. C. King (ed.), Plenum Press, New York (1975).
54. Y. Tazima (ed.), *The Silkworm: An Important Laboratory Tool*, Kodansha, Tokyo (1978).
55. E. R. Terskaya and V. A. Strunnikov, "Methods to activate silkworm eggs to meiotic parthenogenesis," *Dokl. Akad. Nauk SSSR* **219**, 1238–1241 (1974).
56. E. R. Terskaya and V. A. Strunnikov, "Artificial meiotic parthenogenesis in the silkworm," *Genetika* **11**, 54–67 (1975).
57. A. A. Tichomiroff, "Sullo eviluppo delle nova del bombice del gelso sotto l'influenza dell'iccitazioni mecchaniche e chemiche," *Bollettino mensile di Bachicoltura, Ser. II*, Anno **III**, 145–151 (1886).
58. N. M. Tulzeva and B. L. Astaurov, "Increased resistance of the silkworm *Bombyx mori* L. polyploids to radiation damage in connection with the general theory of biological action of ionizing radiation," *Biofizika* **3**, 197–205 (1958).
59. V. N. Vereiskaya, "Meiosis and onset of cleavage in thermoactivated eggs of *Bombyx mori* L.," *Byull. Mosk. Ova. Ispyt. Prir.* **53**(4), 31–40 (1975).
60. V. N. Vereiskaya, "Cytology of maturation of *Bombyx mori* eggs during artificial parthenogenesis," *Ontogenez* **10**, 244–252 (1979).
61. T. Yokoyama, *Silkworm Genetics Illustrated*, Japanese Society for Promoting Science, Tokyo (1959).
62. T. Yokoyama, "Silkworm selection and hybridization," Working Papers, Rockefeller Foundation (1979), pp. 71–83.
63. "The silkworm, a model system," *Biochimie* **61**, 135–320 (1979).
64. "The physiology and biology of spinning in *Bombyx mori*," *Experientia* **39**, 441–473 (1983).

Chapter 10

THE SEA URCHINS *Strongylocentrotus dröbachiensis*, *S. nudus*, AND *S. intermedius*

G. A. Buznikov and V. I. Podmarev

10.1. INTRODUCTION

The important and enduring advantages of sea urchins as objects of developmental studies were discovered as long ago as the last century. These advantages include the possibility of obtaining large numbers of gametes and synchronously developing embryos, the ease of incubation of the embryos under strictly regulated conditions, and the possibility of simple procedures for many kinds of vital observations and treatments of the fixed materials. Subsequent investigations established good permeability of the eggs and embryos to many substances, including amino acids and nucleosides, relative simplicity of isolation of subcellular fractions, and the possibility of obtaining great amounts of parthenogenetic embryos, anuclear embryos, and interspecific hybrids [2, 3, 12, 14, 21, 22, 32, 38, 39, 42, 44, 55, 57, 62, 64, 76]. All this led to the wide utilization of sea urchin gametes and embryos in studies of many aspects of developmental biology, as well as in molecular biology and cytology [3, 18, 21, 26, 35, 41, 44, 48, 49, 60, 63, 65, 67, 69, 76]. In addition, sea urchin embryos are used with increasing frequency for mass toxicological and pharmacological testing of different chemicals (drugs, pollutants, etc.), including potential carcinogens and teratogens [4, 6, 8, 9, 11-13, 36, 37, 50, 53, 59, 75].

The use of sea urchins has certain limitations. In many cases these are purely technical (e.g., difficulties of transportation of sea urchins to scientific centers remote from the sea, absence of necessary scientific equipment at many marine biological stations). Limitations concerned with the difficulties of incubating sea urchin embryos to later developmental stages (beginning with the transition to active feeding) are much more important [42, 43]. Hence, sea urchins are used relatively rarely in studies concerned with analysis of late morphogenesis and in those requiring the development of experimental animals beyond metamorphosis.

Working with sea urchins, Soviet authors most frequently use the following species of the genus *Strongylocentrotus*: *S. dröbachiensis, S. nudus,* and *S. intermedius*. Information on these species will be given below. Data on many other sea urchin species used frequently in studies by Western authors can be found elsewhere [see 20, 39, 42, 44, 49, 62, 76].

10.2. TAXONOMIC DISTRIBUTION AND CONDITIONS OF SPAWNING

The sea urchins *Strongylocentrotus dröbachiensis* (O. F. Müller), *S. nudus* (A. Agassiz), and *S. intermedius* (A. Agassiz) belong to the family Strongylocentrotidae, suborder Camarodonta, order Centechinoida, class Echinoidea, type Echinodermata [39].

10.2.1. *S. dröbachiensis*

This is a widespread species, occurring in the USSR in the Arctic seas, as well as in the Bering Sea [30]. It occurs relatively rarely in the White Sea, and is a common species in the Barents Sea. It is available at marine biological stations on the Atlantic coast of Europe and North America [39, 73]. Females with mature gonads are available in the sublittoral zone from the end of May until the beginning of June (Dalne-Zelenetsky Inlet, White Sea) at water temperatures of 4-8°C. At greater depths egg maturation and spawning proceed 1-2 weeks later. These times are dependent on meteorological conditions. Females with mature gonads may be found in the sublittoral zone in September-October. However, the quality of eggs obtained during this period is usually worse than in the main period of spawning. Males with mature gonads are available in the sublittoral zone for a much longer period, at least from February or March until October.

10.2.2. *S. nudus*

This species occurs in Peter the Great Bay (Sea of Japan), but is absent from the Sea of Okhotsk. Females with mature gonads are available in the sublittoral zone usually from the beginning of July until the second half of August (water temperature 16-24°C); males with mature gonads appear earlier and disappear later [47].

10.2.3. *S. intermedius*

This species occurs frequently with *S. nudus*, but has a more northerly distribution in the Tatar Strait and Sea of Okhotsk. The season of spawning in Peter the Great Bay begins somewhat later than in *S. nudus*, from the end of July or beginning of August, and lasts longer, at least until the first half of October (water temperature 16-24°C); mature sea urchins can be found in autumn at even lower temperatures [47].

The species under consideration show no sexual dimorphism and sex is determined either by dissection or by special methods [39, 73]. To determine the sex of an immature animal, biopsy is used; a piece of gonad is taken by a syringe through the peristome wall and studied under the microscope. To determine the sex of a mature animal, it can be sharply shaken, so that a number of eggs or spermatozoa leak out through the gonopores. The same results are obtained by injection of 0.5 M KCl into the perivisceral cavity (0.5-1 ml if injection through the peristome wall, and 0.2-0.5 ml if injection in one of the gonopores). After shaking or injection, the animal is placed, aboral side downward, on a Petri dish or glass, so that the leaking gametes can be collected. The leakage of gametes from gonopores may also be induced by electrical stimulation of the animal (using an alternating current ~10 V and lead electrodes).

10.3. TRANSPORTATION AND CARE OF ADULT ANIMALS

Sea urchins can be preserved in the cold (0-5°C) without water for 1-2 days after they are caught; if they are kept under these conditions for a longer time, gametes are either spawned or damaged and the animals die. In aquaria with running sea water, sea urchins can live without damage to the gonads from several days to several weeks at water temperatures of 0-8°C. This period can be increased if the animals are fed, in particular with algae (i.e., *Laminaria*).

Ordinary freezer bags are convenient for transporting sea urchins over great distances. A layer of ice made from sea water is placed on the bottom of such a bag, and then sea urchins are placed inside the bag so that they occupy no more than two-thirds of its volume (when loosely packed). Another layer of ice is placed on top and the bag is closed. For periods of up to 2 days small numbers of sea urchins (*S. nudus* or *S. intermedius*) can be transported without cooling, in aerated sea water or without water.

Methods of determining sex and degree of maturity (see above) involving live animals cannot, as a rule, be used on sea urchins reserved for transportation, since these methods can induce premature spawning. The least harmful is electrical stimulation. To establish the condition (sex ratio, percentage of animals with mature gonads) of sea urchins reserved for transportation, selective testing of a certain number of animals should be used.

After transportation, sea urchins are usually kept in aquaria with circulation and constant aeration of the water; 50-75 middle-sized animals can be placed in 150 liters of sea water. Under these conditions we succeeded in keeping sexually mature sea urchins (*S. dröbachiensis*) and using them for experiments for up to 4 weeks; by feeding the animals this period can be increased to 2 months [73], but this requires special systems for water cleaning.

Aquaria can be filled with artificial sea water; there are many recipes [39, 42, 76]. We will give only one of them (Table 10.1), successfully used by many American authors and at the Institute of Developmental Biology in Moscow. Salts are dissolved in distilled or deionized water. Concentrated (stock) solutions of all salts (except NaCl) can be prepared beforehand; concentrations of these solutions are given in Table 10.1. The pH of the prepared water should be between 7.4 and 8.0; all other values suggest poor quality of some components (most often $NaHCO_3$). Besides pH, it is desirable to test the freshly prepared sea water on adult sea urchins, gametes, and embryos. Incapacity of adult animals to attach themselves to the substrate, immobility of spermatozoa, low percentage of fertilization, abnormal cleavage divisions – all these could be due to the poor quality of any component (most often $NaHCO_3$, $MgCl_2$, or $CaCl_2$).

While controlling the temperature at which the adult sea urchins are kept, one can markedly increase the experimental season. It is possible, for example, to accelerate the appearance of sea urchins (*S. dröbachiensis*) with mature gametes by several weeks if the animals are collected at the beginning of April (Dalne-Zelenet-

TABLE 10.1. Composition of Artificial Sea Water

Salts	In grams per liter: prepared sea water	In grams per liter: stock solution	Stock solution (in ml per 50 liters of prepared sea water)
NaCl	26.05	–	–
$MgCl_2 \cdot 6H_2O$	5.20	776	335
$MgSO_4 \cdot 7H_2O$	4.48	448	500
KCl	0.69	276	125
$NaHCO_3$	0.198	39.7	250
$CaCl_2$*	1.43	285.2†	250

*Added last.
†562.7 g/liter if $CaCl_2 \cdot 6H_2O$ is used.

sky Inlet), that is, 1.5-2 months prior to the season of mass spawning, and incubated at 7-8°C, that is several degrees above the water temperature under natural conditions. Conversely, by lowering the temperature of incubation of the animals collected during the period of spawning, one can delay spawning. Stephens [73] incubated adult sea urchins (*S. dröbachiensis*) at 4°C, that is, by 2-5°C below the water temperature during the period of spawning, and obtained mature gametes from them during the 2 months after the end of the normal spawning season in the sublittoral zone.

10.4. OBTAINING MATURE GAMETES AND IMMATURE OOCYTES

Sea urchins are dissected with scissors somewhat above the "equator"; the oral body wall and intestine are removed. The sex of dissected animals is determined by the color of gametes leaking onto the surface of gonads in damaged regions (spermatozoa white, eggs yellowish in *S. dröbachiensis* and *S. nudus*, and orange in *S. intermedius*). After some experience one can determine the sex of animals by the external appearance of intact gonads.

A piece of gonad is taken with forceps and smeared upon the dry surface of a slide. This smear is investigated under the microscope; a drop of sea water is added to a sperm smear prior to investigation; testes and ovaries are taken with different forceps. The external appearance of normal eggs is shown in Fig. 10.5, I; females with damaged eggs (deformation, signs of lysis) and males with immobile spermatozoa are discarded. In those cases when the admixture of oocytes is undesirable, the females with a high percentage of oocytes (identified by a large vesiculiform nucleus, the germinal vesicle) are discarded as well. In many cases the above procedure is unnecessary as a successful experimental fertilization indicates good-quality gametes.

Testes are taken from a selected male and cut in half; drops of sperm leaking from the cuts are diluted with sea water (for use; see below) or collected in a dry glass container (for preservation). The "dry sperm" can be preserved in a refrigerator without any damage for up to 24-48 h and the sperm diluted with sea water up to 12-24 h.

The selected females are placed in glass beakers filled with sea water (the aboral end should be submerged in water) and the open body cavity is filled with 0.5 M KCl. Almost immediately mature eggs begin to leak from the gonopores and settle on the bottom. The majority of the eggs, up to several million from each female, leak out into the water within 15-30 min. The water is then changed 2-3 times. It is desirable to use the washed eggs immediately but, if required, they can be preserved in a refrigerator without stirring for up to 24 h. The addition of sulfa drugs or antibiotics allows this period to be increased somewhat [73]. We find that a mixture of streptomycin and oxytetracycline (50-100 μg/ml of water) gives good results.

Usually under the effect of KCl only mature eggs leak out from the gonopores, while oocytes remain in the ovaries. The above procedure allows, therefore, for a pure suspension of eggs to be obtained, even from those females whose ovaries contain a certain proportion (not more than 5-10%) of oocytes. In the height of the spawning season most females do not have oocytes in the ovaries, and one can simply take out the gonads with forceps and shake out the eggs in water through cotton gauze.

In some cases it is desirable to obtain gametes from an intact animal, and for this purpose the methods used for sex determination and to estimate sexual maturity can be used (see above). To obtain the maximum number of gametes simultaneously from an intact animal, an injection of large quantities (2-15 ml, depending on the animal's size) of 0.5 M KCl into the perivisceral cavity is recommended [73]. To obtain gametes repeatedly from the same animal, electrical stimulation is recommended: the eggs or spermatozoa cease to leak from the gonopores immediately after the cessation of stimulation and begin to leak again once the stimulation is reapplied.

For working with irregular sea urchins (sand dollars) which are available at the Far East marine biological stations (*Echinarachnius parma, Scaphechinus mirabilis, S. griseus*), dissection is not recommended. A small amount of 0.5 M KCl is injected through the peristome wall (0.1-0.2 ml for *S. mirabilis*), and the injected animals are placed, aboral side downward, on penicillin flasks filled with sea water. After the gametes (eggs or spermatozoa) settle on the flask bottom, the sea water is replaced; such washed material is suitable for experiments.

The following practical comments need to be made: 1) Gametes should be obtained from sea urchins at the optimal temperatures 2-8°C (not more than 10°C) for *S. dröbachiensis* and 15-22°C (not more than 25°C) for *S. nudus* and *S. intermedius*; 2) sea water should be filtered, depending on experimental requirements, through cotton, paper, or millipore filters. In many cases (even working at the marine stations) it is recommended to use artificial sea water; 3) to clean the egg suspension from foreign particles, it should be filtered through fine cotton gauze. Filtration of the eggs of sea urchins of the genus *Strongylocentrotus* (but not irregular sea urchins) usually destroys their jelly coats. There are also special methods of removing jelly coats based on various chemical and physicochemical effects [3, 39]. Acidification of the sea water is used most often for this purpose. Stephens [73] recommends the following procedure for *S. dröbachiensis*: the pH of the sea

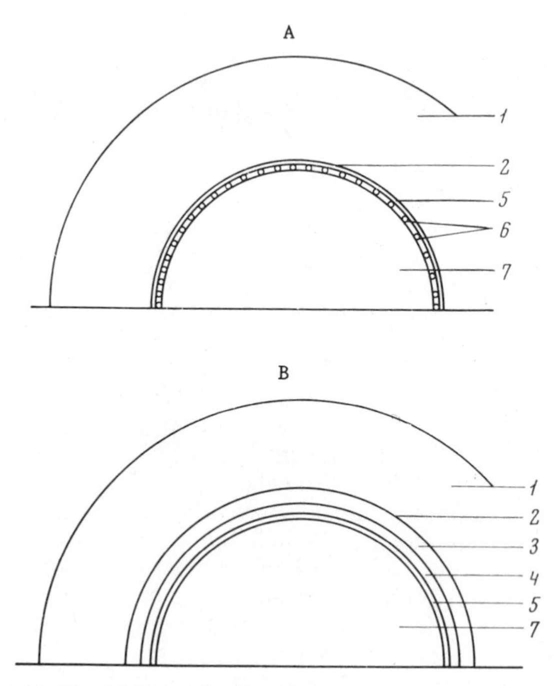

Fig. 10.1. Mature unfertilized (A) and fertilized (B) egg of sea urchin (simplified from [39]). 1) Jelly coat; 2) vitelline membrane; 3) perivitelline space; 4) hyaline layer; 5) cortical layer (cortex) of cytoplasm; 6) cortical granules; 7) endoplasm.

water in which the eggs are suspended is rapidly brought to 5 (not below) by the dropwise addition of 0.1 N HCl, and the eggs are sedimented by means of a hand centrifuge and washed three times with normal sea water. The removal of jelly coats is considered desirable, since their presence increases the probability of infection of the embryos [76]; 4) to prevent contamination of instruments, glassware, millgauze, etc., with sperm, it is necessary to rinse them repeatedly in fresh water. Otherwise, among the unfertilized eggs prepared for experiments a considerable number of embryos at different stages of development can be found.

10.4.1. Obtaining Oocytes

Among the echinoderms, starfish and sea cucumbers are most suitable for studies on oocytes, but in some cases sea urchins are also used. At the beginning of the spawning season females of *S. dröbachiensis* quite frequently occur without mature eggs in their gonads, but with a large number of fully grown oocytes. These oocytes are easily shaken out from the ovaries in water. If the ovaries contain a mixture of mature eggs and oocytes, they can be easily separated using 0.5 M KCl; after the injection the eggs will leak out and the oocytes will remain in the gonads and can be removed separately [33].

10.5. STRUCTURE OF THE EGG

During both natural and induced (electrical stimulation, KCl) spawning, sea urchin females release mature eggs. The maturation divisions and formation of both polar bodies are completed before egg-laying or before they are obtained by the procedures described above. Such eggs are fertilizable and preserve this capacity for quite a long time (times of preservation are indicated above). The form of the eggs can vary greatly (see Fig. 10.5). Their mean diameter is 160 μm in *S. dröbachiensis* and *S. nudus* and 140 μm in *S. intermedius* (without jelly coat).

A simplified scheme of some surface structures of the egg before and after insemination is given in Fig. 10.1. The jelly coat is an external membrane which is easily destroyed both when the eggs are isolated (see above) or preserved for a long time [39]. It overlies a fine vitelline membrane which closely adjoins the plasma membrane. Under the plasma membrane there is a layer of cytoplasm (cortex) which contains numerous cortical granules. The eggs of the sea urchin species discussed here, like those of echinoderms in general, are characterized by a low yolk content (oligolecithal eggs). In the *Strongylocentrotus* species, as in most other sea urchin species (except for *Paracentrotus lividus*), the unfertilized eggs have no morphologically pronounced polarity [29, 44].

10.6. ARTIFICIAL FERTILIZATION

In sea urchins fertilization is external and monospermic; all cases of polyspermy can be considered as abnormal. If the gametes are of good quality and the conditions are optimal, the number of polyspermic eggs does not exceed tenths of a percent [31, 39].

Artificial insemination should be carried out under the optimal temperature conditions which coincide with those of subsequent incubation of the embryos (see below). Eggs suspended in sea water are allowed to sink and most of the water is poured off. A drop of "dry" sperm is diluted in 5 ml of sea water. The diluted sperm is mixed with the eggs (0.1-0.5 ml of sperm per 100 ml of dense suspension of eggs). Within 1-2 min the suspension is diluted 10-15 times with sea water, and after the eggs settle the water is changed. If need be, the eggs are washed with sea water 2 or 3 times using a hand centrifuge. The eggs should be centrifuged with care, since the centrifugation can induce aggregation of the eggs; to prevent aggregation, the eggs should be washed with sea water cooled to 0°C or with calcium-free artificial sea water: there are many recipes for this. The recipe given in Table 10.1 can also be used if $CaCl_2$ is replaced by an equimolar amount of $MgCl_2 \cdot 6H_2O$, or simply removed. The washing of eggs in calcium-free sea water can begin not earlier than 30-60 sec after addition of sperm, since fertilization does not take place in calcium-free sea water. The method of artificial fertilization described is widely accepted and, with some significant modifications, is used by all authors (see, for example, [39, 64, 73]).

The first external sign of successful fertilization is the formation of a perivitelline space between the vitelline and plasma membranes (the separated vitelline membrane is called the "fertilization membrane"). These phenomena are based on the extrusion of cortical granules leading to water pumping from outside into the space between the vitelline and plasma membranes. After a certain latent period, the

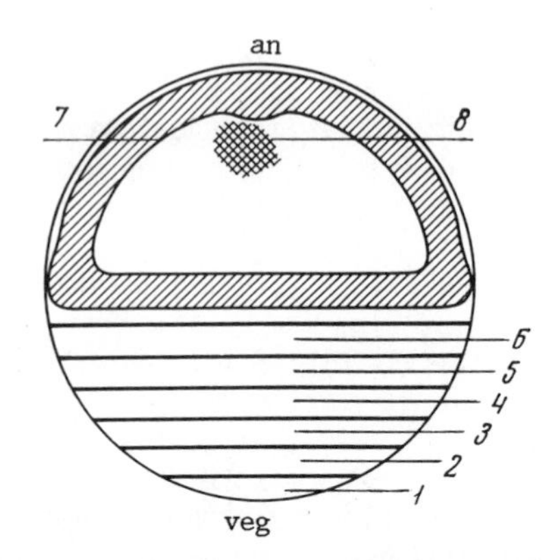

Fig. 10.2. Fate map of the sea urchin egg. Extrapolated from the results of Hörstadius obtained with vital staining [44]. an, animal pole; veg, vegetal pole. 1) Primary mesenchyme; 2) secondary mesenchyme; 3) coelom; 4) pharynx; 5) stomach; 6) proctodeum; 7) ectoderm; 8) stomodeum.

cortical reaction begins at the point of spermatozoon penetration and spreads all over the egg surface. The character of initial changes in the fertilized eggs was studied in sea urchins fairly well [28, 31, 54]. The duration of the latent period in *S. dröbachiensis* is 60-70 sec (at 8°C), the contents of the cortical granules are released within 90 sec all over the egg surface, and the separation of fertilization membrane lags behind by 30-40 sec. Thus, the moment of insemination until the separation of the whole fertilization membrane takes about 3 min in *S. dröbachiensis*: [31] and about 30 sec (at 20°C) in *S. nudus* and *S. intermedius* .

During the formation of the fertilization membrane the egg preserves a spheroid form and suffers practically no changes in volume [39]. As it forms, the perivitelline space looks in section like a crescent, the ends of which embrace the egg to an increasing extent and fuse at the opposite pole. At first the perivitelline space has its maximum width at the place where it started to form, and its minimum width on the opposite pole. Within a certain time (5 min after fertilization in *S. dröbachiensis* at 8°C [73]), these differences disappear.

The formation of a hyaline layer closely adjoining the plasma membrane begins just after the separation of the fertilization membrane. This layer forms at the expense of the contents of the cortical granules and, perhaps, from other substances of cytoplasmic origin [31, 39, 46, 54]. The hyaline layer thickens gradually and attains its definitive size within 75 min of insemination (at 8°C) in *S. dröbachiensis*, but it is already visible by the 15-17th min. It is this layer which connects blastomeres forming as a result of cleavage; the destruction of the hyaline layer leads to disaggregation of the embryonic cells [32, 39, 46]. The arrangement of membranes and surface egg layers after insemination is given in Fig. 10.1 (see also Fig. 10.2).

10.6.1. Some Practical Recommendations

1. Eggs with a percentage of fertilization below 95% (in especially important experiments, below 99%) and with defects in the separation of the fertilization membrane (several poles of separation, deformation of perivitelline space, etc.)

should be discarded. Obviously, one should be certain that these defects are not due to deviations from optimal environmental conditions. In practice, it is advisable to carry out an initial trial fertilization insemination using a small number of eggs; the results can be judged by the separation of the fertilization membranes and, if necessary, by the normality or otherwise of the first developmental stages. Such a trial fertilization allows for rapid and safe estimation of both the quality of gametes and the conditions of incubation.

2. A mixture of sperm from several males can be used for insemination (except in cases where the requirements for homogeneity of the materials are especially high). However, it is advisable not to mix the eggs from several females (although such a mixing frequently proves to be inevitable, especially in biochemical experiments).

3. Artificial fertilization of eggs of good quality by a certain excess of sperm will not lead to polyspermy or other developmental defects. However, if fertilization is preceded by the removal of the vitelline membranes (see below), an excess of sperm becomes deleterious to normal development. In such a case, it is advisable to determine the minimum essential amount of sperm beforehand, using intact eggs of the same female.

10.7. INCUBATION OF EMBRYOS

Sea urchin embryos develop normally in open vessels under optimal temperature conditions and without stirring or aeration if they lie on the bottom of the vessel in a single layer. The thickness of the water layer should not exceed a few centimeters. Small numbers (up to 1000) of embryos of the species in question can be incubated in small containers (1-2 ml in volume) such as penicillin flasks. Large numbers can be incubated in crystallizing dishes, Petri dishes, or the like, provided the "rule of monolayer" is observed (corresponding to a concentration of embryos of about 5000/ml). If the concentration of embryos is higher, water stirring or aeration becomes essential.

Stirrers of different form made of acrylic plastic are employed and connected with electric motors, with a constant rotation rate of 60 rpm, via a rubber tube. At a lower rate than 60 rpm, the eggs settle on the bottom; at a higher rate they are damaged. The flexible connection of the stirrer with the rotor avoids stagnant zones in the incubation vessel. The capacity of the latter can be 5-10 liters, depending on its shape, as well as the form and power of stirrer. To stir large volumes of water, several motors are used simultaneously. Embryonic development in such a suspension is quite normal at a density up to 40,000 embryos per 1 ml. This method not only allows the incubation of large numbers of embryos, but also facilitates the repeated sampling of embryos from the incubation vessel.

Other methods of mixing dense suspensions of embryos are also used, employing magnetic stirrers, shakers, vibrators, microcompressors, etc. [19, 20, 42, 43, 64, 76]. All these methods usually give good results. Our own observations suggest that the mixing of suspensions by air bubbles often results in damage of the embryos and cannot be recommended.

10.7.1. Further Recommendations

1. It is necessary to observe the optimal temperature conditions both during incubation of the embryos and upon artificial fertilization: 4-8°C for *S.*

dröbachiensis (although development proceeds normally but very slowly at 0°C) [73], and 15-22°C for *S. nudus* and *S. intermedius* (our observations). Embryonic development is adversely affected at temperatures over 10°C (*S. dröbachiensis* [73]) or over 25°C (*S. nudus* and *S. intermedius* – our observations). The development of the latter two species is affected, although not so badly, at 10-14°C, but the effect of lower temperatures was not studied by us.

2. While incubating a dense suspension of the embryos, it is necessary to change the water at least once, preferably just after hatching when a large amount of organic matter gets into the water.

3. While working with dense suspensions, one has to bear in mind that the developing sea urchin embryos secrete chemical substances having a certain biological activity in water (see, for example, [5, 17, 78]). These factors can change the sensitivity of embryos to certain external influences.

4. When estimating the density of a suspension of the eggs or of the embryos, the suspension is stirred and a chosen volume (say 0.1 ml) is rapidly taken, diluted with sea water to 25-100 ml, and the number of eggs or embryos is counted in 0.1 ml of the liquid suspension obtained. The counting can be made directly in a micropipette or on a slide placed on a dark background, by the naked eye or with a forehead binocular magnifier. This method is generally used due to its simplicity, ease, and precision.

5. While taking quantitative samples of the eggs and embryos, the suspension should be carefully stirred. With this aim automatic pipettes of different constructions are suitable which allow stirring of the suspension at the moment of sampling.

10.8. NORMAL DEVELOPMENT

10.8.1. General Characteristics

The normal development of sea urchins has been considered in detail in a number of monographs and reviews (see, for example, [34, 39, 44, 49]). Our necessarily brief description is based on these important works.

The organization of the mature egg of sea urchins was studied by Hörstadius (see [44]); the results of these investigations, together with information from other authors, are summarized in Fig. 10.2. In the species under consideration, as in most sea urchin species, the eggs have no morphologically pronounced polarity; the position of the animal–vegetal axis is revealed only after the activation of eggs [39, 44], although in some cases (for example, in *S. dröbachiensis*) it can be determined earlier by a relatively simple technique [58].

Cleavage is complete, radial, and, up to the stage of 8 blastomeres, equal. As in most sea urchin species, desynchronization of cleavage divisions begins from the 5th division, although its first signs appear earlier (there is a very short-term range of 12 blastomeres). The first two cleavage divisions occur in the plane of the animal–vegetal axis, and the third at right angles to these, in the equatorial plane; the eight blastomeres formed as a result of these cleavage divisions are, as a rule, morphologically indistinguishable. The plane of the fourth cleavage division in the animal hemisphere is meridional and, as a result, eight mesomeres arise. In the vegetal hemisphere the planes of cleavage occur at a certain angle to the equatorial one and are directed from the animal–vegetal axis downward and, as a result, four macromeres and, near the vegetal pole, four micromeres arise. Micromeres are

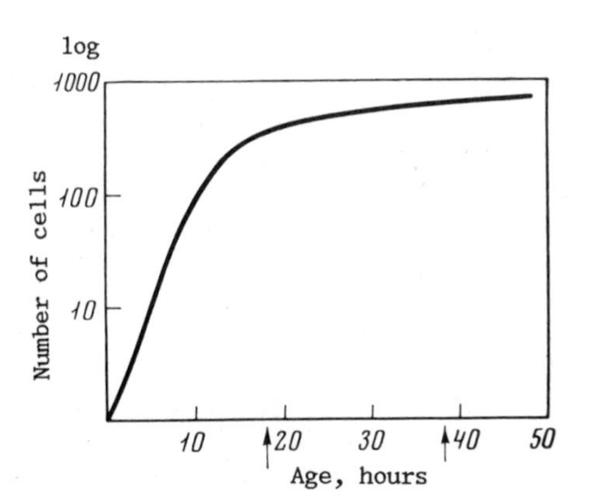

Fig. 10.3. Changes in cell number during early embryonic development of *Strongylocentrotus purpuratus* at 15°C [42].

seen alongside the other embryonic cells at the 32-64 blastomere stages as well and mark the vegetal pole; at the stages of early and mid-blastula the morphological signs of polarity of the embryo disappear.

In some species of sea urchins (e.g., in *Lytechinus variegatus*; see [42]), the mitotic apparatus of the cells during the first cleavage divisions is available for vital observations without any special treatment. In the species considered here, a so-called "strip" structure devoid of visible granules and distinctly contrasting with the granular cytoplasm is clearly seen during the first cleavage division prophase. This "strip" is situated in the equatorial plane of the embryo and coincides topographi-

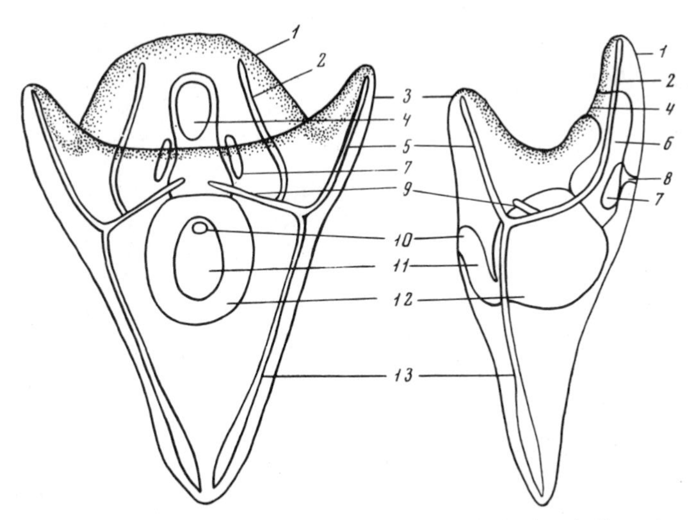

Fig. 10.4. Early echinopluteus [44]. 1) Oral lobe; 2) oral skeletal rod; 3) anal arm; 4) mouth; 5) anal skeletal rod; 6) esophagus; 7) left coelomic sac; 8) hydropore; 9) ventral transverse skeletal rod; 10) anus; 11) intestine; 12) stomach; 13) body skeletal rod.

cally with the mitotic apparatus. With nuclear membrane breakdown the "strip" loses its distinct outlines and broadens, and at anaphase–telophase it acquires a dumbbell-like form. A similar appearance can be seen, although less clearly, during the second mitotic cycle.

At the 16-blastomere stage the embryo already has a blastocoele which increases in size during the subsequent asynchronous cleavage divisions. The increase in cell number during blastulation is almost exponential; the rate of cell divisions decreases somewhat prior to hatching and slows down markedly at the stages of late blastula–gastrula (Fig. 10.3).

At the stage of mid-blastula, prior to hatching, cilia develop on all cells and the embryo begins to move inside its membranes. Soon after hatching, the vegetal side of the embryo flattens and thickens and, as a result, the embryo loses its spheroid form; motile cilia on the animal pole are replaced by a tuft of longer stiff cilia (the so-called apical tuft). Somewhat later cells of the primary mesenchyme originating from the micromeres migrate in the blastocoele from the blastula vegetal wall (mesenchyme blastula stage). Still later, the vegetal blastula wall invaginates, forming the archenteron (gastrula stage). The archenteron grows across the blastocoele to the animal pole. The process of gastrulation in *S. dröbachiensis* has been described by Martynova [52].

Somewhat later the animal pole wall is flattened and the end of the archenteron approaches its internal surface; secondary mesenchyme cells migrate from this region of the archenteron into the blastocoele. Larval skeleton elements (spicules) develop in the primary mesenchyme cells, the ventral and oral zones of ectodermal cilia arise, and the first signs of archenteron differentiation appear. As a result, a bilaterally symmetrical larva (prisma) develops.

During subsequent development, branches of skeletal spicules come into contact with definite regions of ectoderm and induce the formation of outgrowths, the so-called arms. A pair of anal arms forms, then a pair of oral arms. From the time of appearance of the buds of the first pair of arms, the larva is called a pluteus, or echinopluteus. A diagram of the sea urchin pluteus is given in Fig. 10.4.

Some cells of secondary mesenchyme turn into pigment cells; with their appearance the color of the larva changes markedly (it is clearly visible with the naked eye in a dense suspension of larvae). The digestive tract differentiates, and the mouth opening breaks through and the larva begins active feeding.

In conclusion, it should be noted that after fertilization the intensity of macromolecular syntheses in mature unfertilized sea urchin eggs, previously low, is markedly intensified [21, 28, 29, 40]. The increased protein synthesis takes place in *S. dröbachiensis* within 18 ± 2 min after fertilization (at 8°C) and in *S. intermedius* within 12 ± 1 min (at 18°C); it coincides in time with the appearance of a discernible hyaline layer. RNA synthesis is intensified in embryos of different sea urchin species at about the same time, attaining a significant level upon transition from 8 to 16 blastomeres; at the stage of 16 blastomeres it mostly occurs in the micromeres only [23]. DNA synthesis starts after the fusion of pronuclei and intensifies markedly during blastulation [7, 24, 32, 70].

10.8.2. Description of Stages

The characteristics of structure of sea urchin embryos at successive developmental stages coincide, in general, for all the species studied, except for such factors as absolute size, pigmentation pattern, structure of larval skeleton, etc. [39,

42]. Therefore we could use for the species under consideration any of the available tables of normal development [39, 42, 44, 56, 73, etc.], one of which [73] was actually constructed for *S. dröbachiensis*. However, on the basis of many years' studies on sea urchin embryos, we considered it necessary to give a more detailed classification of stages for the periods of blastulation and gastrulation. Moreover, like Harvey [39], and unlike all other authors, we distinguish the stage of "strip" in the course of the first cleavage division. The timing of this stage is a good additional criterion for estimating the results of experimental insemination. The classification of developmental stages given below is, in turn, applicable not only to the species considered here but to the other sea urchin species as well. When describing normal development, we are concerned with the period from fertilization to the onset of active feeding only (Table 10.2).

Table 10.3 (*S. intermedius* and *S. dröbachiensis*) gives the timing of these stages at different temperatures, and Figs. 10.5 and 10.6 (*S. intermedius*) and 10.7 and 10.8 (*S. dröbachiensis*) provide photographs of the embryos and larvae at successive developmental stages. The numbers of photographs in these figures correspond to the numbers of the stages.

S. intermedius. The description of normal development is based on our observations. They were carried out under stable temperature conditions controlled by a thermostat. The morphological state of the embryos and larvae was recorded by microfilming: when taking shots of movable larvae, some of them were immobilized with formaldehyde (final concentration 0.2%). Development at 20°C was studied in most detail: the period of 48 h from the moment of fertilization was studied on embryos from five females. Observations were completed by stage 24; the timing of all subsequent stages was estimated, but only approximately. From the moment of fertilization until stage 4 the embryos were observed and photographed every 2 min; during the formation of the 1st and 2nd cleavage furrows the interval was 1 min; from stage 4 to stage 11, 10-15 min; from stage 11 to stage 23, 30 min; and during the last 6 h of the period studied, 60 min. The development of *S. intermedius* at 15°C was also followed from stage 0 to stage 24; the intervals between observations until stage 4 were 2 min (embryos from two females) and at later stages 30-90 min (embryos and larvae from one female).

These results allowed us to estimate the timing of the above-mentioned developmental stages (Table 10.3), and to obtain microphotographs of the embryos and larvae of *S. intermedius* at successive developmental stages (Figs. 10.5 and 10.6). The time of transition of the embryos and larvae to the next developmental stage was registered when this transition was observed in about 10% of the embryos or larvae in a sample under study.

On the basis of the results obtained, the values of τ_0, the duration of one mitotic cycle during synchronous cleavage divisions [27], were estimated. In sea urchins the duration of the second cell cycle appears to be the most convenient for estimation. We observed 38 min (five estimations) at 20°C, 56 min (two estimations) at 15°C, and 37 min (one estimation) at 23.5°C. At 25°C development is strikingly abnormal: the formation of normal plutei is impossible. Judging by the results of single estimation of τ_0 at 23.5°C, this temperature is already outside the optimal range. Development at this temperature proceeds to the formation of plutei, but is accompanied by some defects during cleavage divisions and gastrulation.

The materials used in Tables 10.2 and 10.3 and Figs. 10.5 and 10.6 were obtained in August-September 1972 and 1973. The sea urchins used in these ob-

TABLE 10.2. Brief Description of Diagnostic Features of the Main Developmental Stages in Sea Urchins

Stage	Diagnostic features of the stages
1	Unfertilized egg
1a	Fertilized egg; fertilization membrane formed
1b	Fertilized egg; hyaline layer formed
2a	Stage of narrow "strip" (prophase of first cleavage division)
2b	Stage of wide and dumbbell-shaped "strip" (metaphase–telophase of first cleavage division
3	2 blastomeres. If need be, stages 3a (narrow "strip" of second cleavage division) and 3b (wide "strip") can be distinguished on the basis of vital observations
4	4 blastomeres
5	8 blastomeres
6	16 blastomeres. End of the period of synchronous cleavage divisions
7	32 blastomeres. Micromeres still clearly discernible among other blastomeres
8	Early blastula 1. Blastomeres preserve spheroid form; micromeres already not discernible
9	Early blastula 2. Blastomeres have partially lost their spheroid form (especially from the side of blastocoele). Diameter of blastocoele is approximately 0.65 of the diameter of the embryo (without fertilization membrane)
10	Middle blastula 1. Blastomeres have lost their spheroid form and the outline of embryo in optical section represents a rather regular circle. Formation of cilia, clearly discernible under phase contrast. Relative diameter of blastocoele 0.7-0.75
11	Middle blastula 2. Hatching. Embryos mobile and preserve spheroid form; thickness of blastocoele wall the same all around. Relative diameter of blastocoele 0.8
12	Late blastula 1. Blastula wall on the vegetal pole somewhat flattened and thickened, embryo slightly elongated along the animal–vegetal axis. First primary mesenchyme cells appear in blastocoele at the vegetal pole
13	Late blastula 2 (mesenchyme blastula). Vegetal wall of blastula markedly flattened. A large number of primary mesenchyme cells in blastocoele
14	Early gastrula 1. Onset of invagination of vegetal wall of embryo: a rudiment of blastopore seen at the vegetal pole. Single primary mesenchyme cells migrate to the animal pole
15	Early gastrula 2. Appearance of a rudiment of archenteron and mass migration of primary mesenchyme cells to the animal pole
16	Middle gastrula 1. Archenteron reaches about the center of blastocoele
17	Middle gastrula 2. Archenteron attains definitive size but does not yet touch the wall of larva in the region of future mouth opening (absence of oral contact). Larva preserves radially symmetrical structure
18	Late gastrula 1. Appearance of oral contact and hence of first signs of bilateral symmetry. Appearance of secondary mesenchyme and onset of formation of ectodermal belts of cilia
19	Late gastrula 2. Animal wall of larva flattened; archenteron displaced from the central axis to the oral contact. Onset of skeleton formation (appearance of two symmetrical spicules)

TABLE 10.2 (continued)

Stage	Diagnostic features of the stages
20	Prisma 1. Larvae lose spheroid shape and assume characteristic angular form. First signs of differentiation of digestive tract
21	Prisma 2. Digestive tract consists of weakly delineated rudiments of esophagus, stomach, and intestine and bends at the boundary between esophagus and stomach. Stomodeum (ectodermal pit in the place of future mouth opening) and elements of larval skeleton clearly seen
22	Early pluteus 1. Mouth opening is broken through, on each side of it (in optical section) a paired rudiment of coelom is clearly seen. Chromatophores appear. A small rudiment of the first pair of arms (anal or ventral) appear
23	Early pluteus 2. Parts of digestive tract delineated by distinct constrictions: stomach markedly increased. Skeletal body rods reach the apex (see scheme in Fig. 10-4). Rudiments of the anal (ventral) pair of arms are as yet small
24	Early pluteus 3. First pair of arms markedly increased; a rudiment of second pair (oral or dorsal) appear. It is at this stage that the color of the larval suspension changes sharply. See scheme of pluteus (Fig. 10-4)
25	Middle pluteus 1. Both pairs of arms developed
26	Middle pluteus 2. Transition to active feeding

servations were collected in the sublittoral zone of the Popov and Putyatin Islands (Peter the Great Bay, Sea of Japan).

In conclusion, we will note some peculiarities which distinguish the embryos of *S. intermedius* from those of the two other species under consideration. The eggs and early embryos of this species are pigmented somewhat more intensively and are of light or bright orange color. With the development of chromatophores, the color of the larvae becomes reddish or light lilac. The embryos of *S. intermedius* are characterized by a somewhat lower degree of synchrony of development than those of the other two species. For example, at 20°C the time of completion of the 1st cleavage division in the embryos from the same female deviates by ±9 min from the mean value of 75 min.

S. dröbachiensis. The microphotographs of the embryos and larvae at successive developmental stages are given in Figs. 10.7 and 10.8 (after Buznikov et al. [10]); the microphotograph of stage 25 has been kindly provided by Dr. R. E. Stephens. The timing of developmental stages is given in Table 10.3, mainly from the data of Stephens [73], obtained at constant temperature of incubation (0, 4, or 8°C).

The eggs and early embryos of *S. dröbachiensis* are weakly pigmented and are of greyish-yellow or light-yellow color. The color of larvae with chromatophores is approximately the same as in the two other species. Embryos obtained from the same female are characterized by a high degree of synchrony of development. The first cleavage division begins practically at the same time in 90% of embryos (at 8°C the first cleavage division begins, on average, within 190 ± 5 min after fertilization) [73].

TABLE 10.3. Developmental Stages of Sea Urchins and Their Timing at Different Temperatures

		S. intermedius				*S. dröbachiensis*		
		15°C		20°C		0°C	4°C	8°C
Stage No.	Stage	h, min	τ_n/τ_0	h, min	τ_n/τ_0	days, hours, minutes		
1a	Fertilization membrane	45 sec	–	30 sec	–	00.00.10	00.00.07	00.00.05
1b	Hyaline layer	00.30	–	00.20	–	00.03.20	00.02.00	00.01.15
2a	Narrow "strip"	00.50	–	00.30	–	–	–	00.01.50*
2b	Wide "strip"	01.20	–	00.50	–	00.05.50	00.03.30	00.02.10
3	2 blastomeres	01.45	1.9	01.07	1.8	00.08.30	00.05.10	00.03.10
4	4 blastomeres	02.40	2.9	01.45	2.8	00.13.30	00.08.00	00.05.10*
5	8 blastomeres	03.35	3.8	02.23	3.8	00.17.00	00.10.30	00.07.00
6	16 blastomeres	04.32	4.9	03.00	4.7	00.23.00	00.14.00	00.08.30
7	32 blastomeres	–	–	03.38	5.7	00.30.00	00.18.00	00.11.00
8	Early blastula 1	–	–	04.20	6.8	–	–	00.14.00*
9	Early blastula 2	07.48	8.6	05.13	8.2	–	–	00.17.00*
10	Middle blastula 1	10.30	10.9	07.15	11.4	02.17.00	01.16.00	01.00.00
11	Middle blastula 2 (hatching)	14.00	15	09.30	15	03.08.00	02.02.00	01.06.00
12	Late blastula 1	–	–	11.00	17.4	–	–	01.11.00*
13	Late blastula 2	17.25	18.7	12.00	18.9	–	–	01.15.00*
14	Early gastrula 1	23.00	24.6	15.45	24.9	05.00.00	03.03.00	01.21.00
15	Early gastrula 2	–		17.00	26.8	–	–	02.03.30*
16	Middle gastrula 1	27.00	28.9	18.25	29	07.00.00	04.04.00	02.12.00
17	Middle gastrula 2	–	–	19.20	30.5	–	–	02.16.00*

18	Late gastrula 1	–	–	20.25	32.2	–	–	02.19.30*
19	Late gastrula 2	36.15	38.8	23.55	37.8	08.00.00	05.00.00	03.00.00
20	Prisma 1	–	–	24.55	39.3	–	–	03.04.00*
21	Prisma 2	42.00	45	27.25	43.3	10.00.00	06.12.00	03.20.00
22	Early pluteus 1	45.30	48.8	29.50	47.1	–	–	04.00.00*
23	Early pluteus 2	–	–	34.25	54.3	–	–	04.13.00*
24	Early pluteus 3	58.00	62	39.35	62.5	14.00.00	08.00.00	05.00.00
25	Middle pluteus 1	–	–	54.00	85.3	18.00.00	11.00.00	07.00.00
26	Middle pluteus 2	–	–	84.00	132.6	30.00.00	20.00.00	12.00.00

Note. The development of *S. intermedius* is given by the original observations of the authors; the timing of development of *S. dröbachiensis* at 0, 4, and (partially) 8°C is given after Stephens [73]. Our data for 8°C are indicated with asterisks.

S. nudus. In size, external appearance, and pigmentation of the eggs, embryos, and larvae, *S. nudus* resembles *S. dröbachiensis* very closely. It differs from *S. intermedius* by a greater degree of synchrony of development of the embryos from the same female and by a greater need for carefully controlled incubation conditions.

At the same temperature, the development of *S. nudus* embryos proceeds somewhat more rapidly than in *S. intermedius*. The timing of individual developmental stages at a constant temperature was not estimated. The duration of the second cell cycle (τ_0) was 36 min at 20°C. From this value one can calculate approximately the time of transition to any developmental stage at the given temperature of incubation (provided that the time of transition to this stage in *S. intermedius* or *S. dröbachiensis*, expressed in τ_0 units, is known). The reliability of such calculations is confirmed by the comparison of the times of the same developmental stages in *S. intermedius* (our data), *S. dröbachiensis*, and *Arbacia punctulata* (data of Harvey [39] and Stephens [73], recalculated by us). The first cleavage division proceeds in all these species within $1.7\tau_0$ after insemination; the relative times of subsequent synchronous cleavage divisions coincide as well. The hatching in all the species takes place within $15\tau_0$, gastrulation starts within $23\text{-}24\tau_0$, skeletal spicules appear within $37\text{-}38\tau_0$, etc.

10.9. TYPICAL DEVELOPMENTAL DEFECTS

Information on some typical developmental defects may be useful for determining the quality of the material used, the conditions of its incubation, and for analyzing the consequences of experiments on developing embryos. Defects of the fertilization membrane have been considered above (Section 10.6.1).

1. Deformation of the embryos prior to the first cleavage division, usually seen as dents on the cell surface, suggests a nonoptimal (too high) temperature of incubation.

2. Polyspermy results in multipolar mitoses, abnormal precocious cleavage, irregular orientation of blastomeres, etc. A typical sign is the formation of 3-4 blastomeres at the same time during the first cleavage division. Polyspermy can be due to the initial poor quality of the eggs, keeping them for too long, or errors in the fertilization procedure. Many chemical influences on the eggs prior to or during insemination will also induce polyspermy [8, 39, 53].

Figs. 10.5–10.8. Developmental stages of the sea urchin. Numbers on photographs correspond to stages.

Fig. 10.5A, B. *Strongylocentrotus intermedius*. There are no photographs for stages 2b, 4, 7, and 11. For these stages, see Fig. 10.7 illustrating *S. dröbachiensis*.

Fig. 10.6A, B. *Strongylocentrotus intermedius* (continued). There are no photographs for stages 12, 14, 16, 23, and 25. For these stages, see Fig. 10.8 illustrating *S. dröbachiensis*.

Fig. 10.7. *Strongylocentrotus dröbachiensis*. There are no photographs for stages 0, 1a, 1b, 2a, and 10. For these stages, see Figs. 10.5A, B illustrating *S. intermedius*.

Fig. 10.8. *Strongylocentrotus dröbachiensis* (continued). There are no photographs for stages 18, 19, 21, 22, and 24. For these stages, see Figs. 10.6A, B illustrating *S. intermedius*.

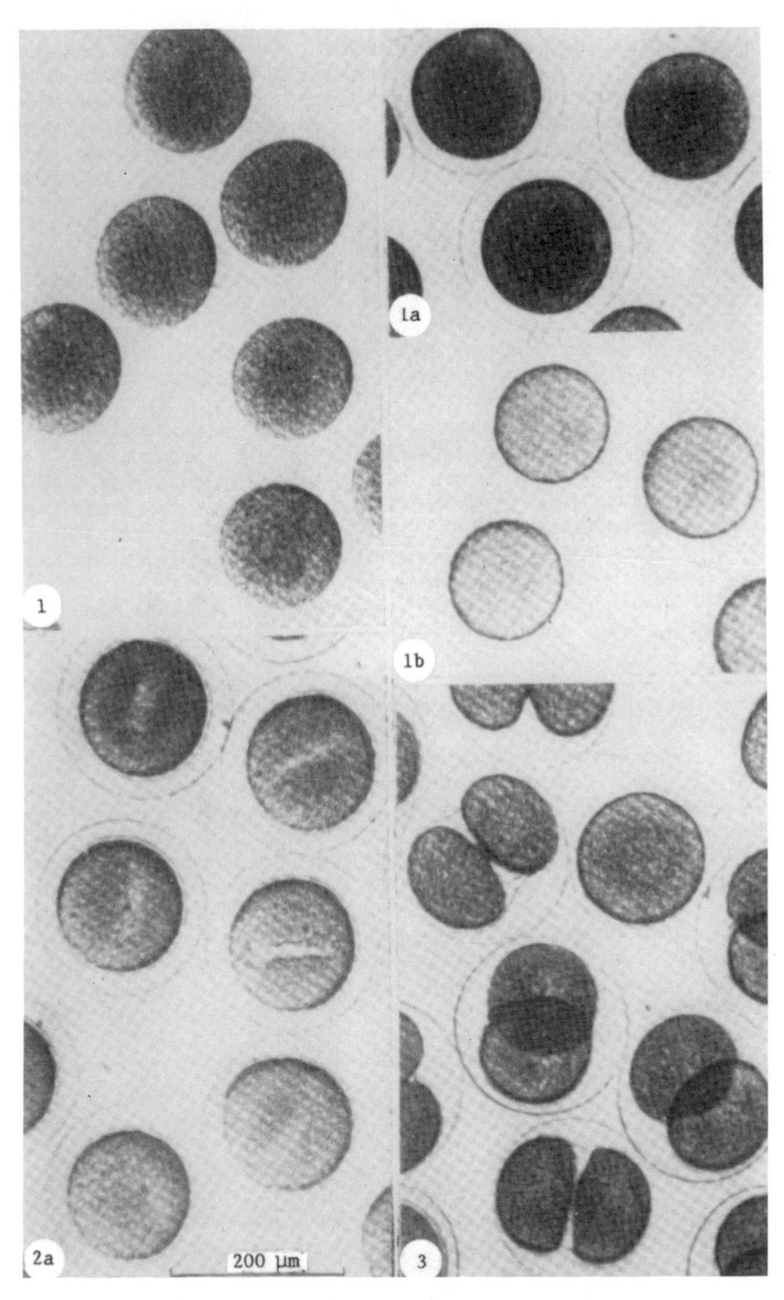

Fig. 10.5A

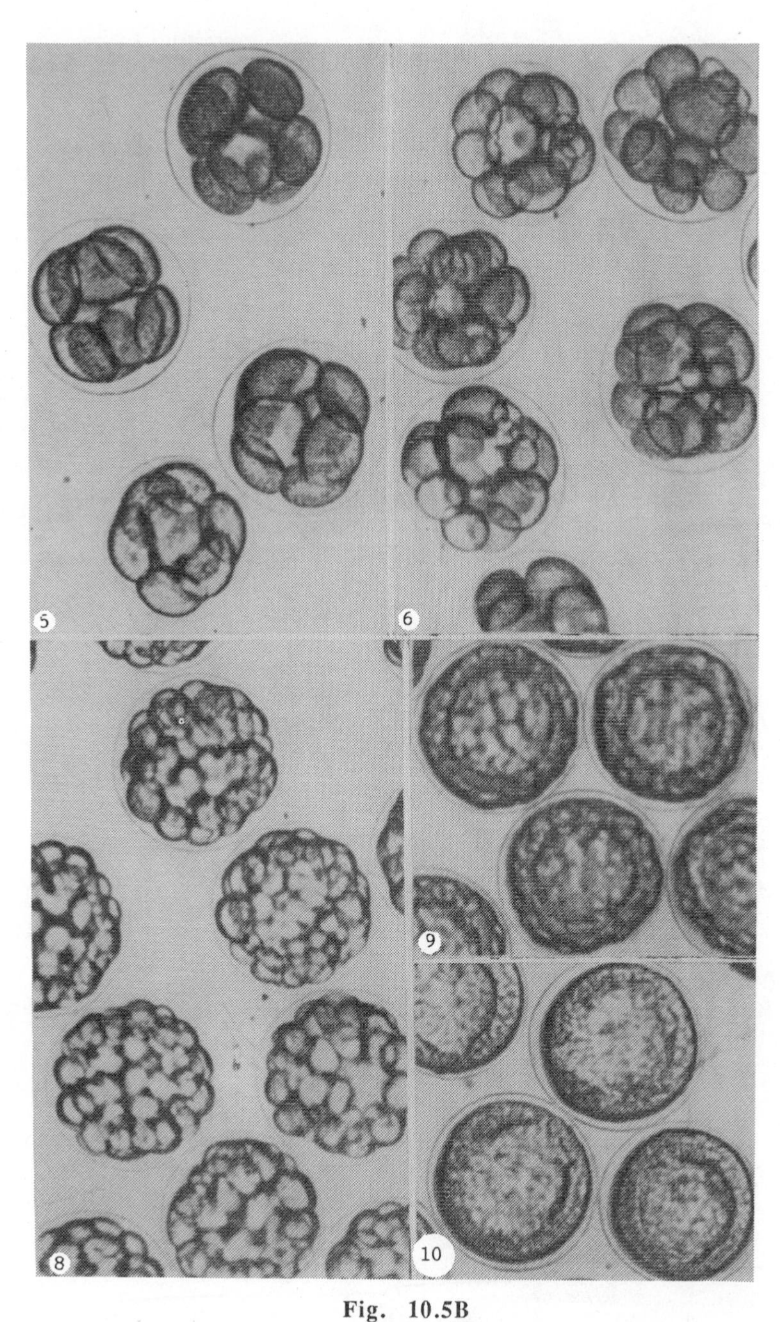

Fig. 10.5B

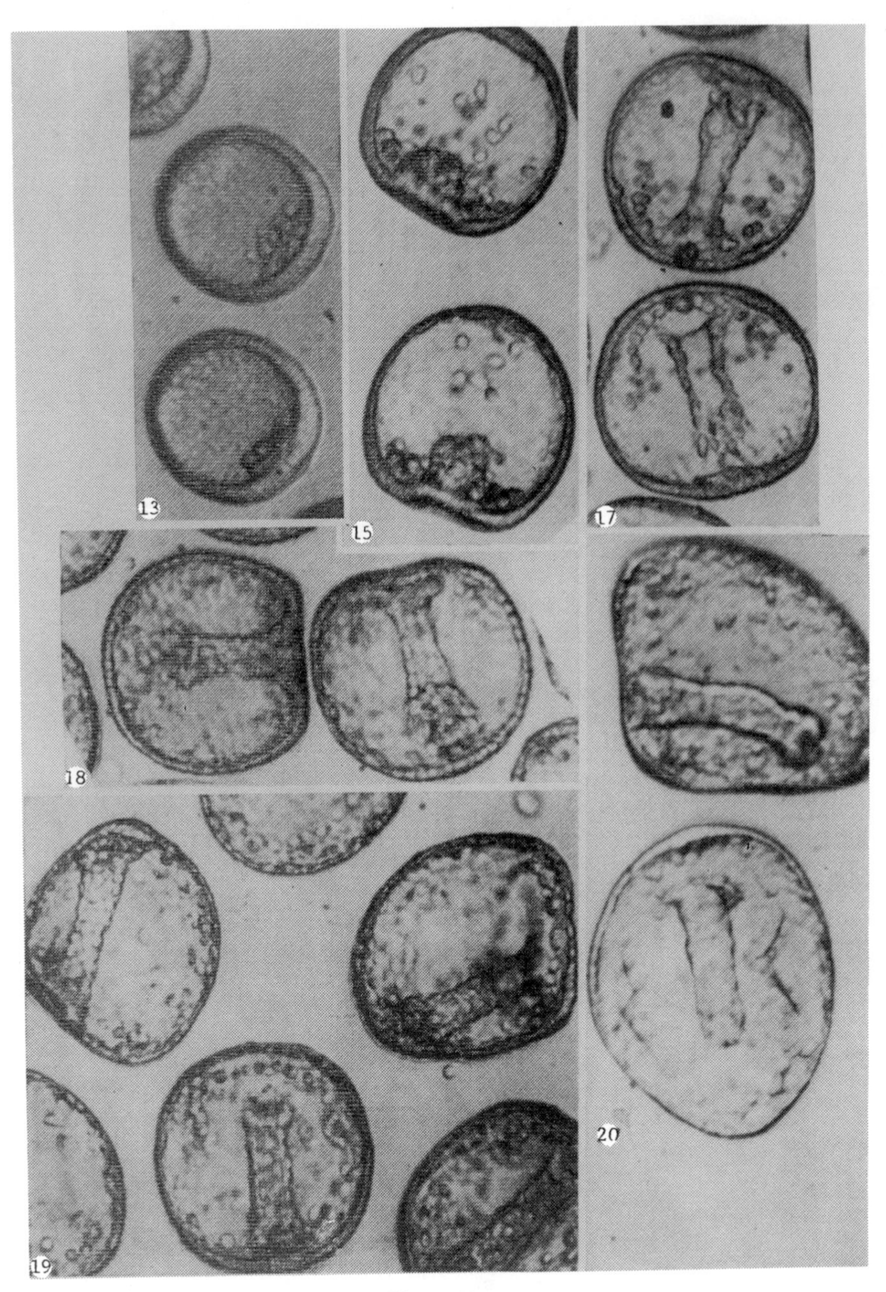

Fig. 10.6A

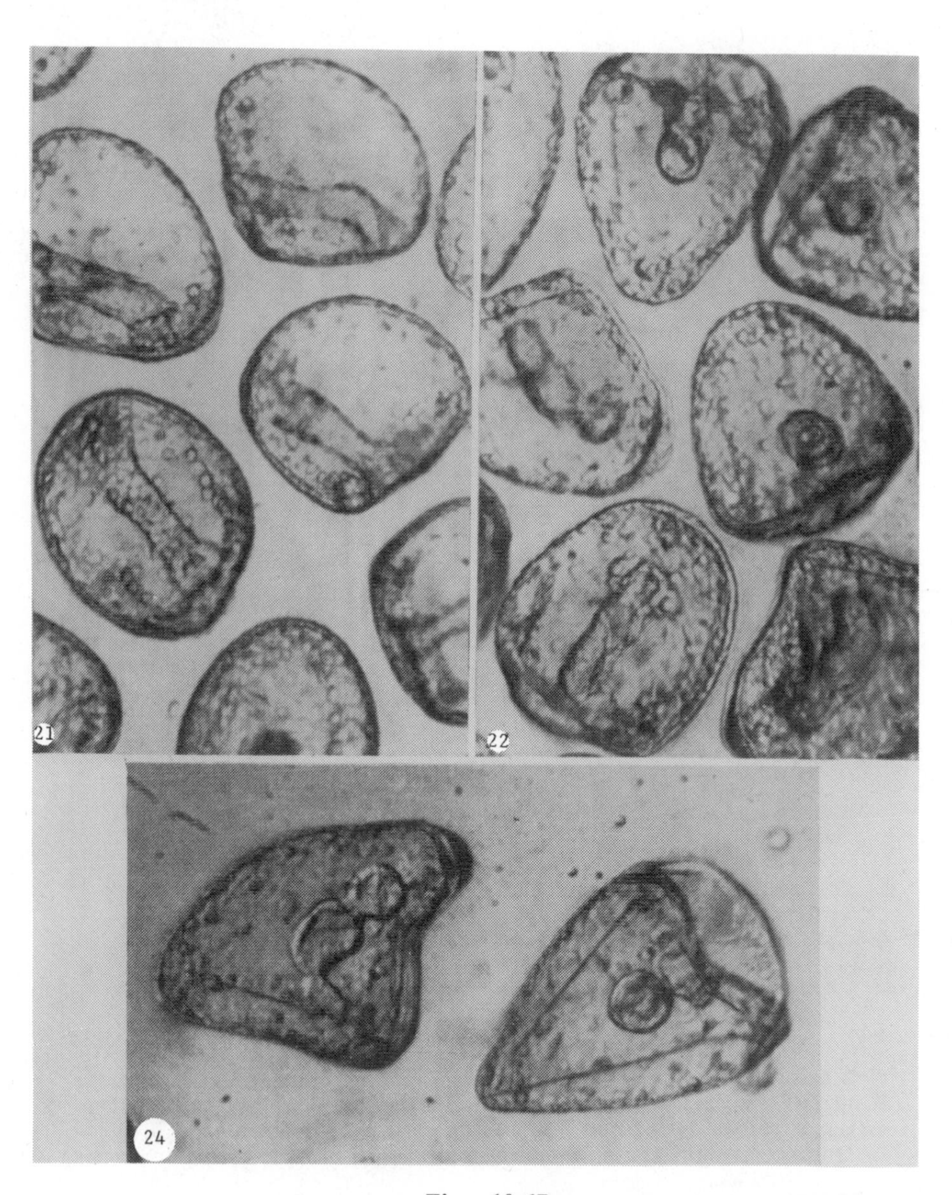

Fig. 10.6B

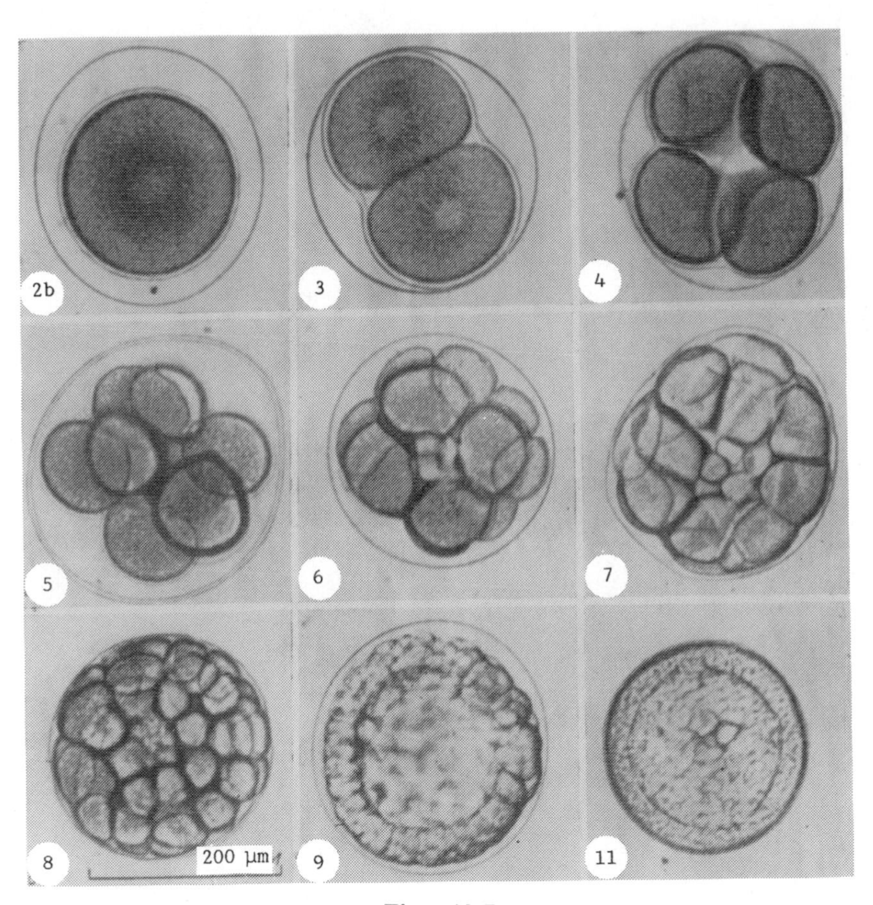

Fig. 10.7

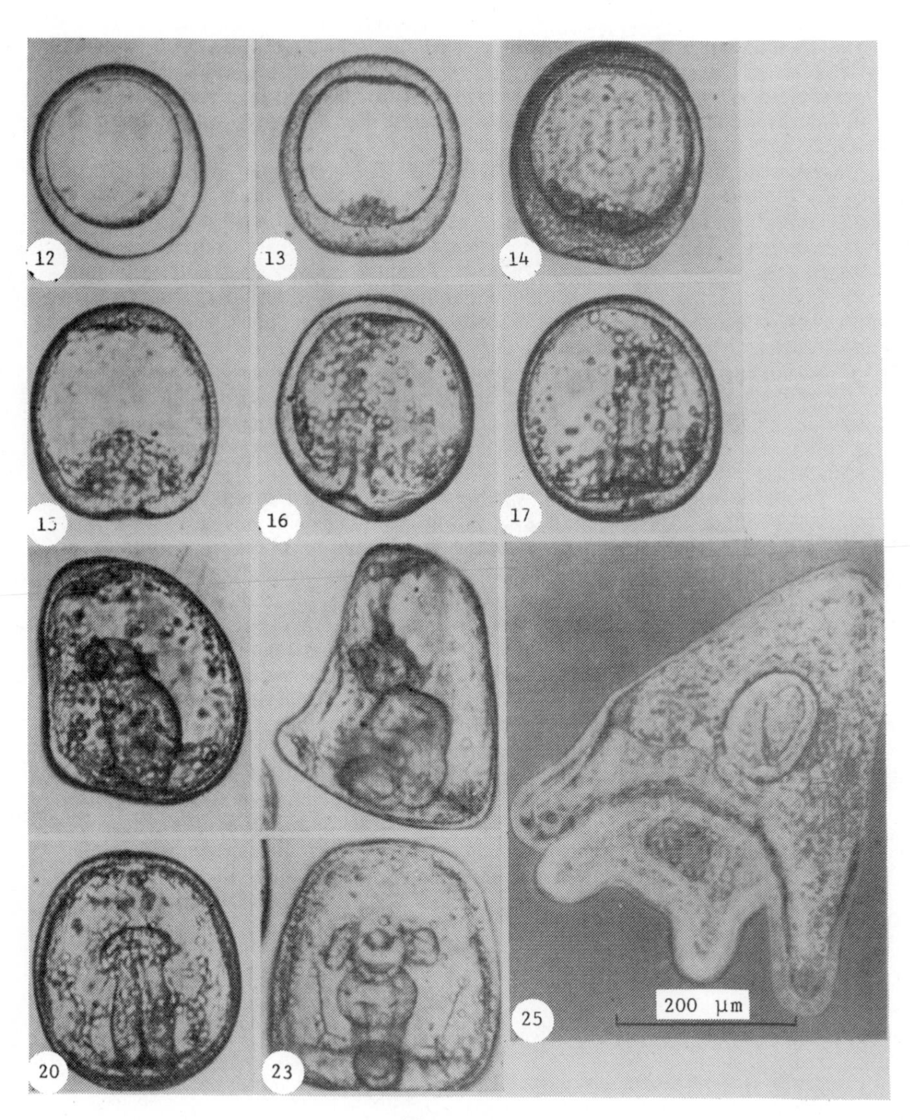

Fig. 10.8

3. Polynuclearity of eggs and blastomeres, not related to polyspermy, differs from that induced by polyspermy in the normal character and rate of mitoses. It occurs rather frequently in eggs obtained at the end of the spawning season, and can also be due to diverse chemical influences on the fertilized eggs [8, 9, 39].

4. Defects may arise due to disturbance or absence of functional intercellular relationships during the first cleavage divisions. Sometimes at stages of 2, 4, or (very rarely) 8 blastomeres, the blastomeres begin to develop totally independently from each other. As a result, each blastomere gives rise to a complete embryo of lesser size: all embryos developed from the same egg (2, 4, or 8) remain mechanically associated with each other. Such a dissociation of the embryos is more often partial and leads to the formation of blastulae with more or less pronounced constrictions. As already mentioned, such defects arise under the influence of some specific agents which disturb or block functional intercellular relationships [21, 32, 58]. At the same time, some females produce eggs which have a tendency to produce such defects. This effect is seen in a large percentage of embryos developing in unfavorable conditions of incubation and is produced by diverse experimental influences bearing no direct relation to functional intercellular connections. The possible character of such results is witnessed by the presence of at least single embryos with such defects in the control sample of embryos.

5. Irregular orientation of blastomeres, which results in the formation of embryos having the shape of a markedly stretched rectangle, is one more example of defects due to the initial quality of the eggs and provoked by nonspecific external influences. This defect occurs rather frequently in *S. intermedius* and can be found by careful examination of the control sample of embryos.

6. Developmental arrest may occur at stage 10; the embryos can in many cases, for a long time (up to several days, Harvey [39]), preserve a normal appearance. This defect can be due to specific switching of the function of cell nuclei and also to the initial poor quality of gametes (in particular, upon insemination with a damaged sperm), and too-low temperatures.

7. Formation of micromeres may occur during the third cleavage division (normally they arise during the fourth division) [25]. The tendency to such a defect, as well as to a number of previously described ones, can be inherent in the eggs of some females and manifest itself under different unspecific influences.

8. Disaggregation of blastomeres soon after hatching is rather frequent when the water is not changed after hatching. This is particularly true of dense suspensions of embryos.

9. Various gastrulation defects (delay of archenteron formation, exogastrulation, etc.) can be due to specific chemical influences (for example, upon animalization and vegetalization of the embryos; see [21, 44]) but can also arise under abnormal conditions of incubation and, especially, in the eggs initially predisposed to these defects.

Of course, this list does not exhaust the possible developmental defects of sea urchin embryos. At the same time it should be stressed once more that if optimal conditions of incubation are observed and good-quality gametes are used, developmental defects in sea-urchin embryos are exceptionally rare.

10.10. SOME EXPERIMENTAL MODELS

One of the main advantages of sea urchins for experimental studies in developmental biology is the possibility of elaboration of diverse experimental models (see [21, 32]). Many such models should be modified for particular species of sea urchins. Therefore, detailed methods for obtaining such models for the sea-urchin species under consideration will be described below.

10.10.1. Eggs and Embryos without the Vitelline Membrane

When working with large numbers of the eggs and embryos, chemical methods of removal of the vitelline membrane (fertilization membrane) are recommended. We will describe here one such method [45] which gives good results on all species of the genus *Strongylocentrotus*. A dense suspension of eggs is placed for 5 min in a solution of papain (0.04%) and cysteine (0.2%) prepared in sea water. The sperm is suspended in the same solution and artificial fertilization is carried out. The eggs are washed in sea water within 5-10 min after fertilization for the first time in calcium-free artificial sea water. The membranes are removed more easily and more completely if artificial fertilization is carried out in lowered calcium ion concentration. With this aim calcium-free artificial sea water is mixed with normal sea water (4:1) [66]. While working with small numbers of eggs, the embryos can be mechanically shaken just after fertilization to remove the fertilization membranes. The removal of the membrane does not affect normal development and is a prerequisite for various experimental procedures on the embryos prior to stage 11, in particular for the disaggregation of embryonic cells and for testing different high-molecular-weight compounds.

10.10.2. Activated Eggs

The eggs of the species under investigation can be activated by a somewhat modified method elaborated initially on *Paracentrotus lividus* [51]. The unfertilized eggs are placed for 40-60 sec in thymol/water (a mixture of sea water previously saturated with thymol and normal sea water, 1:4), washed in sea water, and transferred for 20 min in sea water and for 20 min in 0.27 M KCl prepared in sea water. The eggs are then washed and resuspended in sea water. For *S. dröbachiensis* eggs, these times should be increased 1.5-2.5 times. The separation of fertilization membranes and activation of macromolecular syntheses are observed in 60-80% of the activated eggs; a certain number of eggs (5-20%) start to cleave, and their development leads sometimes to the formation of plutei, although it is more often arrested prior to gastrulation. Good results are obtained with calcium ionophore A23127, which induces egg activation in various groups of animals [29]. The calcium–magnesium salt of this ionophore (2-5 μg/ml for 2-3 min) induces activation and hence stimulation of protein synthesis in practically all eggs, as well as in anuclear egg fragments (see Section 10.10.6). This procedure should be of interest for molecular biology and biochemistry studies of development.

A partial activation of the eggs is also possible when neither cleavage nor cortical reaction are observed (no separation of fertilization membrane, in particular) but macromolecular syntheses are markedly intensified. Such an activation is induced, for example, by ammonium salts [61]. Good results on the eggs of *Strongylocentrotus* species are obtained with NH_4Cl (final concentration 10 mM, pH 8, incubation of the unfertilized eggs at the corresponding optimal temperatures with constant stirring). The time of activation varies in different experiments, and it is necessary to determine it afresh every time the method is used.

10.10.3. Embryos without the Hyaline Layer

An excellent method to remove the hyaline layer was proposed by Kane [46]. The eggs, just after sperm addition, are sedimented in a hand centrifuge and resuspended in 1 M glycine (10 min). The removal of the hyaline layer results in defects of early intercellular relationships; the development of experimental embryos is, as a rule, abnormal, and ends in the formation of a disorderly accumulation of semidisaggregated cells [15]. This model can be more widely used on some other sea urchin species, for example *Echinus esculentus*, where the removal of the hyaline layer does not adversely affect normal development and, at the same time, makes the embryos sensitive to substances blocking intercellular interactions (see below). The hyaline layer is capable of a relatively rapid regeneration [46] and, if need be, it should be removed repeatedly by a 10-min treatment of the embryos with 1 M glycine.

10.10.4. Disaggregated Cells of Embryos and Larvae

While working on the species considered here, most often we used Kane's method, described above [46]. The larvae, hatched embryos, or early embryos previously denuded are washed once in 1 M glycine with 0.001-0.002 M EDTA and resuspended by stirring in the same solution. Disaggregation of cells is practically complete within 10-20 min. The cells are washed several times with calcium-free sea water and resuspended in the same water. Their reaggregation starts upon their transfer to the normal sea water. The embryos can also be incubated in calcium-free sea water with 0.001-0.002 M EDTA; in this case, an additional mechanical effect is necessary for disaggregation: the suspension of embryos is shaken or pipetted several times [32]. The methods described above, or similar ones, are often applied to 16-cell embryos to obtain pure fractions of macro-, meso-, and micromeres [65, 72]. We disaggregated 16-cell embryos by Kane's method [46], without preliminary removal of fertilization membranes, to compare the sensitivity of macro-, meso-, and micromeres of the same embryo to some cytotoxic drugs [12].

To obtain a relatively small number of isolated blastomeres, microsurgery (separation of blastomeres with a fine glass needle) may be used. This method is recommended when it is necessary to break intercellular relationships at a strictly defined moment of development [71].

10.10.5. Dwarf and Twin Embryos

If the blastomeres are isolated during the first cleavage division up to the 8-cell stage, each blastomere can give rise to a complete dwarf embryo [44]. Some blastomeres (not micromeres) are capable of normal development if isolated at the 16-cell stage [15, 16]. Individual blastomeres can develop into complete dwarf embryos without preliminary mechanical isolation, as a result of treatment with some chemical substances such as sulfhydryl drugs, some detergents, and some structural analogs of serotonin [15, 16, 77]. In the sea urchin species considered here, such a functional separation of cells leading to the formation of twin embryos is impossible, but it is quite possible in irregular sea urchins, *S. mirabilis* in particular. The best results are obtained with dithiothreitol dissolved in sea water (0.05-

0.067 M) and applied 2-3 min prior to the appearance of the 1st or 2nd cleavage furrow for 15-25 min [15]. This method can be applied for the embryos of a regular sea urchin *E. esculentus*, but only after the removal of the hyaline layer. The experiments described are suitable for studying early intercellular relationships.

10.10.6. Stratified and Fragmented Eggs and Embryos

For stratification of the unfertilized eggs of *S. intermedius* a small volume of their suspension in sea water is layered on 0.95 M sucrose and centrifuged at 0°C (K-24, 8000-9000 rpm, 15 min). An increase of speed up to 10,000-12,000 rpm and/or of time up to 15-20 min results in the division of the unfertilized eggs into animal and vegetal halves; to obtain pure fractions of these halves, a sucrose gradient is used. In experiments on *S. intermedius* good results were obtained with a stepwise gradient of six layers. Sucrose (0.95 M) was poured on a test-tube bottom and mixtures of 0.95 M sucrose with sea water were layered successively on top: 3:1 (2 ml); 2:1 (2 ml); 1.5:1 (2 ml); 1.25:1 (2 ml), and 1:1 (1.5 ml); 0.5 ml of a dense suspension of the eggs in sea water were layered on top. After the centrifugation, the fraction of upper halves is found at the boundary between layers "1:1" and "1.25:1" and the fraction of lower halves at the boundary between layers "1.5:1" and "2:1." The fractions obtained are collected by a syringe through a puncture in a plastic test-tube wall or by a Pasteur pipette with a tip bent at a right angle. The stratified eggs or egg fragments are washed and resuspended in sea water; if need be, they can be inseminated or activated. The upper halves of the egg of *S. intermedius*, like those of most sea urchin species studied [32, 39], contain the nucleus and most mitochondria. After fertilization these halves can develop normally until the formation of plutei. The stratified eggs usually cleave abnormally after fertilization, but their development can thereafter be normalized and result in the formation of plutei. The lower halves are devoid of nucleus and impoverished in mitochondria: the embryos obtained from them (androgenetic haploids) are usually incapable of gastrulation. The embryos from upper and lower halves and intact embryos can differ markedly from each other in sensitivity to many inhibitors of development [12].

By centrifugation in a sucrose gradient of a fraction of upper or lower halves, pure fractions of quarter cells can be obtained. Conditions for obtaining these minicells vary greatly and should be determined anew for every portion of the eggs.

10.10.7. Animalized Embryos

The embryos and larvae of sea urchins with predominant development of animal or vegetal rudiments (animalized or vegetalized) have been used as experimental models for a long time. Many methods have been used to obtain such embryos [21]. We will describe here only one of them: animalization by zinc salts, which gives an almost 100% success rate in the species considered here. Prior to the onset of cleavage, or during the first cleavage division, the embryos are placed in $ZnCl_2$ or $ZnSO_4$ solution in sea water (0.007-0.035 M) and incubated with constant stirring. By stage 22 in the control embryos (early pluteus 1), the experimental embryos look like motile transparent globes with markedly underdeveloped or absent archenteron, without skeleton and arm rudiments, but with chromatophores.

10.10.8. Removal and Regeneration of Cilia

Two methods of removing cilia were tested on the motile embryos and larvae of the species in question: treatment with hypertonic sea water (32 g of NaCl per liter of water) [74] or 1 M sodium acetate [1]. In the first case, treatment lasts 2 min at 0°C and, in the second case, 30 sec at room temperature. The cilia disappear completely after these treatments, and their regeneration takes 1-6 h (according to temperature), and is accompanied by restoration of motility. The cilia can be removed several times, and the development of embryos is not affected adversely. It has been suggested [1] that the disturbance of intercellular interactions makes the regeneration of cilia impossible. A special test has shown that even the complete disaggregation of embryos of the species under consideration does not prevent the regeneration of cilia (Shmukler, personal communication). Therefore, the "deciliation–regeneration" model cannot be used for the functional estimation of intercellular relationships. Nevertheless this model has been successfully used in developmental molecular biology and biochemistry studies, in particular for studying the synthesis and transport of cell skeleton proteins [74].

REFERENCES

1. S. Amemiya, "Relationship between cilia formation and cell association in sea urchin embryos," *Exp. Cell Res.* **64**, 227–229 (1971).
2. J. Bennett and D. Mazia, "Interspecific fusion of sea urchin eggs. Surface events and cytoplasmic mixing," *Exp. Cell Res.* **131**, 197–207 (1981).
3. W. E. Berg, "Some experimental techniques for eggs and embryos of marine invertebrates," in: *Methods in Developmental Biology*, F. H. Wilt, N. K. Wessels, and T. Y. Crowell (eds.), Academic Press, New York (1967).
4. A. Berisha, G. A. Buznikov, L. A. Malchenko, and L. Rakić, "The action of heavy metal salts on the development of sea urchin embryos and on the protein synthesis of mouse transplantable tumor cells," *Ontogenez* **14**, 173–179 (1983).
5. J. Brachet and J. Aimi, "Effects of conditioned media on sea urchin egg development," *Exp. Cell Res.* **72**, 46–55 (1972).
6. H. Bresch and H. Ockenfels, "The influence of Tween surfactants on the development of the sea urchin embryo," *Naturwissenschaften* **64**, 593 (1977).
7. J. W. Brookbank, "DNA and RNA synthesis by fertilized, cleavage-arrested sea urchin eggs," *Differentiation* **6**, 33–40 (1976).
8. G. A. Buznikov, *Low-Molecular-Weight Regulators of Embryonic Development* [in Russian], Nauka, Moscow (1967).
9. G. A. Buznikov, "Neurotransmitters in early embryogenesis," in: *Problems of Developmental Biology*, Mir, Moscow (1981).
10. G. A. Buznikov, I. V. Chudakova, and N. D. Zvezdina, "The role of neurohumours in early embryogenesis. I. Serotonin content of developing embryos of sea urchin and loach," *J. Embryol. Exp. Morphol.* **12**, 563–573 (1964).
11. G. A. Buznikov, L. A. Malchenko, T. M. Turpaev, and Tien Vu Dung, "Embryos of echinoderms from the Vietnamese coastal waters as an object for testing cytotoxic neuropharmacological drugs," *Biol. Morya* **6**, 24–29 (1982).

12. G. A. Buznikov, B. N. Manukhin, and L. Rakić, "The sensitivity of whole, half, and quarter sea urchin embryos to cytotoxic neuropharmacological drugs," *Comp. Biochem. Physiol., Pt. C.* **64**, 129–135 (1979).
13. G. A. Buznikov, L. Rakić, T. M. Turpaev, and L. N. Markova, "Sensitivity of sea urchin embryos to antagonists of acetylcholine and monoamines," *Exp. Cell Res.* **86**, 317–324 (1974).
14. G. A. Buznikov, L. Rakić, N. I. Kudryashova, T. R. Ovsepyan, and N. V. Khromov-Borisov, "Stratified embryos of sea urchin *Arbacia lixula* as a model for studying intercellular interactions," *Ontogenez* **11**, 411–416 (1980).
15. G. A. Buznikov and Y. B. Shmukler, "Effect of antimediator preparations on intercellular interactions in early sea urchin embryos," *Ontogenez* **9**, 141–145 (1978).
16. G. A. Buznikov and Y. B. Shmukler, "The possible role of 'prenervous' neurotransmitters in cellular interactions of early embryogenesis: a hypothesis," *Neurochem. Res.* **6**, 55–69 (1981).
17. G. A. Buznikov, N. D. Zvezdina, B. N. Manukhin, N. V. Prokazova, and L. D. Bergelson, "Protection of sea urchin embryos against the action of some neuropharmacological agents and some detergents by endogenous gangliosides," *Experientia* **31**, 902–904 (1975).
18. A. D. Cooper and W. R. Crain, "Complete nucleotide sequence of a sea urchin actin gene," *Nucleic Acids Res.* **10**, 4081–4106 (1982).
19. D. P. Costello, M. E. Davidson, A. Eggers, M. H. Fox, and C. Henley, *Methods for Obtaining and Handling Marine Eggs and Embryos*, Marine Biol. Lab., Woods Hole (1957).
20. D. P. Costello and C. Henley, *Methods for Obtaining and Handling Marine Eggs and Embryos*, (2nd edn.), Marine Biol. Lab., Woods Hole (1971).
21. G. Czihak (ed.), *The Sea Urchin Embryo. Biochemistry and Morphogenesis*, Springer, New York (1973).
22. G. Czihak, "Kinetics of RNA-synthesis in 16-cell stage of sea urchin *Paracentrotus lividus*," *Wilhelm Roux's Arch. Dev. Biol.* **182**, 59–68 (1977).
23. G. Czihak and S. Hörstadius, "Transplantation of RNA-labeled micromeres into animal halves of sea urchin embryos. A contribution to the problem of embryonic induction," *Dev. Biol.* **22**, 15–30 (1970).
24. G. Czihak and E. Pohü, "DNS Synthese in frühen Furchungstadien von Seeigelembryonen," *Z. Naturforsch.* **25b**, 1047–1052 (1970).
25. K. Dan and M. Ikeda, "On the system controlling the time of micromere formation in sea urchin embryos," *Dev. Growth Differ.* **13**, 285–302 (1971).
26. E. H. Davidson, B. R. Hough-Evans, and R. J. Britten, "Molecular biology of the sea urchin embryo," *Science* **217**, 17–24 (1982).
27. T. A. Dettlaff and A. A. Dettlaff, "On relative dimensionless characteristics of development duration in embryology," *Arch. Biol.* **72**, 1–16 (1961).
28. D. Epel, "Protein synthesis in sea urchin eggs: a 'late' response to fertilization," *Proc. Natl. Acad. Sci. USA* **57**, 899–906 (1967).
29. D. Epel, "The physiology and chemistry of calcium during the fertilization of eggs," in: *Calcium and Cell Function*, Vol. II, Academic Press, New York (1982).
30. N. S. Gayevskaya, *Guide to Northern Seas Fauna and Flora in the USSR* [in Russian], Sovjetskaya Nauka, Moscow (1948).
31. A. S. Ginsburg, *Fertilization in Fishes and the Problem of Polyspermy*, Israel Program for Scientific Translations, IPST Catalog No. 600418 (1972).

32. G. Guidice, *Developmental Biology of the Sea Urchin Embryo*, Academic Press, New York (1973).
33. G. Guidice, G. Sconzo, A. Bono, and J. Albanese, "Studies on sea urchin oocytes. I. Purification and cell fractionation," *Exp. Cell Res.* **72**, 90–94 (1972).
34. T. Gustafson, "Cell recognition and cell contacts during sea urchin development," in: *Cell Recognition*, R. T. Smith and R. A. Good (eds.), Meredith Corp., Mexico (1969).
35. T. Gustafson, "The cellular basis of morphogenesis and behavior during sea urchin development," in: *Biology of Normal Human Growth*, Raven Press, New York (1981).
36. T. Gustafson and M. Toneby, "On the role of serotonin and acetylcholine in sea urchin morphogenesis," *Exp. Cell Res.* **62**, 102–117 (1970).
37. B. E. Hagström and S. Lönning, "The sea urchin egg as a testing object in toxicology," *Acta Pharmacol. Toxicol.* **32**, Suppl. 1, 1–83 (1973).
38. M. A. Harkey and A. H. Whiteley, "Isolation, culture, and differentiation of echinoid primary mesenchyme cells," *Wilhelm Roux's Arch. Dev. Biol.* **189**, 111–122 (1980).
39. E. B. Harvey, *The American Arbacia and Other Sea Urchins*, Princeton University Press, Princeton (1956).
40. M. B. Hille and A. A. Albers, "Efficiency of protein synthesis after fertilization of sea urchin eggs," *Nature (London)* **278**, 469–470 (1979).
41. M. B. Hille, D. C. Hall, Z. Yablonka-Reuveni, M. V. Danilchik, and R. T. Moon, "Translational control in sea urchin eggs and embryos: initiation is rate limiting in blastula stage embryos," *Dev. Biol.* **86**, 241–249 (1981).
42. R. T. Hinegardner, "Echinoderms," in: *Methods in Developmental Biology*, F. H. Wilt and N. K. Wessels (eds.), T. Y. Cromwell, New York (1967).
43. R. T. Hinegardner, "Growth and development of the laboratory cultured sea urchin," *Biol. Bull.* **137**, 465–475 (1969).
44. S. Hörstadius, *Experimental Embryology of Echinoderms*, Clarendon Press, Oxford (1973).
45. R. O. Hynes and P. R. Gross, "A method for separating cells from early sea urchin embryos," *Dev. Biol.* **21**, 383–402 (1970).
46. R. E. Kane, "Hyaline release during normal sea urchin development and its replacement after removal at fertilization," *Exp. Cell Res.* **81**, 301–311 (1973).
47. V. L. Kasyanov, L. A. Medvedeva, S. N. Yakovlev, and Y. M. Yakovlev, *Reproduction of Echinodermata and Lamellibranchiata* [in Russian], Nauka, Moscow (1980).
48. R. Lallier, "Morphogenetic action of lactins on the development of sea urchin eggs," *Biol. Cell.* **39**, 295–300 (1980).
49. R. Lallier, P. Chang, and C. Sardet, "La polarité de l'oeuf d'Oursin et les inducteurs de la différentiation," *Ann. Biol.* **20**, 201–225 (1981).
50. M. A. Landau, G. A. Buznikov, A. S. Kabankin, N. A. Teplitz, and P. R. Chernilovskaya, "The sensitivity of sea urchin embryos to cytotoxic neuropharmacological drugs; the correlations between activity and lipophilicity of indole and benzole derivatives," *Comp. Biochem. Physiol., Pt. C* **69**, 359–366 (1981).
51. W. Ledebur-Villiger, "Cytology and nucleic acid synthesis of parthenogenetically activated sea urchin eggs," *Exp. Cell Res.* **72**, 285–308 (1972).

52. L. E. Martynova, "Gastrulation of sea urchin in normal development and after action of various drugs," *Ontogenez* **12**, 310–315 (1981).
53. G. M. Mateyko, "Developmental modifications in *Arbacia punctulata* by various metabolic substances," *Biol. Bull.* **131**, 184–228 (1957).
54. D. R. McClay and R. D. Fink, "Sea urchin hyaline: appearance and function in development," *Dev. Biol.* **92**, 285–293 (1982).
55. D. R. McClay and R. B. Marchase, "Separation of ectoderm and endoderm from sea urchin pluteus larvae and demonstration of germ layer-specific antigens," *Dev. Biol.* **71**, 289–296 (1979).
56. W. E. Müller, W. Forster, G. Zahn, and R. K. Zahn, "Morphologische und biochemische Charakterisierung der Entwicklung befruchteter Eier von *Sphaerechinus granularis*. I. Aufzucht, Morphologie, und elektronmikroskopische Stadienbestimmung," *Wilhelm Roux's Arch. Entwicklungsmech. Org.* **167**, 99–117 (1971).
57. S. Nakashima and M. Ishikawa, "Cytological changes and DNA and protein synthesis in parthenogenetically activated sea urchin eggs," *Wilhelm Roux's Arch. Dev. Biol.* **185**, 323–332 (1979).
58. A. Nicotra, "On the fine structure of *Paracentrotus lividus* permanent blastulae obtained with 5-bromodeoxyuridine," *Dev. Growth Differ.* **24**, 39–54 (1982).
59. G. Pagano, A. Esposito, and G. G. Giordano, "Fertilization and larval development in sea urchins following exposure of gametes and embryos to cadmium," *Arch. Environ. Contam. Toxicol.* **11**, 47–55 (1982).
60. E. Parisi, S. Filosa, B. de Petrocellis, and A. Monroy, "Pattern of cell division in early development of sea urchin *Paracentrotus lividus*," *Dev. Biol.* **65**, 38–49 (1978).
61. M. Paul and R. N. Johnston, "Absence of a calcium response following ammonia activation of sea urchin eggs," *Dev. Biol.* **67**, 330–335 (1978).
62. G. Reverberi (ed.), *Experimental Embryology of Marine and Freshwater Invertebrates*, American Elsevier, New York (1971).
63. J. V. Ruderman and M. R. Schmidt, "RNA transcription and translation in sea urchin oocytes and eggs," *Dev. Biol.* **81**, 220–228 (1981).
64. G. D. Ruggieri, "Echinodermata," in: *Culture of Marine Invertebrate Animals*, Plenum Press, New York (1975).
65. K. Sano, "Changes in cell surface charges during differentiation of isolated micromeres and mesomeres from sea urchin embryos," *Dev. Biol.* **60**, 404–415 (1977).
66. K. Sano and N. Usui, "Changes in cell morphology during differentiation of micromere-derived cells of the sea urchin embryo," *Dev. Growth Differ.* **22**, 179–186 (1980).
67. G. Schatten, "Sperm incorporation, the pronuclear migrations, and their relation to the establishment of the first embryonic axis: time-lapse videomicroscopy of the movements during fertilization of the sea urchin *Lytechinus variegatus*," *Dev. Biol.* **86**, 426–437 (1981).
68. T. E. Schroeder, "Expressions of the prefertilization polar axis in sea urchin eggs," *Dev. Biol.* **79**, 427–443 (1980).
69. T. E. Schroder, "Interrelations between the cell surface and the cytoskeleton in cleaving sea urchin eggs," in: *Cytoskeletal Elements and Plasma Membrane Organization*, North Holland, Amsterdam (1981).
70. D. R. Senger and P. R. Gross, "Macromolecule synthesis and determination in sea urchin blastomeres at the 16-cell stage," *Dev. Biol.* **65**, 404–415 (1978).

71. Yu. B. Shmukler, L. M. Chailakhyan, V. V. Smolyaninov, Z. L. Bliokh, A. L. Karpovich, E. V. Gusareva, T. D. Medvedeva, Z. K.-M. Khashaev, and T. K. Naidenko, "Cellular interactions in the early embryos of sea urchins. II. Dated mechanical separation of blastomeres," *Ontogenez* **12**, 298–403 (1981).
72. E. Spiegel and M. Spiegel, "The internal clock of reaggregating embryonic sea urchin cells," *J. Exp. Zool.* **213**, 271–281 (1980).
73. R. E. Stephens, "Studies on the development of the sea urchin *Strongylocentrotus dröbachiensis*. I. Ecology and normal development," *Biol. Bull.* **142**, 132–144 (1972).
74. R. E. Stephens, "Differential protein synthesis and utilization during cilia formation in sea urchin embryos," *Dev. Biol.* **61**, 311–329 (1977).
75. Y. Tanaka, "Effects of surfactants on the cleavage and further development of sea urchin embryos. I. The inhibition of micromere formation at the fourth cleavage," *Dev. Growth Differ.* **18**, 113–122 (1976).
76. A. Tyler and B. S. Tyler, "The gametes: some procedures and properties. Physiology of fertilization and early development," in: *Physiology of Echinodermata*, R. A. Boolootian (ed.), Wiley, New York (1966).
77. V. D. Vacquier and D. Mazia, "Twinning of sand dollar embryos by means of dithiothreitol. The structural basis of blastomere interactions," *Exp. Cell Res.* **52**, 209–221 (1968).
78. V. D. Vacquier, M. J. Terner, and D. Epel, "Protease released from sea urchin eggs at fertilization alters the vitelline layer and aids in preventing polyspermy," *Exp. Cell Res.* **80**, 111–119 (1973).

Chapter 11

THE STARFISH *Asterina pectinifera* (Müller et Troschel, 1842)

P. V. Davydov, O. I. Shubravyi, and S. G. Vassetzky

11.1. INTRODUCTION

Asterina pectinifera has a number of advantages which make this species very suitable for studying various problems of developmental biology. It occurs commonly in shallow zones of the sea and can easily be collected and transported. It also adapts well to artificial conditions, and its feeding is nonspecialized. The embryos and larvae, as well as the adults, can be maintained in a wide range of temperatures. The body structure is fairly simple, and this allows the isolation of gonad fragments and even whole gonads from live starfish by surgery: wounds are healed and the animals can be repeatedly used in experiments. The ovaries and/or fragments of them can be kept in sea water for many hours, and the oocytes retain their ability for maturation and fertilization [37]. Among the invertebrates, the starfish are often used when the oocytes are needed for experimentation [31]. The large and transparent oocytes of *A. pectinifera* are quite suitable for such studies. These and some other advantages provide the basis for the ever-increasing use of *A. pectinifera* in studies of oocyte maturation, fertilization, and embryonic and larval development.

This species was used to study the mechanisms underlying oocyte maturation and the role of 1-methyladenine in this process (for reviews, see [35, 36, 56]) and, more recently, the role of gonad-stimulating substance (GSS) [34, 37, 75, 79, 87]. Polar body formation [27], changes in the rigidity of the egg surface during cleavage [69], the oocyte's competence to the effect of 1-methyladenine [6], the inhibiting effects of cysteine and its derivatives on 1-methyladenine production by the follicle cells [60], and the fine structure of oocytes at different stages of oogenesis and after 1-methyladenine-induced maturation of the oocytes [1] are among the notable investigations in this area. In addition, studies have been made on the effects of steroids upon gonads [85, 86], control of sperm motility [78], early development after insemination of the oocytes at different stages of maturation with different concentrations of sperm [89], and initiation of cleavage by injecting Triton X-100-treated sperm into the eggs [95]. Other investigations include those of the structure and localization of DNA polymerases in the oocytes [28, 70, 71], artificial parthenogenesis [68], and studies of early morphogenetic processes in tetraploid embryos [61]. Many experimental studies deal with the spatial organization and cell interactions during embryonic and early larval development [12-18, 33, 50, 59, 61, 62, 73, 97].

This list of references, which is far from being complete, is in itself a recommendation for using *A. pectinifera* as an experimental species for developmental biology studies.

11.2. TAXONOMIC STATUS, DISTRIBUTION, AND BREEDING IN NATURE

Asterina (*Patiria*) *pectinifera* (Müller et Troschel, 1842) belongs to the class Asteroidea (order Spinulosa, family Asterinidae). In Soviet scientific literature this species is usually referred to as *Patiria pectinifera* (see, for example, [39]), whereas in scientific literature of other countries (mostly Japan), it is referred to as *Asterina pectinifera* (see, for example, [14, 88]). At present the genus *Patiria* is included into the genus *Asterina* [10] and the species in question should be referred to as *Asterina pectinifera* (Müller et Troschel, 1842).

This species occurs in the Yellow Sea and the Sea of Japan, along the far east coast of the USSR from the Posieta Bay to the Tatar Strait, near the southern coast of Sakhalin Island, and in Aniva Bay of the Sea of Okhotsk [23, 39]. It is the most common starfish in Peter the Great Bay of the Sea of Japan [39].

A. pectinifera occurs predominantly on stony ground, but can also occur on various substrata down to 40 m [2, 23].

Studies in Peter the Great Bay have shown that two generations of the oocytes are produced in the ovaries during the reproductive cycle: one in the autumn and another in the spring [39]. Spawning takes place from the second half of July to the beginning of September at surface water temperatures of 18-22°C [39]. According to Novikova [67], the oocytes of the autumn generation are shed during the second spawning in November. At the end of autumn to the beginning of winter the gonads are reorganized and, from midwinter until August, they are characterized by the accumulation and differentiation of gametes [39].

In Sagami Bay (Pacific coast of Honshu Island), *A. pectinifera* spawns from April to May [52], and in Mutsu Bay (northern coast of Honshu Island), from August to September [89].

A. pectinifera females produce, on the average, 500,000 eggs per year [5].

11.3. TRANSPORT AND KEEPING ADULTS IN CAPTIVITY

Before transportation, the collected animals are kept for a few days in basins with running sea water. The starfish should not be fed during this period. They are best transported in 50-liter plastic bags containing one-third of sea water and two-thirds of oxygen. For safety, the plastic bags are placed into paper bags or wooden boxes. The starfish are best transported over long distances at low temperatures (5-10°C) in foam plastic boxes or portable refrigerator bags with a layer of ice on the bottom. Under these conditions the starfish can be kept for 2-3

TABLE 11.1 Composition of Artificial Sea Water of 35.00% [80] (all values in g/100 liters)

Basic constituents	
NaCl	2640
$MgCl_2 \cdot 6H_2O$	491
$MgSO_4 \cdot 7H_2O$	674
KCl	65.5
H_3BO_3	2.48
$NaHCO_3$	13.6
KBr	9.7
Na_2CO_3	3.34
$SrCl_2 \cdot 6H_2O$	2.38
NaF	0.27
$CaCl_2$	111
Complexing agent	
EDTA (Na_4-salt)	1
Trace elements	
KIO_3	0.009
$MnSO_4 \cdot 5H_2O$	0.397
$NaH_2PO_4 \cdot 2H_2O$	0.397
LiCl	0.099
$Na_2SiO_3 \cdot 9H_2O$	0.099
$Na_2MoO_4 \cdot 2H_2O$	0.099
$Al_2(SO_4)_3 \cdot 18H_2O$	0.172
$CoSO_4 \cdot 7H_2O$	0.0089
$CuSO_4 \cdot 7_2O$	0.0009
$ZnSO_4 \cdot 7H_2O$	0.0096
Rb_2SO_4	0.0149
$BaCl_2$	0.009

days. Small numbers of starfish can be transported in portable refrigerator bags on a layer of ice and under a cover of moist algae without water.

In the laboratory the starfish are kept in closed-cycle aquaria with artificial sea water (see [20, 81]). There are many recipes for making artificial sea water [42]. *A. pectinifera* is an undemanding species and can live for a few months in sea water made of a limited number of components (only basic constituents). But in order to keep starfish for a long time (e.g., a few years) and obtain gonad maturation, a more sophisticated water is needed. We prepare artificial sea water according to the "Instant Ocean" recipe with a somewhat altered ratio of basic constituents [80]. This recipe proved to be universally suitable for many species of marine invertebrates and fish (Table 11.1). The water in the aquaria should be changed every month (by 20% in volume). It should be made 2-3 weeks before change and kept in darkness with slight aeration; the pH should be in the range of 7.8-8.3.

A. pectinifera can withstand a temperature rise to 26°C and fall to 8°C. At temperatures of 18-24°C (that is, in an aquarium without water cooling and heating) the starfish normally live several years. But at high temperatures spontaneous spawning may occur; this is especially true of starfish which live in captivity for more than a year and never spawn. Therefore, the starfish are better kept at moderate temperatures (up to 16°C) in aquaria with water cooling (see [20]).

The starfish can feed on algae (*Ulva, Porphyra*), various microphytes, fish and molluscs, polychaetes, and remains of various marine organisms. If they are kept in captivity for a long time and their food is not diverse enough, avitaminosis may be observed: the dark-blue color converts to a pale-blue color. In order to avoid this, we give the starfish the following food mixture: squid meat, 500 g; sweet pepper powder, 2 g; calcium gluconate (glycerophosphate), 1.5 g. This mixture is ground and kept at –10°C.

Since direct sunlight is detrimental for most marine invertebrates [43], only moderate illumination should be used when keeping the starfish in captivity.

11.4. SEX IDENTIFICATION, OBTAINING OF GAMETES, AND ARTIFICIAL FERTILIZATION

A. pectinifera is a dioecious species but has no external sexual dimorphism. The sex of starfish can be identified by several methods; one of them is the gonad biopsy [11, 72]. A hollow needle is introduced into the arm base closer to the central region, thus obtaining enough gametes to identify the sex. The advantage of this method is that the starfish is practically uninjured. However, it is rather difficult to isolate gametes from immature gonads and it is therefore more useful to cut the starfish open. It is placed on a Petri dish with its aboral side up and a cut is made across the interradius (1.5-2 cm) (Fig. 11.1a). A piece of gonad is then cut off. The mature ovaries are orange, and the testes cream colored. Under normal conditions the cut is usually healed within 5-6 months. Unlike in some other starfish species [26], no autotomy of injured arms takes place.

A. pectinifera has short arms; therefore, in order to obtain a large number of oocytes, a circular cut should be made and whole gonads should be isolated (Fig, 11.1b). An oocyte suspension can be obtained as follows [58]: the ovaries are rapidly washed with an ice-cold Ca^{2+}-free sea water and they are delicately eased apart for 1-2 min with fine forceps. The suspension is then filtered through a double layer of cheesecloth and centrifuged at low speed for 0.5-1 min. As a result, the

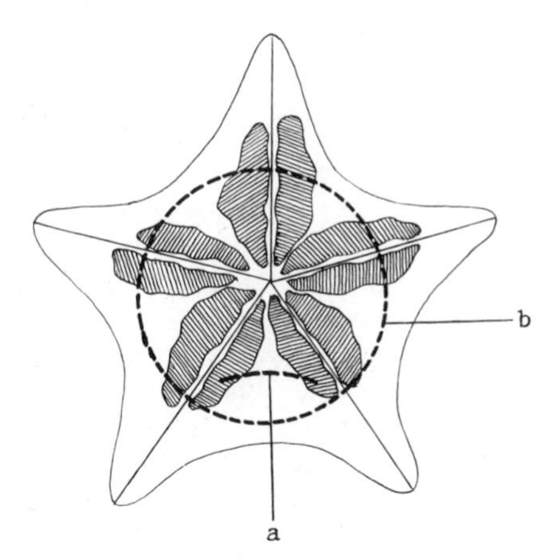

Fig. 11.1. Sketch of a starfish section when isolating a piece of gonad (a) or the whole gonad (b). Aboral body side.

follicle cells remain in the supernatant and are removed. The suspension is washed until it becomes free of follicle cells. The use of Ca^{2+}-free water prevents the secretion of 1-methyladenine by the follicle cells and the adhesion of the follicle cells to the oocyte's surface; the low temperature increases the necessary contact time for 1-methyladenine to induce maturation. Various recipes for making Ca^{2+}-free water are provided by Meijer et al. [58].

Isolation of gonads, by the method described above, will inevitably cause the death of the starfish; therefore, this method is used mostly at the marine stations. However, a part of the ovary, or even the whole gonad, can be isolated through a small cut between the arms. The operated animal is then placed back into the aquarium.

Mature gametes can also be obtained by the hormonal stimulation of spawning [8, 24, 38]. The starfish are injected in each arm with 0.5-1 ml of $1 \cdot 10^{-4}$ M solution of 1-methyladenine in the filtered sea water and placed in separate vessels. The males start spawning first (10-20 min after the injection), and they are immediately placed into a refrigerator in dry Petri dishes. The spawning of females starts 30-60 min after the injection. The intensity of spawning varies and depends, above all, on the state of ovaries.

The fertilized eggs can be obtained by placing the injected males and females into the same vessel. For artificial fertilization, a piece of testis is cut off, minced on a dry glass, and kept in a refrigerator. When needed, the "dry" sperm is diluted with sea water and the activity of spermatozoa is monitored under the microscope. The method of artificial fertilization proposed by Busnikov and Podmarev [7] (see also Chapter 10) for sea urchins can be used: the diluted sperm is mixed with the oocytes (0.1-0.3 ml of sperm per 50-100 ml of dense suspension of oocytes), the volume is brought to 1 liter within 1-2 min, and, after the oocytes have settled, the water is changed. Fertilization can be monitored by the rate of membrane elevation.

Although the starfish oocytes are fertilizable at any stage of maturation after germinal vesicle breakdown, the period from the germinal vesicle breakdown until the extrusion of the first polar body is considered to be optimal for artificial fertilization [11, 25, 51, etc.]. When the spawning is induced, the oocytes shed by the females are at this stage of oocyte maturation [52]. Therefore, oocytes are best fertilized 30-45 min after the beginning of spawning (with respect to its intensity). The first polar body is extruded 60-70 min, and the second polar body 90-105 min after 1-methyladenine treatment of the oocytes at 20°C [14, 51]. A depression is formed on the egg surface at the site of the polar body extrusion some 10 min before its appearance [14].

It is advisable to gradually increase the temperature, by 0.5-1.0°C a day, in the aquarium for 1-2 weeks before spawning to 18-20°C.

We induce spawning of *A. pectinifera* by hormonal stimulation during the autumn–winter period of its gonad cycle, when no spawning is observed in nature. The embryos and larvae developed quite normally [20, 21, 81].

In addition to 1-methyladenine, dithiothreitol [45], methylglyoxal-2-(guanylhydrazone) [55], arachidonic acid [57], and some other substances have also been used to induce oocyte maturation (see [58]).

11.5. EGG STRUCTURE, REMOVAL OF EGG ENVELOPES, AND ISOLATION OF GERMINAL VESICLES

The mature eggs of *A. pectinifera* are transparent, of brownish color, about 170 μm in diameter; their shape is typically spherical (Figs. 11.2 and 11.3), but it can be irregular when the oocytes are densely packed in the ovaries. The eggs are heavier than the sea water. As in most echinoderms, the egg is surrounded by a thin hyaline membrane, adjacent vitelline membrane, and outer jelly coat [40]; the eggs are liberated from the follicular layer during spawning.

The membranes can be removed by the methods used for many starfish species (see [58]). The jelly coat is dissolved by washing the eggs in acidified sea water (pH 4.5-5.5) [65, 77]. The vitelline membrane can be removed by short-term treatment with pronase (0.05-0.1% w/v) or trypsin (0.05% w/v) with subsequent washing in 1 M urea [65]. Chiba and Hoshi [9] removed the vitelline membrane by 5-min treatment with trypsin (7 mg/70 ml of Ca^{2+}- and Mg^{2+}-free sea water). The trypsin effect was arrested with diisopropylfluorophosphate (1 ml of 1 mM solution).

A simple and effective method has been developed for mass isolation of germinal vesicles from the oocytes of *A. pectinifera* [9]. After the vitelline membranes have been removed, the oocytes are washed in Ca^{2+}- and Mg^{2+}-free sea water and suspended in 100 ml of 0.2 M sucrose in the presence of 50 mM Tris-HCl (pH 7.5). If the oocytes are not lysed within 5 min they are passed through a glass column (1.5 mm × 5.0 cm) three times. Within 5 min, 80 ml of 0.2 M sucrose is added dropwise to the homogenate and gently mixed. Aliquots (30 ml) of the homogenate are layered with a pipette on the sucrose gradient (10 ml of 1.7 M + 5 ml of 1.4 M) and centrifuged at 750 g for 8 min. The germinal vesicles settle to the bottom. The isolated germinal vesicles are gently suspended in 1.7 M sucrose, and they are ready for further treatment.

Some other methods used on the oocytes and embryos of *A. pectinifera* and other starfish species (obtaining nucleate and anucleate fragments, fusion of oocytes, microinjection and cytoplasmic transfer, parthenogenetic activation, surface staining, etc.) are briefly described by Meijer et al. [58]. More detailed infor-

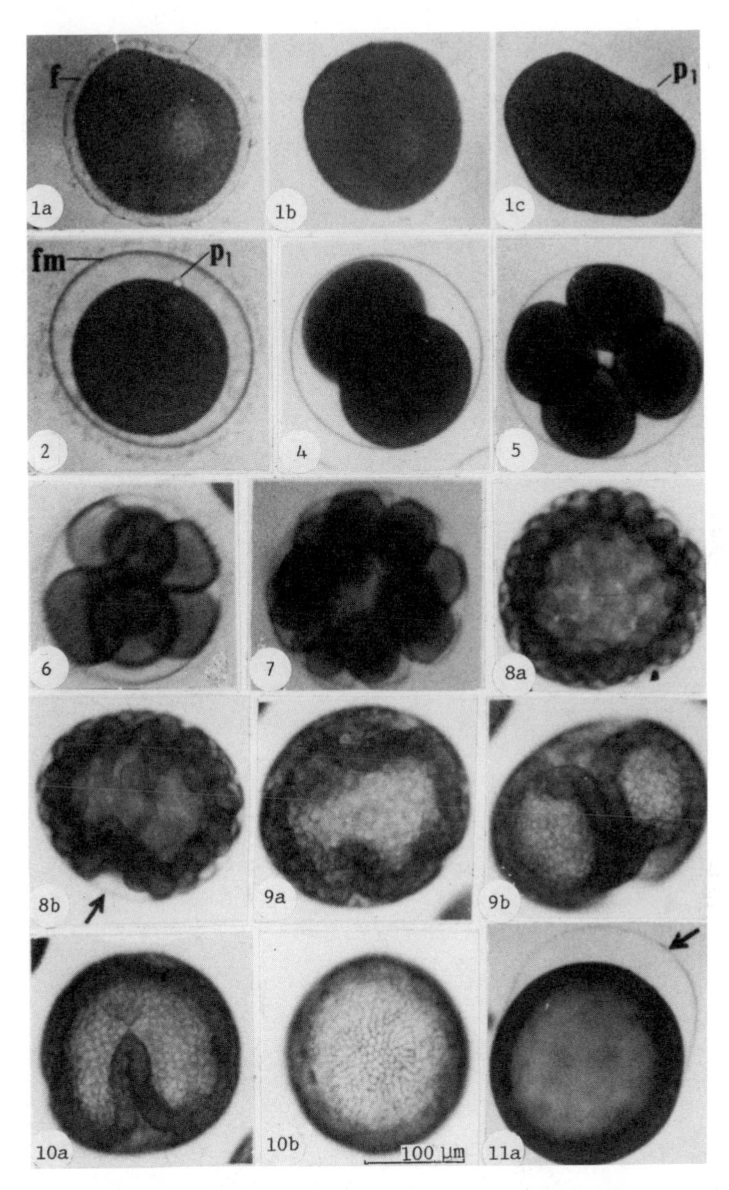

Fig. 11.2. Normal development of the starfish *Asterina pectinifera*. Numbers correspond to stage numbers in Table 20. (No photograph is given for stage 3). 1a) Immature egg; 1b) mature egg; 1c) extrusion of the 1st polar body; 2) formation of fertilization membrane; 4) 2 blastomeres; 5) 4 blastomeres; 6) 8 blastomeres; 7) 16 blastomeres; 8) early blastula before (a) and after the onset of wrinkling (b); 9) mid-blastula; 10) late blastula; 11a) hatching. f, follicular envelope; p_1, 1st polar body; fm, fertilization membrane. All photographs are from Teshirogi and Ishida [88].

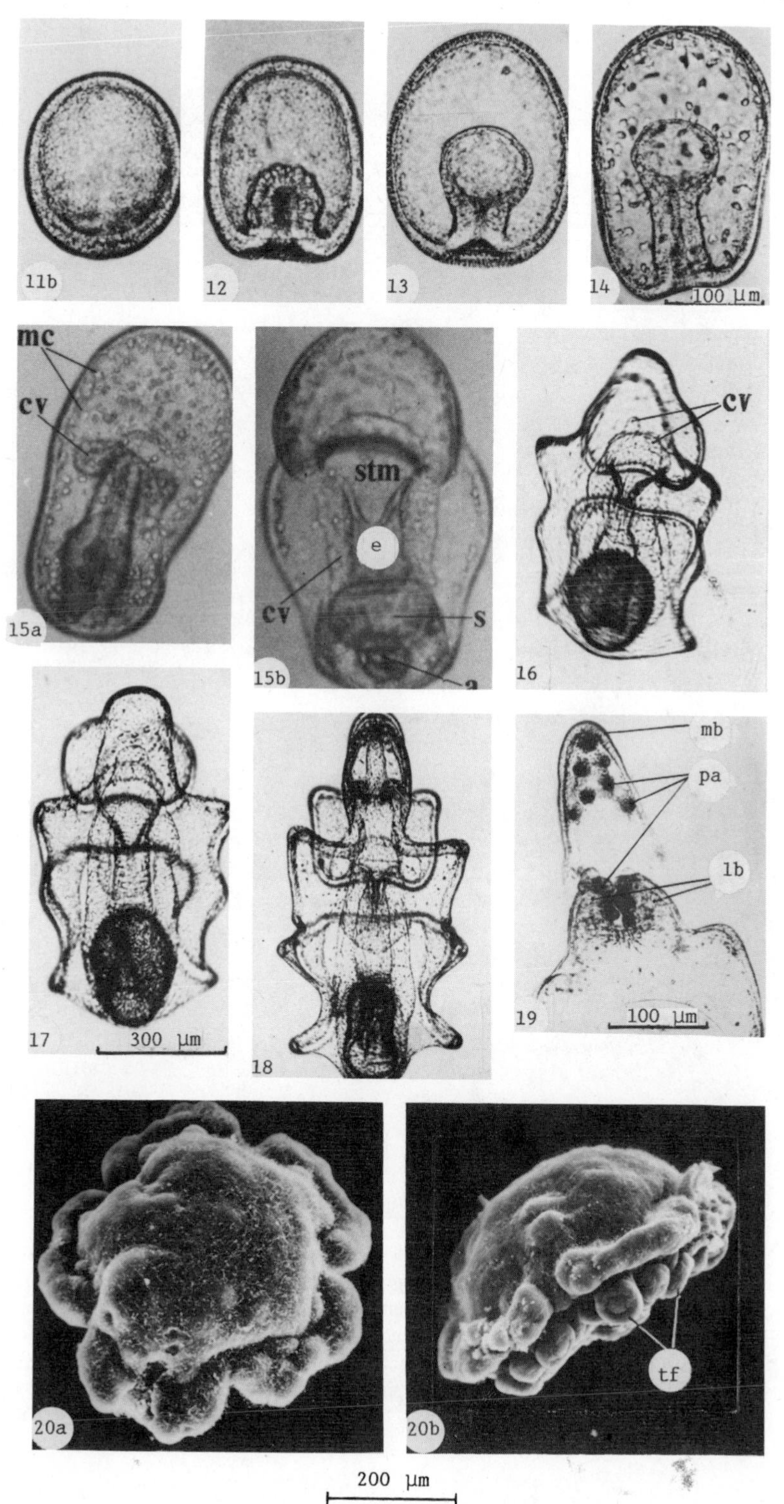
11b
12
13
14
100 µm
mc
cv
15a
stm
e
cv
s
a
15b
cv
16
17
300 µm
18
mb
pa
lb
19
100 µm
20a
20b
tf
200 µm

mation about these methods is also available [44, 66, 68, 74, 76, 92-94, 96, 98, etc.].

11.6. INCUBATION OF EMBRYOS

The embryos can normally develop in shallow vessels with a relatively large base (e.g., crystallizing dishes, Petri dishes). If they are placed on the bottom in one layer no disturbance of water is required. Increase in density will aggravate the oxygen regime in the vessel; in this case the water should be intensively mixed. Water flow can be created by stirrers made of a plastic pipette and a cover of the Petri dish. The cover is glued to the pipette at an angle of 30° and the pipette is connected via a rubber tube to the spindle of an electric motor set at about 60 rpm [7] (see also Chapter 10). The embryos are best incubated by mixing in high vessels, for example in 1-liter beakers. The water in the vessels should be changed 2 or 3 times a day.

The light regime appears not to affect markedly the embryonic development in *A. pectinifera* but the developing embryos are best kept shadowed. The blastulae are characterized by positive phototaxis, and this can be used for their catching and transferring to another vessel.

When incubating the embryos for experimental studies, an artificial sea water composed of 5-6 basic constituents can be used (see [58]). But in order to obtain the late larval stages, settling, and metamorphosis, we recommend the recipe given in Table 11.1.

11.7. REARING OF LARVAE

The technique of rearing the planktonic larvae of *A. pectinifera* is much more complicated than that of incubating the embryos. This is related to the ever-increasing requirements of the larvae kept under artificial conditions, including their active feeding on phytoplankton.

One-liter beakers are suitable for rearing the larvae. They are compact; the high water column ensures the vertical distribution of the larvae. The development and behavior of the larvae can be easily observed, and they are easily caught. The permissible density of larvae is determined first of all by the stage of their development [43]. At the early larval stages (to middle bipinnaria) the larvae can be reared at high concentrations (up to 10,000/liter) using a plastic stirrer. Later, after the larvae have started active feeding, a shaker is best used for water mixing since it does not interfere with the movement of the larvae in the vessels. Strathmann [82] suggests

Fig. 11.3. Normal development of the starfish *Asterina pectinifera*. 11b) Hatching; 12) early gastrula; 13) mid-gastrula; 14) late gastrula; 15) dipleurula; 16) early bipinnaria; 17) mid-bipinnaria; 18) late bipinnaria; 19) brachioles of the brachiolaria; 20) juvenile starfish. a, anus; cv, coelomic vesicles; lb, lateral brachiole; mb, medial brachiole; mc, mesenchyme cells; e, esophagus; pa, papillae; s, stomach; stm, stomodeum; tf, tube feet. Scale: 11b-15, 100 μm; 16-18, 300 μm; 19, 100 μm; 20, 200 μm. Photographs 11b-14 are from Dan-Sohkawa et al. [17], and 15a, b are from Teshirogi and Ishida [88].

TABLE 11.2. Duration of Development at Different Temperatures, Expressed in Absolute (hours) and Relative (τ_n/τ_0) Time Units

t°C	τ_0, min	To rotation stage				To hatching stage				To gastrulation stage			
		τ_1 hours	τ_1/τ_0	τ_1* hours	τ_1*/τ_0	τ_2 hours	τ_2/τ_0	τ_2* hours	τ_2*/τ_0	τ_3 hours	τ_3/τ_0	τ_3* hours	τ_3*/τ_0
11.8	83	30.0	21.7			32.8	23.7						
14.8	57	22.8	24.0			21.5	24.7			25.5	26.8		
15.7	52	20.8	24.0			21.3	24.6			23.3	26.9		
16*	49			19.5	23.9			20.5	25.1			23.5	29.8
19.3	38	15.3	24.2			15.8	25.0			16.7	26.3		
20*	33			13.8	25.1			14.5	26.4			16.5	30.0
20.4	35	13.3	22.9			14.0	24.0			14.3	24.6		
23	28	12.3	26.4			12.3	26.4			13.3	28.6		

*Data from Kominami and Satoh [51].

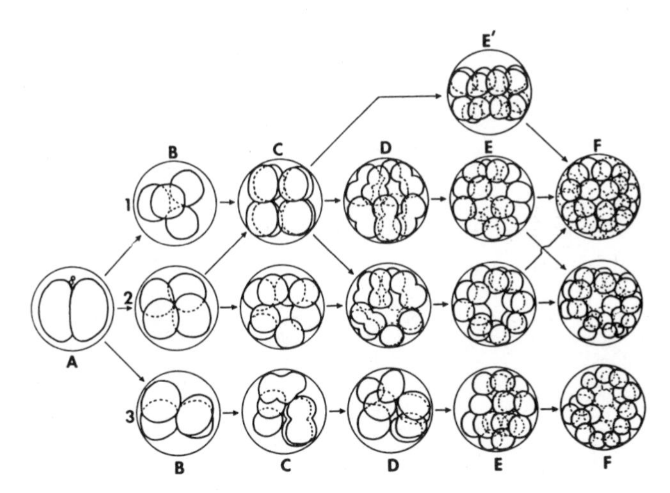

Fig. 11.4. Various cleavage patterns in *Asterina pectinifera*. Side view. A) 2-Cell stage; B) 4-cell stage; 1C, 2C) 8-cell stage; 3C) beginning of 8-cell stage; 1D, 2D) beginning of 16-cell stage; 3D) 8-cell stage; 1E-3E, E') 16-cell stage; 1F-3F) 32-cell stage (after Teshirogi and Ishida [88]).

that the concentration at the later stages should be reduced to 100/liter. We have reared *A. pectinifera* larvae to the brachiolaria stage at a concentration of 100-150 larvae per liter. The late brachiolariae should be isolated from the less-advanced larvae and reared separately.

There are different recommendations as to the frequency of water change when rearing the starfish larvae (see, for example, [4, 11, 84]). We obtained the best results while changing the water in the containers by one-third every 2 days and, at the later larval stages, every 4-5 days. The *A. pectinifera* blastulae and larvae are characterized by positive phototaxis; therefore, when the water is changed, the glass containers are covered with black paper or dense tissue to shield the upper two-thirds and the source of light is left to one side. Within 1.5-2 h the larvae sink to the illuminated part of the vessel, and the upper part of water can be removed using a rubber tube.

Water mixing is desirable for all larval stages since it helps to disperse the larvae and reduces the possibility of bacterial infections [4]. There are different methods for creating a water flow in the vessels (see, for example, [4, 30, 32]). Water mixing by air bubbles leads to the death of sea urchin [7] (see also Chapter 10) and starfish larvae, and thus cannot be used.

The feeding of the echinoderm and, in particular, starfish larvae, has been studied by many authors (for review, see [53]). At the early stages the larvae can survive without feeding, but their settling and metamorphosis are possible only if suitable food is available [43]. For the larvae of some species of the family Asterinidae, including *A. pectinifera*, microalgae *Nitzchia* sp., *Dunaliella tertiolecta, Isochrysis galbana*, and *Phaeodactylum tricornutum* are suitable [53]. Strathmann [82] suggests algal species as food for the larvae of many starfish species: *Phaeodactylum tricornutum, Dunaliella tertiolecta, Isochrysis galbana, Cyclotella nana, Thalassiosira fluviatilis, Crisosphaera carterae, Amphidinium carteri, Ditylum brightwelli*. We have fed larvae on microalgae *Platymonas viridis, Exuviella* sp.,

TABLE 11.3. Normal Development of *Asterina pectinifera*

Stage No.	Stage	t°C	Time after insemination: day	hour	min	τ_n/τ_0	Size, μm	Observations
1a	Immature egg (oocyte)	25 ± 2					170	Germinal vesicle with a nucleolus is seen; the egg is surrounded by the follicular envelope
1b	Mature egg	25 ± 2						Germinal vesicle and follicular envelope disappear within 30-40 min after 1-methyladenine treatment
1c	Extrusion of the 1st polar body	25 ± 2						Takes place at the animal pole within 80 min after 1-methyladenine treatment
Insemination								
2	Formation of fertilization membrane	20 ± 1			2-3			The envelope is elevated at the site of sperm entry. The oval egg assumes a regular spherical shape
		23			3	0.1		
		25 ± 2			2			
3	Fusion of pronuclei	25 ± 2			40			Zygote nucleus enlarges
4	2 blastomeres	20 ± 1		1	40			1st cleavage is meridional, equal. The furrow is laid down in the animal pole region
		23		1	30	3.2		
		25 ± 2		1	25			
5	4 blastomeres	23		2	00	4.3		2nd division. Blastocoele appears
		25 ± 2		1	40			
6	8 blastomeres	25 ± 2		2	99			3rd cleavage is more often equal. Unequal blastomeres can form
		23		2	35	5.5		
7	16 blastomeres	23		3	10	6.8		4th cleavage is equal. Blastomeres of spherical shape. Blastocoele increases
		25 ± 2		2	20			
8	Early blastula	25 ± 2		4	30			Blastomeres surrounding the blastocoele are located in one layer. Blastoderm starts to invaginate into the clastocoele

9	Midblastula	25 ± 2	6-8	30			As the number of cells increases, the invaginating folds thicken and reach the blastocoele center
10	Late blastula	25 ± 2	9	30			The blastoderm folds return to the initial position; they are superimposed and fuse – in these places the blastoderm becomes multilayered. The egg surface is smoothed out. A large blastocoele is again seen in the center
11	Hatching	20-22 25 ± 2 23	11 10 11 12	30 50 20 20	 26.4		The embryo starts to rotate inside the fertilization envelope, the direction of rotation varies. In the vegetal pole region the blastula wall thickens. Within 10-40 min after the beginning of rotation, the envelope is attenuated and breaks down; the envelope is not shed at once but remains as a cap on the embryo for some time. Folds can be seen in some embryos at the moment of hatching
12	Early gastrula	25 ± 2 23 20	12 13 17	00 20	 28.6		The body surface is densely covered with cilia. The embryos rapidly swim while rotating counterclockwise. Invagination starts in the vegetal pole region, the blastopore rudiment can be seen. There are still embryos with the fertilization envelope as a cap
13	Midgastrula	25 ± 2 23 20	17 20	00		260 × 210	The body elongates along the animal–vegetal axis. The archenteron invagination continues; the primary mesenchyme cells emigrate from its distal end into the blastocoele cavity

TABLE 11.3 (continued)

Stage No.	Stage	t°C	Time after insemination day	hour	min	τ_n/τ_0	Size, µm	Observations
14	Late gastrula	25 ± 2 23 20		21 30			280	A thin-walled spherical expansion forms at the archenteron's distal end. Stomodeum forms on the lateral surface of the larval body. Archenteron bends toward stomodeum
15	Dipleurula	25 ± 2 23	1	3			320	The expansion on the archenteron's distal end becomes fungiform. Two symmetrical coelomic rudiments are pinched off the archenteron's distal end. The intestine parts start to form. The anterior part connects with stomodeum where the mouth opening forms. A peristomal ring of cilia forms. Esophagus forms in the anterior part. In the middle part archenteron starts to expand in the transverse direction. Stomach forms in the middle part. Blastopore shifts toward stomodeum and transforms into anal orifice by which the posterior gut opens outside. In front of stomodeum and behind anal orifice protrusions form, frontal and anal regions. Ciliary bands start to form. The cilia are concentrated forming the anterior–transverse, posterior–transverse, and longitudinal bands
16	Early bipinnaria	25 ± 2 23	 3				350	The anterior body part is elongated. The preoral ciliary band is closed into a ring, which limits the frontal field of the preoral lobe. It is separated from the rest of the band. The body shape becomes more

					complicated, lateral protrustions appear. The larva starts to feed on microalgae. Coeloms grow in the anterior–posterior direction
17	Mid-bipinnaria	23	7	650	Coeloms fuse in the preoral lobe before pharynx
18	Late bipinnaria	23	12-14	1000	Rudiments of transient attachment organs of brachiolaria, brachioles, and attachment disk appear in the preoral region. The medioventral protrusion transforms into the medial brachiole. The ciliary band configuration becomes more complicated, lateral protrusions of the larval body are elongated
19	Brachiolaria	23	16	1000-1100	Three brachioles form (two lateral and one medial) which carry attachment papillar Five to six days before settlement the attachment disk becomes distinctly seen between the lateral brachioles (24-25 days of development). The developing coelom gets into the medioventral protrusion and lateral brachioles. In the coelom region, which corresponds to the left hydrocoele, five lobe-shaped coelomic protrusions appear on the 20th day: rudiments of the ambulacral system's radial channels
20	Juvenile starfish	23	30	410-485	The attachment stalk is first clearly seen, and then gradually (in 3-4 days) disappears. The primary tube feet touch the substrate. The developing starfish have five primary tube feet. Lobe-shaped radial channels of ambulacral system get into the arms

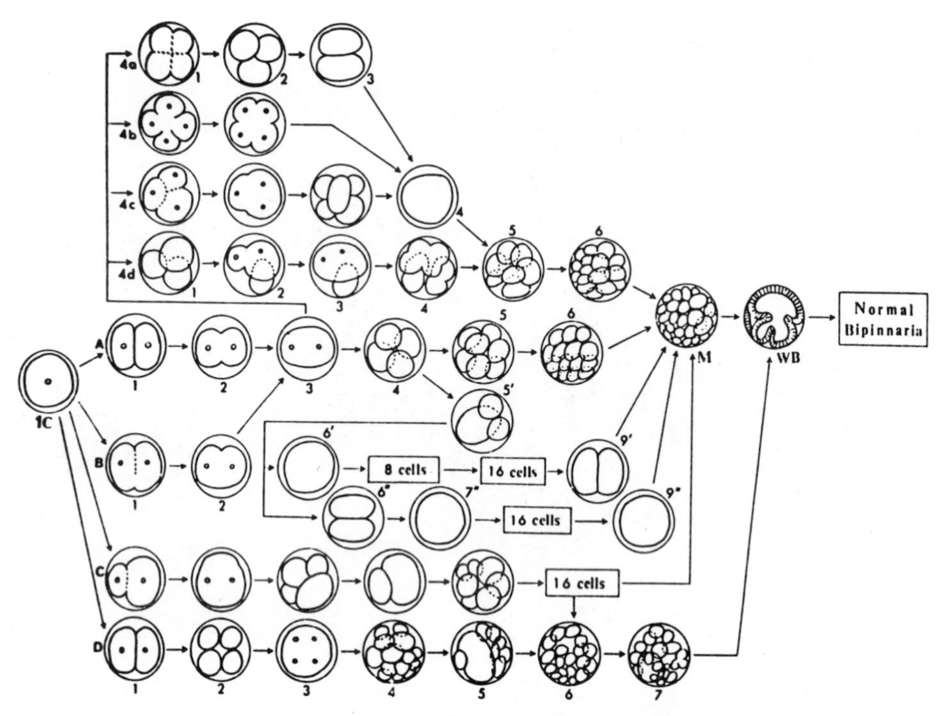

Fig. 11.5. Development of *Asterina pectinifera* in the case of full blastomere fusion. A–D) Various developmental patterns; 1C) 1-Cell stage. 4 in A-4a pattern; A3, A6', A7", A9", C2, D3 – apparently 1 cell. 3 in A-4a pattern; A9', A6", C4 – apparently 2 cells. M, morula; WB, wrinkled blastula (after Teshirogi and Ishida [88]).

Glenodinium sp., *Pavlova lutheri,* and *Bodo marina*; the late brachiolariae were given diatomeans a few days before settling. The food was given during water change. According to Strathmann [83], starfish larvae in a suspension of algae with a concentration of 5000 cells/ml cease to feed within a few minutes: the ciliary band becomes saturated with algal cells; therefore, an excess of microalgae in the vessels for rearing the larvae should be avoided. For feeding, we prepare a dense mixture of the five above-mentioned microalgal species and introduce it into water on a pipette tip.

In addition to planktonic algae, the larvae require bacteria for successful rearing: in their absence the growth of larvae is delayed [54]. That is why it is better to take water for change from an acting closed-cycle aquarium, since fresh water is practically sterile (it is better to add this to the aquarium). The role of bacteria and dissolved organic matter in the feeding of starfish larvae has as yet been insufficiently studied [40].

When settling, some starfish species are characterized by substrate selectivity: for example, *Stichaster australis* brachiolaris settle only on algae of the genus *Mesophyllum* [3]. The larvae of other starfish species, such as *Coscinasterias calamaria* and *Acanthaster placi*, settle on practically all substrates covered with a primary bacterial film [3, 84]; substrate selectivity also appears to be absent in the *A. pectinifera* larvae. A few days before settling, the larvae are placed into clean vessels,

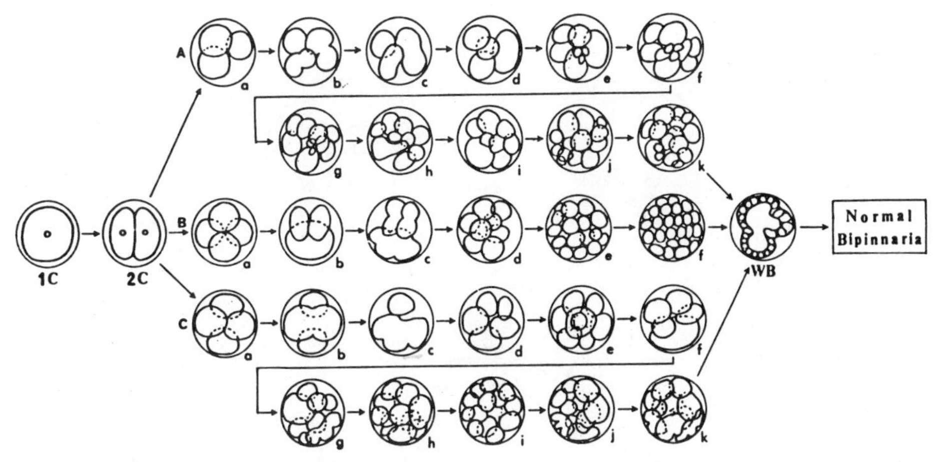

Fig. 11.6. Development of *Asterina pectinifera* in the case of partial blastomere fusion. A–C) Various developmental patterns. 1C) 1-cell stage; 2C) 2-cell stage. WB, wrinkled blastula (after Teshirogi and Ishida [88]).

with mussel or oyster shells covered with a bacterial-algal film (for this film to be formed, the shells are placed for a few days into an illuminated aquarium). This suffices to stimulate the settling of brachilariae on the shells.

Continuous illumination inhibits the metamorphosis of the starfish larvae [19]; therefore, the larvae are better kept in darkness during the night.

It is advisable to rear the larvae at a temperature of no less than 16°C, since the development is markedly delayed at lower temperatures. We obtained settling and metamorphosis at a temperature of 23-24°C.

11.8. DEVELOPMENT OF EMBRYOS AND LARVAE

The normal development of *A. pectinifera* is poorly documented. There are some incomplete descriptions of embryonic development [14, 47, 51] and Teshirogi and Ishida [88] compiled the normal tables for *A. pectinifera* development until the early bipinnaria stage. The embryonic and larval development of *A. pectinifera* has been briefly described by Kasyanov et al. [40] using materials collected from Peter the Great Bay. We have also studied the development of embryos and larvae, but under artificial conditions [20, 21, 81]; the gonad development in juvenile (6-12 months old) starfish has been outlined [41].

In *A. pectinifera* the fertilization membrane is elevated 3 min after fertilization at 23°C and 2-3 min at 20 ± 1°C [14]; the elevation lasts 45 sec. The membrane is elevated unequally starting from the fertilization cone at the site of the sperm entry. As a result of cortical reaction a 30-µm-wide perivitelline space forms [14].

The cleavage is holoblastic, radial, to a weak extent bilateral, and somewhat asynchronous [40, 47]. At the same time, according to Teshirogi and Ishida [75], marked deviations from this scheme can be observed. For example, during cleavage the furrows may disappear and the blastomeres may completely (Fig. 11.4) or

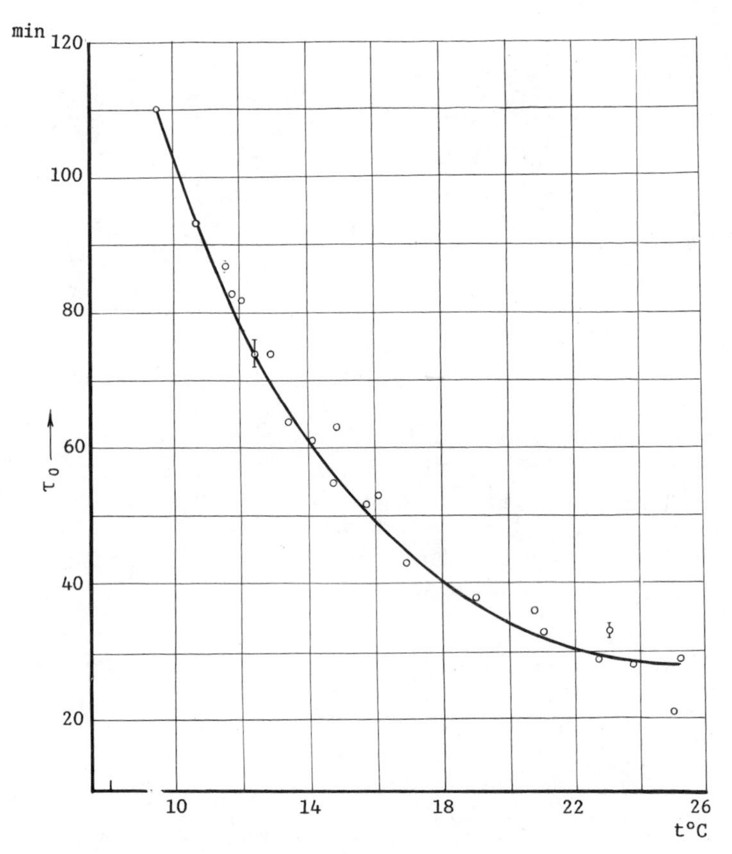

Fig. 11.7. Duration (τ_0) of one mitotic cycle during synchronous cleavate divisions in *Asterina pectinifera* as a function of temperature.

partially (Fig. 11.5) fuse. In addition, there are three possible patterns of cleavage at the 4-, 8-, 16-, and 32-cell stages (Fig. 11.6). Not all these variants result in the formation of normal larvae. Teshirogi et al. [89] demonstrated that the frequency of blastomere fusion does not depend on the sperm concentration during fertilization, but increases when the oocytes are fertilized at the later stages of maturation (after the 1st polar body extrusion). The development of the eggs with fully fused blastomeres is arrested at the blastula or gastrula stage and of those with partially fused blastomeres results in the formation of normal bipinnariae [89].

Synchronous cleavage divisions in *A. pectinifera* lasts to the 8th [51] or 10th cleavage division [14]. The intervals between the cleavage divisions are constant until the 7th division (about 33 min at 20.0 ± 0.5°C) and gradually increase from the 8th division (more than 1 h between the 10th and 11th divisions [51]). According to Dan-Sohkawa and Satoh [14], the intervals between the cleavage divisions until the 5th division are roughly 35 min (at 20 ± 1°C). There were no published data on the cell cycle duration during cleavage at any other temperatures.

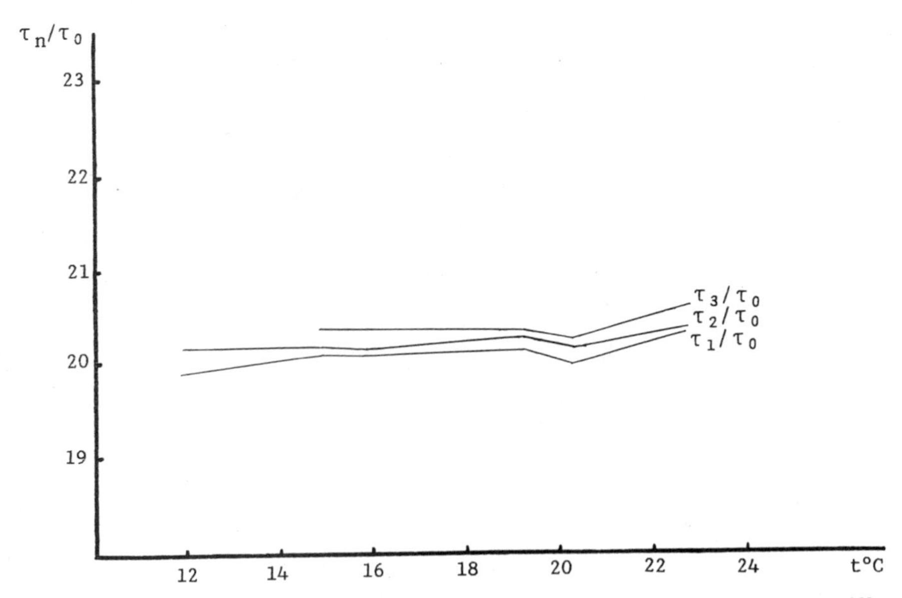

Fig. 11.8. Duration of certain developmental periods in relative time units (τ_n/τ_0) at different temperatures. τ_1, rotation; τ_2, hatching; τ_3, gastrulation.

We measured the values of τ_0, the duration of one cell cycle during synchronous cleavage divisions [22], at different temperatures and plotted the τ_0 against temperature curve (Fig. 11.7). The duration of the 2nd cell cycle is the easiest to measure. Using the plotted curve (Fig. 11.7), we can determine the value of τ_0 at any temperature in the zone of optimal temperatures. According to our data, the range of temperatures at which the cleavage of *A. pectinifera* from Peter the Great Bay is possible, is limited by 10-11 and 24°C. At a temperature above 25°C, the mass death of embryos is observed; at 25°C the amount of abnormal embryos rapidly increases to 80-90% at the 8- and 16-cell stages. At low temperatures the development proceeded in one of the experiments at 9°C, although at a very slow rate and with a high percentage of defects. No cleavage was observed at 7.6-7.8°C. The embryos of *A. pectinifera* from Mutsu Bay (Honshu Island) could develop at higher temperatures: the cleavage was arrested at temperatures above 28 and below 12°C [51].

We determined the time from fertilization to the stages of rotation (τ_1), hatching (τ_2) at six temperatures, and gastrulation (τ_3) at five temperatures (Table 11.2). In addition, Table 11.2 includes the data obtained by Kominami and Satoh [51] for the stages of rotation, hatching, and gastrulation (τ_1*, τ_2*, τ_3*) (from the moment of 1-methyladenine treatment) at 16 and 20°C. Using the τ_n/τ_0 ratio, we can determine approximately the limits of the zone of optimal temperatures for the embryonic and larval development of *A. pectinifera*. It can be seen from Table 11.2 that, in the range of temperatures from 14.8 to 20.4°C, the τ_n/τ_0 values differ by about unity; outside this zone the difference amounts already to 2-3τ_0 (we calculated the data by Kominami and Satoh using our τ_0 versus temperature curve). Figure 11.8 shows three τ_n/τ_0 curves plotted at different temperatures. It can be seen that the τ_n/τ_0 ratio is stable in the range of 15-20°C and changes at higher and lower temperatures (inflections appear).

The wrinkled blastula stage is observed during cleavage in *A. pectinifera* [47, 88], as well as in some other starfish species (see, for example [29, 48, 49]). Its formation starts within 5.5 h after insemination at 20-22°C: folds appear on the surface of the spherical coeloblastula and multiple blastoderm invagination takes place. The invaginating regions reach the blastocoel center. In a few hours the adjacent folds fuse and the blastula's surface is smoothed out; in some places it becomes multilayered. There are three different types of blastula wrinkling [88]: 1) Invaginating folds increase in size during a few hours and fill almost the whole blastocoel. The blastula's surface is then smoothed out. 2) The folds invaginate to a small extent and the blastula develops into the gastrula without the typical wrinkled stage. 3) The wrinkled blastula develops from the wrinkled morula. Table 11.3 shows the normal development of *A. pectinifera* . In compiling the normal table we gave ordinal numbers to successive stages (1-20). Stages 1-16 are described mainly according to Teshirogi and Ishida [88]. Our changes concern only the stage numeration (the number of stages is diminished since some stages are divided into designated substages, for example, as 1a, 1b, 8a, 8b, etc.), and additions to the description of some stages. The timing of stages after fertilization at different temperatures is given (where possible) according to Teshirogi and Ishida [88] (25 ± 2°C), Dan-Sohkawa and Satoh [14] (20 ± 1°C), Komatsu [47] (20-22°C), and our data (23°C). Stages 17-20 are described on the basis of our observations. We have also calculated the relative timing of developmental stages (where possible) using our τ_0 versus temperature curve (Fig. 11.7).

REFERENCES

1. T. B. Aisenshtadt and S. G. Vassetzky, "Fine structures of oocytes and accessory cells of the ovary in the starfish *Patiria* (*Asterina*) *pectinifera* at different stages of oogenesis and after 1-methyladenine-induced maturation," *Dev. Growth Differ.* **28**(5), 449–460 (1986).
2. Z. I. Baranova, "Echinoderms of Posieta Bay, Sea of Japan," in: *Fauna and Flora of Posieta Bay, Sea of Japan* [in Russian], Nauka, Leningrad (1971).
3. M. F. Barker, "Observations on the settlement of the brachiolaria larvae of *Stichaster australis* (Verrill) and *Coscinasterias calamaria* (Gray) (Echinodermata: Asteroidea) in the laboratory and on the shore," *J. Exp. Mar. Biol. Ecol.* **30**(1), 95–108 (1977).
4. M. F. Barker, "Descriptions of the larvae of *Stichaster australis* (Verrill) and *Coscinasterias calamaria* (Gray) (Echinodermata: Asteroidea) from New Zealand obtained from laboratory culture," *Biol. Bull.* **154**, 32–46 (1978).
5. M. G. Biryulina, "Starfish of Peter the Great Bay. Their influence on the numbers of food invertebrates," in: *Problems of Hydrobiology of Some Regions of the Pacific Ocean* [in Russian], Vladivostok (1972).
6. P. Bulet, T. Kishimoto, and H. Shirai, "Oocyte competence to maturation-inducing hormone: 1. Breakdown of germinal vesicles of small oocytes in starfish, *Asterina pectinifera*," *Dev., Growth Differ.* **27**(3), 243–250 (1985).
7. G. A. Buznikov and V. I. Podmarev, "The sea urchins *Strongylocentrotus dröbachiensis, S. nudus,* and *S. intermedius*," in *Objects of Developmental Biology* [in Russian], Nauka, Moscow (1975).

8. A. B. Chaet and R. A. McConnaughy, "Physiologic activity of nerve extracts," *Biol. Bull.* **117**, 407–408 (1959).
9. K. Chiba and M. Hoshi, "Mass isolation of germinal vesicles from starfish *Asterina pectinifera* oocytes," *Dev. Growth Differ.* **27**(3), 277–282 (1985).
10. A. M. Clark, "Notes on Atlantic and other Asteroidea. 3. The families Ganeriidae and Asterinidae, with description of a new asterinid genus," *Bull. Br. Mus. (Nat. Hist.) (Zool.)* **45**(7), 359–380 (1986).
11. D. P. Costello and C. Henley, *Methods for Obtaining and Handling Marine Eggs and Embryos* (2nd edn.), Mar. Biol. Lab., Woods Hole (1971).
12. M. Dan-Sohkawa, "A 'normal' development of denuded eggs of the starfish *Asterina pectinifera*," *Dev. Growth Differ.* **18**, 439–445 (1976).
13. M. Dan-Sohkawa, "Formation of joined larvae in the starfish *Asterina pectinifera*," *Dev. Growth Differ.* **19**(3), 233–239 (1977).
14. M. Dan-Sohkawa and N. Satoh, "Studies on dwarf larvae developed from blastomeres of starfish *Asterina pectinifera*," *J. Embryol. Exp. Morphol.* **46**, 171–185 (1978).
15. M. Dan-Sohkawa and H. Fujisawa, "Cell dynamics of the blastulation process in the starfish *Asterina pectinifera*," *Dev. Biol.* **77**, 328–339 (1980).
16. M. Dan-Sohkawa, G. Tamura, and H. Mitsui, "Mesenchyme cells in starfish development: effect of tunicamycin on their differentiation, migration, and function," *Dev. Growth Differ.* **22**(3), 495–502 (1980).
17. M. Dan-Sohkawa, H. Yamanaka, and K. Watanabe, "Reconstruction of bipinnaria larvae from dissociated embryonic cells of the starfish *Asterina pectinifera*," *J. Embryol. Exp. Morphol.* **94**, 47–60 (1986).
18. M. Dan-Sohkawa, Y. Komiya, H. Kaneko, and H. Yamanaka, "Morphogenetic capability of epithelial and mesenchymal cells dissociated from swimming starfish embryos," in: *Progress in Developmental Biology*, Part A, H. C. Slavkin (ed.), Alan R. Liss, New York (1986).
19. S. Sh. Dautov, "Changes in behavioral reactions during ontogenesis of the larvae of starfish from the family Asteriidae," in: *Materials of the 4th All-Union Colloquium on Echinoderms* [in Russian], Tbilisi Univ. (1979).
20. P. V. Davydov, O. I. Shubravyi, A. G. Tsetlin, S. Sh. Dautov, and S. G. Vassetzky, "Cultivation of the starfish *Patiria pectinifera* in a closed cycle aquarium," *Biol. Morya,* No. 5, 67–71 (1986).
21. P. V. Davydov, O. I. Subravyi, and S. G. Vassetzky, "Larval development of starfish as revealed by long-term culture of embryos," *Dev., Growth Differ.* **30**(5), 463–469 (1988).
22. T. A. Dettlaff and A. A. Dettlaff, "On relative dimensionless characteristics of the development duration in embryology," *Arch. Biol. (Liege)* **72**, 1–16 (1961).
23. A. M. Dyakonov, "Sea stars from the seas of the USSR," in: *Key to Fauna of the USSR* [in Russian], Vol. 34, Leningrad (1950).
24. A. Fraser, J. Gomes, B. Hartwick, and J. Smith, "Observation on the reproduction and development of *Pisaster ochraceus* (Brandt), *Can. J. Zool.* **59**, 1700–1707 (1980).
25. T. Fujimmori and S. Hirai, "Differences in starfish oocyte susceptibility to polyspermy during the course of maturation," *Biol. Bull.* **157**, 249–254 (1979).
26. J. W. Fuseler, "Repetitive procurement of mature gametes from individual sea stars," *J. Cell Biol.* **57**, 879–881 (1973).

27. M. S. Hamaguchi and Y. Hiramoto, "Protoplasmic movement during polar body formation in starfish oocytes," *Exp. Cell. Res.* **112**(1), 55–62 (1978).
28. T. Haraguchi and H. Nagano, "Isolation and characterization of DNA polymerases from mature oocytes of the starfish *Asterina pectinifera*," *J. Biochem. (Tokyo)* **93**(3), 687–698 (1983).
29. R. Hayashi, M. Komatsu, and C. Oguro, "Wrinkled blastula of the sea-star *Acanthaster planci* (L.)," *Proc. Jpn. Soc. Syst. Zool.* **9**, 59–61 (1973).
30. J. A. Henderson and J. S. Lucas, "Larval development and metamorphosis of *Acanthaster planci* (Asteroidea)," *Nature (London)* **232**, 655–657 (1971).
31. T. R. Hinegardner, "Echinoderms," in: *Methods in Developmental Biology*, F. H. Wilt and N. K. Wessels (eds.), T. Y. Cromwell (1967).
32. R. T. Hinegardner, "Growth and development of the laboratory cultured sea urchin," *Biol. Bull.* **137**(3), 465–475 (1969).
33. H. Honda, M. Dan-Sohkawa, and K. Watanabe, "Geometrical analysis of cells becoming organized into a tensile sheet, the blastular wall in the starfish *Asterina pectinifera*," *Differentiation* **25**(1), 16–22 (1983).
34. N. Hondo, H. Shirai, P. Bulet, M. Isobe, K. Imai, and T. Goto, "Effective desalting techniques for a hormonal peptide, gonad-stimulating substance of starfish," *Biomed. Res.* **7**(2), 89–95 (1986).
35. H. Kanatani, "Maturation-inducing substance in starfish," *Int. Rev. Cytol.* **35**, 253–298 (1973).
36. H. Kanatani, "Hormones in echinoderms," in: *Hormones and Evolution*, E. J. W. Barrington (ed.), Vol. 1, Academic Press, New York (1979).
37. H. Kanatani, "Nature and action of the mediators inducing maturation of the starfish oocyte," in: *Molecular Biology of Egg Maturation*, Pitman Books, London (1983).
38. H. Kanatani, H. Shirai, K. Nakanishi, and T. Kurokawa, "Isolution and identification of meiosis-inducing substance in starfish *Asterias amurensis*," *Nature (London)* **221**, 273–274 (1969).
39. V. L. Kasyanov, L. A. Medvedeva, U. M. Yakovlev, and S. N. Yakovlev, *Breeding of Echinoderms and Marine Clams* [in Russian], Nauka, Moscow (1980).
40. V. L. Kasyanov, G. A. Kryuchkova, V. A. Kulikova, and L. A. Medvedeva, *Larvae of Marine Clams and Echinoderms* [in Russian[, Nauka, Moscow (1983).
41. V. L. Kasyanov and N. K. Kolotukhina, "The gonad development in a starfish *Patiria pectinifera*," *Zool. Mag. (Russ.)* **64**(10), 1591–1594 (1985).
42. O. Kinne, "Cultivation of marine organisms: water-quality management and technology," in: *Marine Ecology*, O. Kinne (ed.), Vol. 3, Books Demand, Univ. of Michigan (1976).
43. O. Kinne, "Echinodermata," in: *Marine Ecology*, O. Kinne (ed.), Vol. 3, Books Demand, Univ. of Michigan (1977).
44. T. Kishimoto, "Microinjection and cytoplasmic transfer in starfish oocytes," in: *Methods in Cell Biology*, Vol. 27 (1986).
45. T. Kishimoto and H. Kanatani, "Isolation of starfish oocyte maturation by disulfide-reducing agents," *Exp. Cell Res.* **82**, 296–302 (1973).
46. T. Kishimoto and H. Kanatani, "Cytoplasmic factor responsible for germinal vesicle breakdown and meiotic maturation in the starfish oocyte," *Nature (London)* **260**, 321–322 (1967).

47. M. Komatsu, "On the wrinkled blastula of the sea-star *Asterina pectinifera*," *Zool. Mag.* **81**, 227–231 (1972).
48. M. Komatsu, "A preliminary report on the development of the sea-star *Leiaster leachii*," *Proc. Jpn. Soc. Syst. Zool.* **9**, 55–58 (1973).
49. M. Komatsu, "Development of the sea-star, *Asterina coronata japonica Hayashi*," *Proc. Jpn. Soc. Syst. Zool.* **11**, 42–47 (1975).
50. T. Kominami, "Allocation of mesendodermal cells during early embryogenesis in the starfish *Asterina pectinifera*," *J. Embryol. Exp. Morphol.* **84**, 177–190 (1985).
51. T. Kominami and N. Satoh, "Temporal and cell-numerical organization of embryos in the starfish *Asterina pectinifera*," *Zool. Mag.* **89**(3), 244–251 (1980).
52. M. Kume and K. Dan, *Invertebrate Embryology*, Nolit, Belgrade (1968).
53. J. M. Laurence and M. Jangoux, "Food used to maintain postmetamorphic echinoderms in the laboratory," in: *CRC Handbook Series in Nutrition and Food*, M. Rechcigl (ed.), Vol. 1, Section D, CRC Press, Cleveland (1977).
54. J. M. Laurence and M. Jangoux, "Nutrition of larvae," in *Echinoderm Nutrition*, A. A. Balkema, Rotterdam (1982).
55. L. Meijer and P. Guerrier, "Immobilized methylglyoxal-bis(guanylhydrazone) induces starfish oocyte maturation," *Dev. Biol.* **100**, 308–317 (1983).
56. L. Meijer and P. Guerrier, "Maturation and fertilization in starfish oocytes," *Int. Rev. Cytol.* **86**, 129–196 (1984).
57. L. Meijer, P. Guerrier, and J. Maclouf, "Arachidonic acid, 12- and 15-hydroxy-eicosatetraenoic acid, eicosapentaenoic acid, and phospholipase A_2 induce starfish oocyte maturation," *Dev. Biol.* **106**, 368–378 (1984).
58. L. Meijer, P. Pondaven, P. Guerrier, and M. Moreau, "A starfish oocyte user's guide," *Cah. Biol. Mar.* **35**, 457–480 (1984).
59. I. Mita, "Studies on factors affecting the timing of early morphogenetic events during starfish embryogenesis," *J. Exp. Zool.* **225**, 293–299 (1983).
60. I. Mita, "Effect of cysteine and its derivatives on 1-MeAde production by starfish *Asterina pectinifera* follicle cells," *Dev. Growth Differ.* **27**(5), 563–572 (1985).
61. I. Mita and C. Obata, "Timing of early morphogenetic events in tetraploid starfish *Asterina pectinifera* embryos," *J. Exp. Zool.* **229**(1), 215–222 (1984).
62. I. Mita and N. Satoh, "Timing of gastrulation in fused double embryos formed from eggs with different cleavage schedules in the starfish *Asterina pectinifera*," *J. Exp. Zool.* **223**(1), 67–74 (1982).
63. M. Morisawa and H. Kanatani, "Oocyte-surface factor responsible for 1-methyladenine-induced oocyte maturation in starfish," *Gamete Res.* **1**, 157–164 (1978).
64. S. I. Nemoto, "Nature of the 1-MeAde-requiring phase in maturation of starfish *Asterina pectinifera* oocytes," *Dev. Growth Differ.* **24**(5), 429–442 (1982).
65. S. I. Nemoto, M. Yoneda, and I. Uemura, "Marked decrease in the rigidity of starfish oocytes induced by 1-methyladenine," *Dev. Growth Differ.* **22**, 315–325 (1980).
66. S. I. Nemoto, S. Washitani-Nemoto, and A. Hino, "DNA synthesis relating tetraploidy of parthenogenetic starfish eggs," *Dev. Growth Differ.* **28**(4), 385 (1986).

67. G. P. Novikova, "Gonad cycles of the sea-stars *Asterias amurensis* and *Patiria pectinifera* from Peter the Great Bay, Sea of Japan," *Biol. Morya,*, No. 6, 33–40 (1978).
68. C. Obata and S. I. Nemoto, "Artificial parthenogenesis in starfish eggs: Production of parthenogenetic development through suppression of polar body formation by methylxanthines," *Biol. Bull.* **166**(3), 525–536 (1984).
69. M. Ohtsubo and Y. Hiramoto, "Regional difference in mechanical properties of cell surface in dividing echinoderm eggs," *Dev. Growth Differ.* **27**(3), 371–384 (1985).
70. N. Oishi and H. Shimada, "Intracellular localization of DNA polymerases in the oocyte of the starfish *Asterina pectinifera*," *Dev. Growth Differ.* **25**(6), 547–552 (1984).
71. N. Oishi and H. Shimada, "Intracellular localization and sedimentation coefficient of DNA ligase in oocytes of the starfish *Asterina pectinifera*," *Dev. Growth Differ.* **26**(6), 571–574 (1984).
72. S. J. Ruggieri, "Echinodermata," in: *Culture of Marine Invertebrate Animals*, Plenum Press, New York (1975).
73. T. E. Schroeder, "Cortical expressions of polarity in the starfish oocyte," *Dev. Growth Differ.* **27**(3), 311–322 (1985).
74. G. G. Sekirina, M. N. Skoblina, S. G. Vassetzky, and A. A. Bilinkis, "The fusion of oocytes of the starfish *Aphelasterias japonica*. 1. Formation of cell hvbrids," *Cell Differ.* **12**, 67–71 (1983).
75. H. Shirai, "Gonad-stimulating and maturation-inducing substance," in: *Methods in Cell Biology*, Vol. 27 (1986).
76. H. Shirai and H. Kanatani, "Effect of local application of 1-methyladenine on the site of polar body formation in starfish oocyte," *Dev. Growth Differ.* **22**, 555–560 (1980).
77. H. Shirai and H. Kanatani, "Effect of 1-methyladenine on responses of spermatozoa to egg jelly in starfish *Asterina pectinifera*: A convenient method for counting the rate of acrosome reaction and for measuring sperm motility," *Zool. Mag. (Tokyo)* **91**(3), 272–280 (1982).
78. H. Shirai, S. Ikegami, H. Kanatani, and H. Mohri, "Regulation of sperm motility in starfish *Asterina pectinifera*. 1. Initiation of movement," *Dev. Growth Differ.* **24**(5), 419–428 (1982).
79. H. Shirai, N. Hondo, P. Bulet, M. Isobe, K. Imai, and T. Goto, "Separation by preparative electrofocusing of several components of gonad-stimulating substance," *Biomed. Res.* **7**(2), 97–102 (1986).
80. O. I. Shubravyi, "An aquarium with artificial sea water for keeping and breeding of a primitive multicellular organism *Trichoplax* and other marine invertebrates," *J. Zool. (Moscow)* **62**(4), 618–621 (1983).
81. O. I. Shubravyi, H. J. Herrmann, and P. V. Davydov, "Über Haltung und Zucht des Kammseesternes (*Patiria pectinifera* Müller et Troschel) im Meeresaquarium," *Aquarien Terrarienz*, No. 7, 240–243 (1985).
82. R. Strathmann, "The feeding behavior of planktotrophic echinoderm larvae: Mechanisms, regulation, and rates of suspension feeding," *J. Exp. Mar. Biol. Ecol.*, No. 6, 109–160 (1971).
83. R. Strathmann, "Larval feeding in Echinoderms," *Am. Zool.* **15**, 717–730 (1975).
84. R. Strathmann, "Larval settlement in Echinoderms," in: *Settlement and Metamorphosis of Marine Invertebrate Larvae*, Elsevier, New York (1977).

85. N. Takahashi, "The relation between injection of steroids and ovarian protein amounts in the starfish *Asterina pectinifera*," *Bull. Jpn. Soc. Sci. Fish.* **48**(4), 509–512 (1982).
86. N. Takahashi, "Effect of injection of steroids on the starfish testes *Asterina pectinifera*," *Bull. Jpn. Soc. Sci. Fish.* **48**(4), 513–516 (1982).
87. N. Takahashi, "Release of a gonad-stimulating substance from the radial nerves of the cushion star *Asterina pectinifera* in response to excess potassium," *Dev. Growth Differ.* **27**(4), 447–452 (1985).
88. W. Teshirogi and S. Ishida, "Early development of the starfish *Asterina pectinifera*, especially on the cleavage patterns, blastomere fusion, wrinkled morula, and wrinkled blastula," *Rep. Fukaura Mar. Biol. Lab.*, No. 7, 2–24 (1978).
89. W. Teshirogi, S. Ishida, and M. Kodama, "Early development of the starfish eggs (*Asterina pectinifera*) inseminated with various sperm concentrations in oocyte maturation process," *Rep. Fukaura Mar. Biol. Lab.*, No. 8, 2–13 (1980).
90. N. Usui and I. Takahashi, "Quantitative analysis of modulations in intramembrane particles during maturation of starfish *Asterina pectinifera* oocytes," *J. Ultrastruct. Res.* **87**(1), 86–96 (1984).
91. S. G. Vassetzky, G. G. Sekirina, M. N. Skoblina, and A. A. Bilinkis, "The fusion of oocytes of the starfish *Aphelasterias japonica*. I. The capacity of cell hybrids for maturation and cleavage," *Cell. Differ.* **12**, 73–76 (1983).
92. S. G. Vassetzky, G. G. Sekirina, B. L. Veisman, M. N. Skoblina, and A. A. Bilinkis, "The fusion of oocytes of the starfish *Aphelasterias japonica*. II. Reconstruction of oocytes from cells and cell fragments (cytoplasts)," *Cell Differ.* **14**, 47–52 (1984).
93. S. G. Vassetzky, M. N. Skoblina, and G. G. Sekirina, "Induced fusion of echinoderm oocytes and eggs," in: *Methods in Cell Biology*, Vol. 27, Academic Press, New York (1986).
94. S. Washitani-Nemoto, A. Hino, and S. Nemoto, "Production of artificial parthenogenesis by premature activation of starfish oocytes," *Dev. Growth Differ.* **28**(4), 395 (1986).
95. H. Yamada and S. Hirai, "Initiation of cleavage in starfish *Asterina pectinifera* eggs by the injection of Triton-treated spermatozoa," *Gamete Res.* **13**(2), 135–142 (1986).
96. K. Yamamoto and M. Yoneda, "Cytoplasmic cycle in meiotic division of starfish oocytes," *Dev. Biol.*, **96**, 166–172 (1983).
97. H. Yamanaka, Y. Tanaka-Ohmura, and M. Dan-Sohkawa, "What do dissociated embryonic cells of the starfish *Asterina pectinifera* do to reconstruct bipinnaria larvae?," *J. Embryol. Exp. Morphol.* **94**, 61–72 (1986).
98. M. Yoneda, "The compression method for determining the surface force," in: *Methods in Cell Biology*, Vol. 27, Academic Press, New York (1986).

INDEX